ACING PHYSICS 12

Review of Physics 12 for Students in BC

Second Edition

June Chung
Patricia Grindlay

유한문화사

ACING PHYSICS 12

Review of Physics 12 for Students in B. C.

Dedicated to All the Scientifically Inquisitive Young Minds

Direct comments, suggestions, questions, and requests to:

JC Publishing
#1004 - 933 Hornby St.
Vancouver, BC V6Z 3G4
e-mail: acingphysics@hotmail.com

Acing Physics 12- Review of Physics 12 for Students in B.C.

ISBN 0-9781249-0-1

Author: June Chung

- Ph. D., University of Illinois at Urbana-Champaign, 2002
- M.S., Carnegie Mellon University, 1997
- B.S., Carnegie Mellon University, 1996

Associate Author & Editor: Patricia Grindlay

- Head of Physics & Chemistry, Little Flower Academy, Vancouver, B.C., 1985-2006
- Ed. Diploma, University of London, England, 1970
- B.Sc. (Tech), University of Manchester, England, 1961

Acknowledgement: Adele Coady

- Head of Science, Little Flower Academy, Vancouver, B.C.

Printed in S. Korea

PREFACE

While teaching physics, we often lamented the shortage of excellent and appropriate workbooks customized especially for secondary school students in British Columbia. *Acing Physics 12* is a workbook for those students who want to review the major concepts in Physics 12 and prepare themselves for school tests and government exams.

In this second edition, we explained each concept more thoroughly and provided more figures to help the readers. Each explanation is followed by examples that show how the concepts are applied to the types of problems given in school tests and other exams. The students will find those examples particularly helpful. We also added more exercise problems so that students can apply the concepts themselves and practice on their own. The review problems at the end of each chapter consist of vocabulary, true-false, multiple-choice, and written response questions, and will test the student's understanding from many angles.

Although this edition is 50 percent longer than the previous one, it is relatively brief and should not overwhelm the readers – unlike some textbooks! Thus we recommend that students go over this book again and again until they become fully comfortable with all the given concepts and problem-solving techniques demonstrated in the book. Of course, this book cannot be (and is not meant to be) a substitute for physics textbooks or notes taken in class. But we believe that, when used as a complement, this book will definitely help students improve their performance in school, and help them prepare for future physics courses in colleges and universities.

Sincerely,

June Chung

Patricia Grindlay

TABLE OF CONTENTS

1

CHAPTER 1

VECTORS

This chapter is an essential preparation for the work you are going to do in Physics 12. In Physics 11 you studied changes taking place in one dimension - an important introduction to the subject but rather limited in its application to real life. In this course we shall work in a two-dimensional world, where vectors play a most important role. You must understand what a vector is and how to resolve a vector into two component parts that act independently of each other. Resolving vectors, and adding and subtracting vectors using their components is essential to understanding and solving many problems in Physics 12. Once you have mastered vectors, you are well on your way to mastering this course.

In this chapter, you will:

- ✓ review vectors and scalars from Physics 11
- ✓ test your knowledge of vectors from Physics 11
- ✓ learn how to resolve two-dimensional vectors into components
- ✓ add and subtract two-dimensional vectors by components
- ✓ solve exercise problems on resolving vectors and adding vectors
- ✓ test your knowledge of vectors from Physics 12

Review of Physics 11: Vectors

1. Vectors and Scalars

- **Vector** – a quantity with both magnitude and direction

  Examples: displacement, velocity, acceleration, force, momentum, impulse, etc.

- **Scalar** – a quantity that has magnitude only

  Examples: mass, time, distance, speed, work, energy, etc.

- To determine whether a given quantity is a scalar or a vector, think about whether it has a direction. If it does, it is a vector. Otherwise, it is a scalar.

2. Graphical Representation of a Vector

- A vector is represented by an arrow, as shown in Fig. 1-1. The arrow's length and direction represent the vector's magnitude and direction, respectively. The longer the arrow, the larger the magnitude of the corresponding vector.

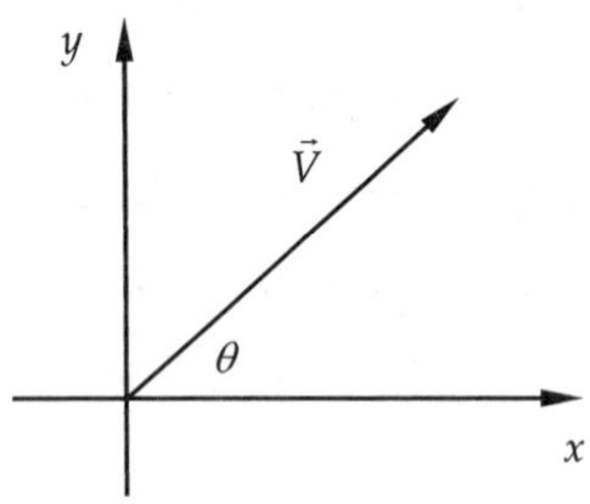

Since the length of an arrow cannot be negative, the magnitude of a vector can never be negative!

Figure 1-1: Graphical representation of a vector

3. Operations Involving Vectors

- When a vector is multiplied or divided by a positive scalar quantity, its magnitude changes accordingly, but there is no change in direction.

  Example: If you multiply a vector, say $\vec{A}$, by 2, the product is a vector whose magnitude is twice that of $\vec{A}$, with no change in direction, as shown in Fig. 1-2.

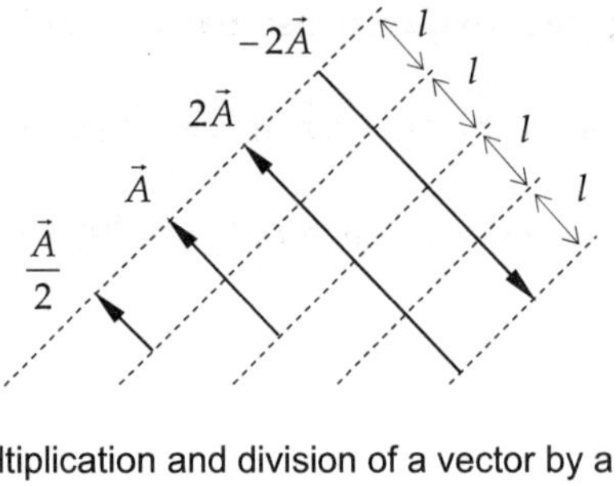

Figure 1-2: Multiplication and division of a vector by a scalar

- When a vector is multiplied or divided by -1, its direction is reversed with no change in magnitude.

  Example: Multiplying $\vec{A}$ by -2 can be interpreted as doubling its magnitude and reversing its direction (see Fig. 1-2).

$$\vec{A}\times(-2)=\vec{A}\times(-1)\times 2$$

multiply magnitude by 2 (refers to $\times 2$)

reverse direction (refers to $\times(-1)$)

- When two vectors are added, the result is another vector called the **resultant** vector. It can be obtained graphically by the **tip-to-tail method**, shown in Fig. 1-4. When adding vectors algebraically, pay attention to their directions.

  Example: Adding 2.7 m, east, with 1.5 m, east, yields a resultant vector of 4.2 m, east. However, adding 2.7 m, east, with 1.5 m, west, produces a resultant vector of 1.2 m, east. These additions are illustrated in Fig. 1-3.

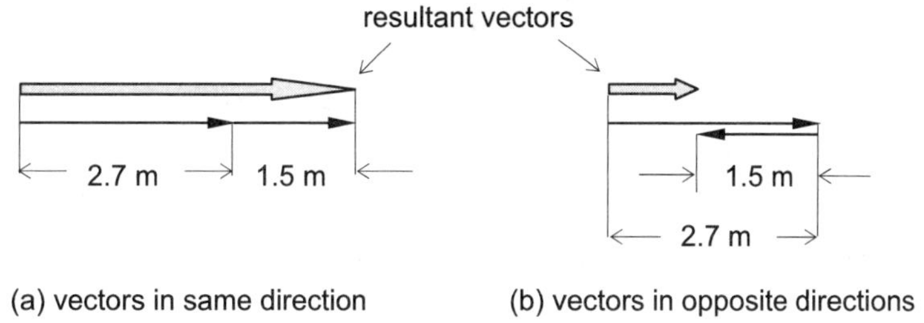

Figure 1-3: Addition of two parallel vectors

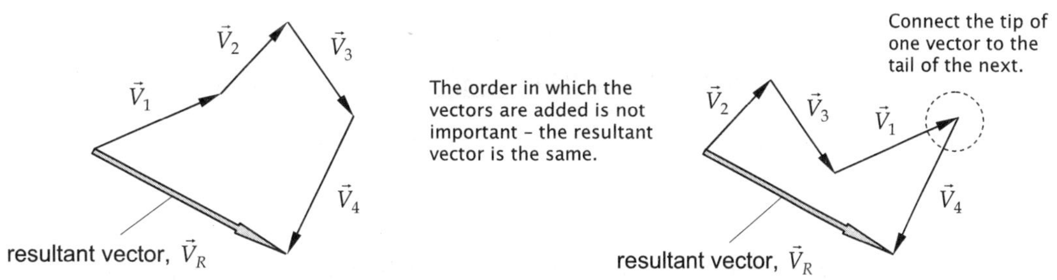

$$\vec{V}_R=\vec{V}_1+\vec{V}_2+\vec{V}_3+\vec{V}_4 \qquad \vec{V}_R=\vec{V}_2+\vec{V}_3+\vec{V}_1+\vec{V}_4$$

Figure 1-4: Addition of two-dimensional vectors using the tip-to-tail method

- When subtracting a vector (say, $\vec{A}$), reverse its direction first (to get $-\vec{A}$) and then add it to the other vector.

$$\vec{B}-\vec{A}=\vec{B}+\left(-\vec{A}\right)$$

reverse its direction and then add! (refers to $+\left(-\vec{A}\right)$)

When subtracting a vector... (refers to $-\vec{A}$)

Test Your Knowledge of Physics 11: Vectors

(Answer Key on p. 365)

1 – 5. True/False questions

1. _____ Temperature is a scalar quantity.
2. _____ Momentum is a vector quantity.
3. _____ Two vectors with the same magnitude and the same direction are identical.
4. _____ When a vector is multiplied by a scalar quantity, the product is another vector quantity.
5. _____ If $\vec{A}+\vec{B}=\vec{0}$, the vectors have the same magnitude but point in opposite directions.

6 – 10. Multiple choice questions

6. Which of the following quantities is <u>not</u> a vector?

   a) displacement b) force
   c) acceleration d) work

7. Which of the following quantities is a vector?

   a) 750 W b) -50 °C
   c) 13 N·s, 45° E of S d) 80 km/h

8. Two vectors whose magnitudes are 3.5 m and 4.5 m, respectively, are added together. Which of the following <u>cannot</u> be the magnitude of the resultant vector?

   a) 0.5 m b) 1.0 m
   c) 3.5 m d) 7.0 m

9. If the vector sum of two vectors is zero, which of the following is <u>false</u>?

   a) Their magnitudes are equal.
   b) Their directions are opposite.
   c) Multiplying one of them by -1 results in the other vector.
   d) They are identical vectors.

10. For the two vectors shown below, which of the following vectors represents the resultant of $\vec{A}$ and $\vec{B}$?

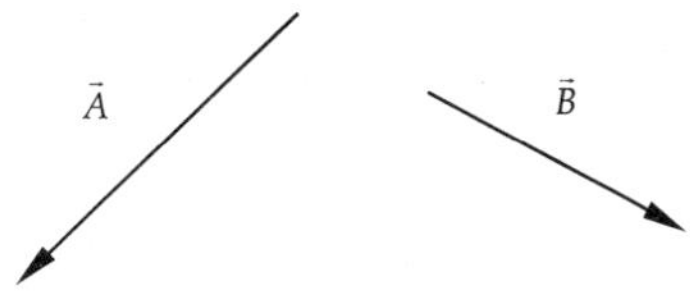

a) 

b) 

c) 

d) 

11 - 15. Written-response questions

11. Given that $\vec{A} = 32$ m, north, and $\vec{B} = 12$ m, south, list the following vectors in the order of increasing magnitude:

$\vec{P} = 3.0 \times \vec{A}$ $\qquad$ $\vec{Q} = -2.5 \times \vec{B}$ $\qquad$ $\vec{R} = \vec{A} + \vec{B}$ $\qquad$ $\vec{S} = \vec{A} - (2.0 \times \vec{B})$

12. Given that $\vec{A} = 3.7$ m, north, and $\vec{B} = 1.5$ m, south, $2.0\vec{A} - x\vec{B}$ is 9.8 m, north. Find the value of the scalar quantity, x.

13. Determine the resultant of the following four vectors: 3.2 km, north; 2.7 km, east; 5.1 km, south; and 2.7 km, west.

14. If $\vec{A} = 5.00$ m, south, and $\vec{B} = 12.0$ m, west, find the magnitude of their vector sum.

15. The vector sum of four vectors is 1.5 m, north. If three of the four vectors are 2.5 m, west; 7.5 m, north; and 3.0 m, south, find the magnitude of the fourth vector.

What's New in Physics 12?

1-1. Resolving a Vector into Components

We know that adding two vectors produces another vector (for example, $\vec{A}+\vec{B}=\vec{C}$), and the reverse is also true; any vector can be broken up or resolved into two other vectors which add to it (for example, $\vec{C}=\vec{A}+\vec{B}$). As you can see in Fig. 1-5, there are numerous possible combinations. However, if the direction of one vector is given,

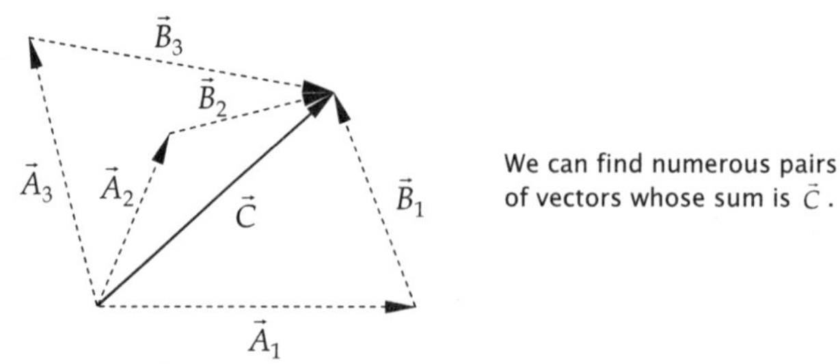

Figure 1-5: A vector represented as the sum of two vectors

and the second vector is to be perpendicular to the first, there is only one possible combination. Then the vector is said to be resolved into two orthogonal (that is, perpendicular) components. Figure 1-6 shows an identical vector ($\vec{v}$) resolved into three different sets of components. Notice in each set that the components (drawn in dashed lines) are perpendicular to each other, and thus form a right triangle with the original vector ($\vec{v}$) as the hypotenuse. Then denoting the components' magnitudes as v_x and v_y, we find the following relationship among the magnitudes of vector $\vec{v}$ and its components:

$$v^2 = v_x^2 + v_y^2 \qquad \ldots (1.1)$$

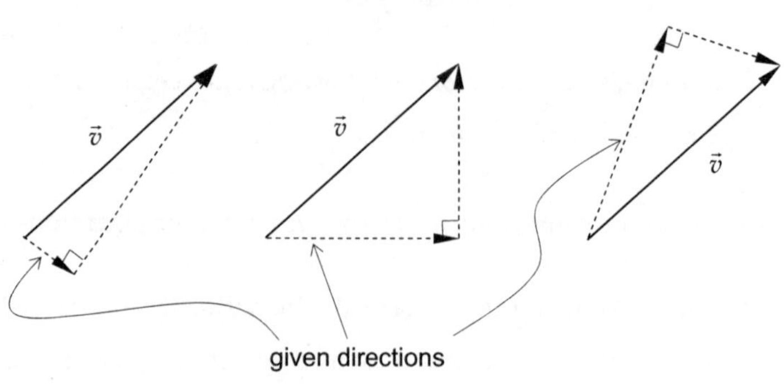

Figure 1-6: Resolving a vector into orthogonal components

It should be clear from Fig. 1-6 and Eq. (1.1) that the magnitude of neither component can be greater than that of the vector itself.

The step-by-step procedure to obtain the components of a vector, say $\vec{V}$, is as follows:

1. Set up the +*x*- and +*y*-directions. In later chapters, when resolving vectors such as velocity and force, you will choose the *x*-axis to be along the surface and the *y*-axis perpendicular to the surface. But it does not really matter as long as the chosen axes are perpendicular to each other.
2. Find the angle between the vector and the *x*-axis, say θ. Then the magnitude of the *x*-component is always $V{\cdot}\cos\theta$, and that of the *y*-component is $V{\cdot}\sin\theta$.
3. Assign proper signs to the components based on the +*x*- and +*y*-directions you set up in Step 1. For example, an *x*-component is positive if it follows the +*x*-direction, and a *y*-component is negative if it points in the opposite direction to the +*y*-direction. Figure 1-7 shows vectors in four different quadrants and the signs of their components.

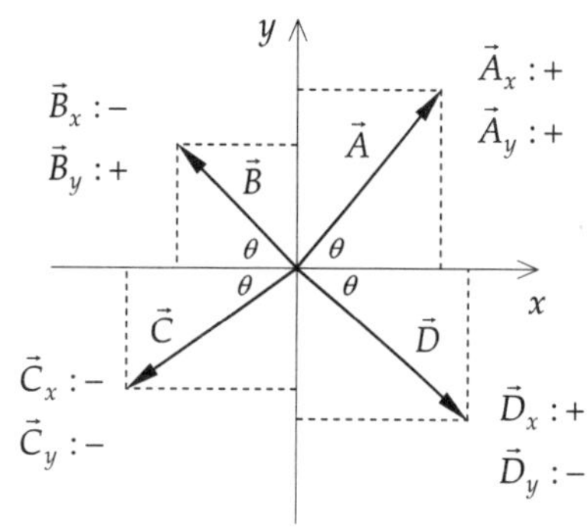

Figure 1-7: Signs of vector components

4. After calculating the components, check your answers using Eq. (1.1).

Let's apply this method to resolve two vectors into their components. Consider the following vectors: $\vec{A}$ = 13 m, 62° N of E, and $\vec{B}$ = 11 m, 53° E of S. The +*x*- and +*y*-directions are already given as shown below.

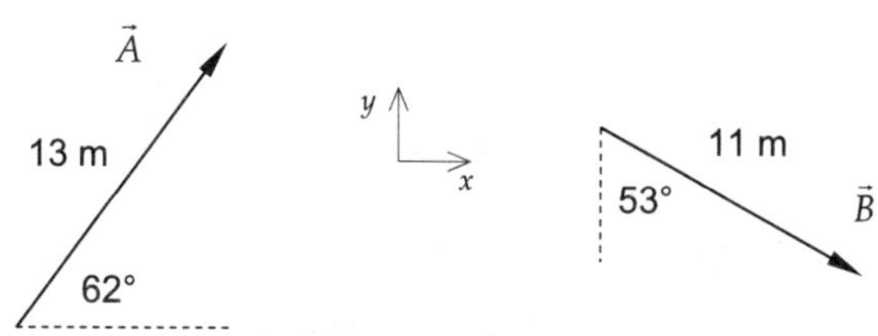

With the +*x*- and +*y*-directions set, we can find the angle between each vector and the *x*-axis. For $\vec{A}$ the angle is 62°, and for $\vec{B}$ it is 90° − 53° = 37°. Hence, the magnitudes of each vector's components are calculated as

x-comp. of $\vec{A}$: $A_x = 13\,\text{m}\cdot\cos 62° = 6.10\text{ m}$ x-comp. of $\vec{B}$: $B_x = 11\,\text{m}\cdot\cos 37° = 8.78\text{ m}$

y-comp. of $\vec{A}$: $A_y = 13\,\text{m}\cdot\sin 62° = 11.5\text{ m}$ y-comp. of $\vec{B}$: $B_y = 11\,\text{m}\cdot\sin 37° = 6.62\text{ m}$

According to the above coordinate system, all the components are positive except the y-component of $\vec{B}$ (which is negative because it points towards south). By attaching proper signs, we can express the components as

x-comp. of $\vec{A}$: $\vec{A}_x = +6.1\text{ m}$ x-comp. of $\vec{B}$: $\vec{B}_x = +8.8\text{ m}$

y-comp. of $\vec{A}$: $\vec{A}_y = +12\text{ m}$ y-comp. of $\vec{B}$: $\vec{B}_y = -6.6\text{ m}$

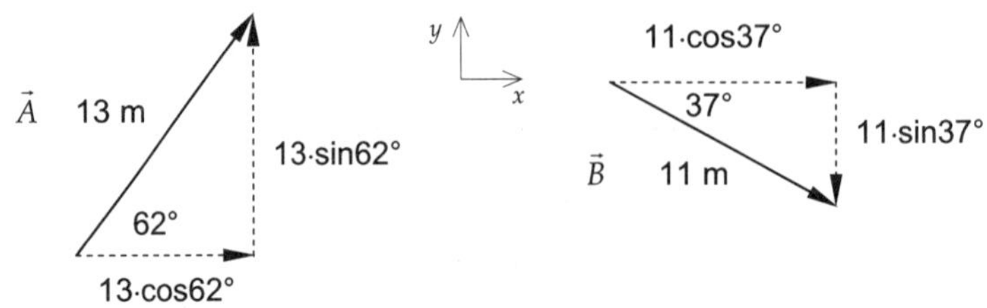

Finally, we make sure that Eq. (1.1) is satisfied for each vector:

$$A = \sqrt{(6.10\,\text{m})^2 + (11.5\,\text{m})^2} = 13.0\text{ m}$$

$$B = \sqrt{(8.78\,\text{m})^2 + (-6.62\,\text{m})^2} = 11.0\text{ m}$$

> Note that the calculations were done using three significant digits. Sometimes, if you do not include a sufficient number of digits, the results may be different from the actual values because of the round-off error.

As you will see throughout this course, a lot of problems involve resolving vectors into their components. So, if you have trouble doing this, you will have trouble solving many types of problems later on. It is therefore absolutely critical that you practice this until you become proficient. Take this advice seriously!

1-2. Addition of Vectors by Components

Adding vectors graphically requires drawing a vector diagram (such as the ones in Fig. 1-4), and it is not easy to get an accurate solution for the vector sum of two-dimensional vectors. Using components to add vectors is a more systematic and accurate method. To understand how it works, consider the addition of the two vectors, $\vec{A}$ and $\vec{B}$, in Fig. 1-8. According to the vector diagram, $\vec{R}$ is the sum of $\vec{A}$ and $\vec{B}$. When all three vectors are resolved into their x- and y-components, we find that the x-component of $\vec{R}$ is the sum of the x-components of $\vec{A}$ and $\vec{B}$; and that the y-component of $\vec{R}$ is the sum of the y-components of $\vec{A}$ and $\vec{B}$. With the components of $\vec{R}$, we can then find its magnitude from Eq. (1.1) and its direction θ by using the tangent ratio:

$$\tan\theta = \frac{R_y}{R_x} \qquad \text{... (1.2)}$$

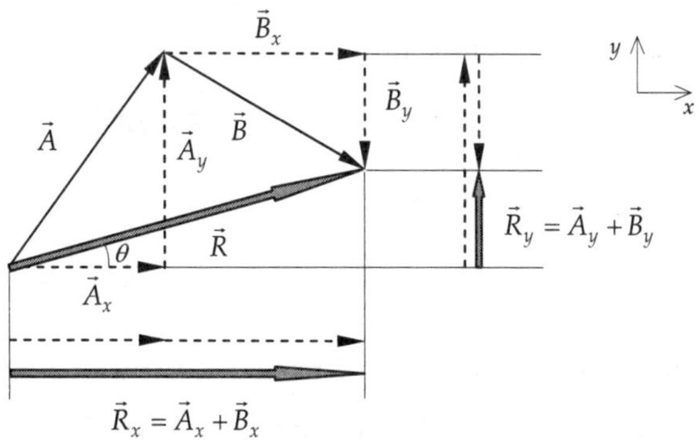

Figure 1-8: Addition of vectors by components

Let's apply this method to add the two vectors ($\vec{A}$ and $\vec{B}$) that we considered in the previous section (p. 11). As you resolve each vector into the x- and y-components, it helps to organize them in a table as shown below:

	x-component	y-component
$\vec{A}$	6.10 m	11.5 m
$\vec{B}$	8.78 m	-6.62 m
$\vec{R}$	6.10 m + 8.78 m = 14.88 m	11.5 m + (-6.62 m) = 4.88 m

The magnitude of the resultant vector ($\vec{R}$) is obtained from Eq. (1.1):

$$R = \sqrt{R_x^2 + R_y^2} = \sqrt{(14.88\,\text{m})^2 + (4.88\,\text{m})^2} = 15.7\ \text{m}$$

The direction (θ) is calculated by using inverse tangent of the ratio of components:

$$\theta = \tan^{-1}\left(\frac{R_y}{R_x}\right) = \tan^{-1}\left(\frac{4.88\,\text{m}}{14.88\,\text{m}}\right) = 18.2°$$

Thus, the vector sum of $\vec{A}$ and $\vec{B}$ is stated as "16 m, 18° north of east."

As you can see from the above example, vector addition is fairly straightforward if you can resolve vectors into their components properly. When adding any number of vectors, you can follow the three-step procedure described in the next page:

1. Resolve <u>each</u> vector into its x- and y-components. Construct a table as shown above to organize the components. Make sure the components have proper signs.

2. Add all the x-components together. The resulting number is the x-component of the resultant vector, R_x. Do the same for the y-components to calculate the y-component of the resultant vector, R_y.

3. The magnitude of the resultant vector is obtained from

$$R = \sqrt{R_x^2 + R_y^2} \quad \ldots (1.3)$$

and the direction θ from

$$\theta = \tan^{-1}\left(\frac{R_y}{R_x}\right) \quad \ldots (1.4)$$

The full description of the direction must include the reference direction as well as the angle from it. For instance, if you say "18° north of east," east is the reference direction, and the direction of the vector is 18 degrees north from that reference direction. Other examples are "25° below the +x-axis" and "33° above the horizontal."

(the +x-axis: reference direction; the horizontal: reference direction)

PHYSICS 12 ANALYZED: VECTORS

1-1. Resolving a Vector into Components

- Two-dimensional vectors can be represented as a vector sum of two orthogonal components.
- To resolve a vector into two orthogonal components,
  1. Set up the +*x*- and +*y*-directions.
  2. Find the magnitude of each component.
  3. Determine the sign of each component.

1-2. Addition of Vectors by Components

- To add vectors by components,
  1. Resolve each vector into its *x*- and *y*-components. ← Organize them in a table!
  2. Add all the *x*-components together and all the *y*-components together.
  3. Find the magnitude and direction of the resultant vector from

$$R = \sqrt{R_x^2 + R_y^2} \text{ and } \theta = \tan^{-1}\left(\frac{R_y}{R_x}\right)$$

EXERCISES ON CHPATER 1: VECTORS

Exercise 1-1

An object is thrown 42.5° above the horizontal at 11.3 m/s as shown right. Resolve the velocity vector using the coordinate system given.

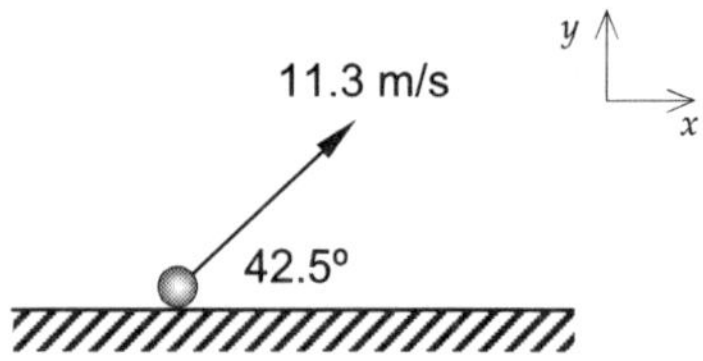

Exercise 1-2

An object traveling in mid-air has the velocity of 7.50 m/s, 35.2° below the horizontal at the position shown. Resolve the velocity vector using the coordinate system given.

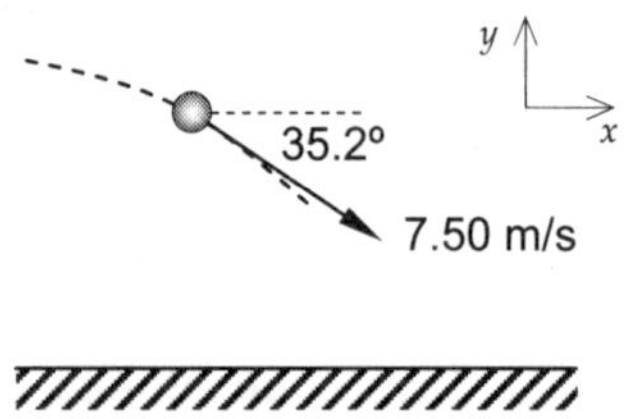

Exercise 1-3

The object in Exercise 1-2 is later seen to be traveling 58.0° below the horizontal. If the x-component of its velocity has remained the same, find the magnitude of the velocity vector.

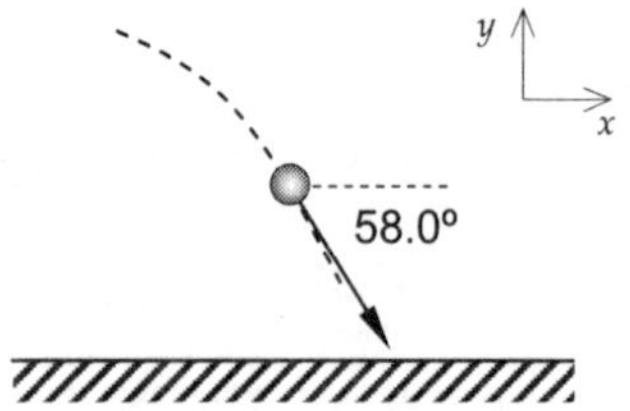

Exercise 1-4

An object is sliding down a 30.0° slope as shown right. Resolve the force of gravity, $\vec{F}_g$, into its components using the coordinate system given. (Note that the x-axis is parallel to the surface, and the y-axis is perpendicular to it.)

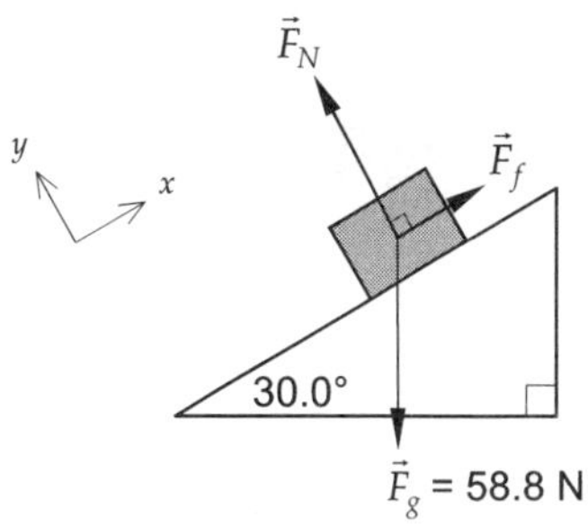

Exercise 1-5

If the vector sum of the three forces in Exercise 1-4 is parallel to the surface, what is the magnitude of the normal force, $\vec{F}_N$?

Exercise 1-6

If the resultant of the three forces in Exercise 1-4 is zero, what is the magnitude of the friction force, $\vec{F}_f$?

Exercise 1-7

Three forces are acting on an object as shown. Resolve all three force vectors using the coordinate system given.

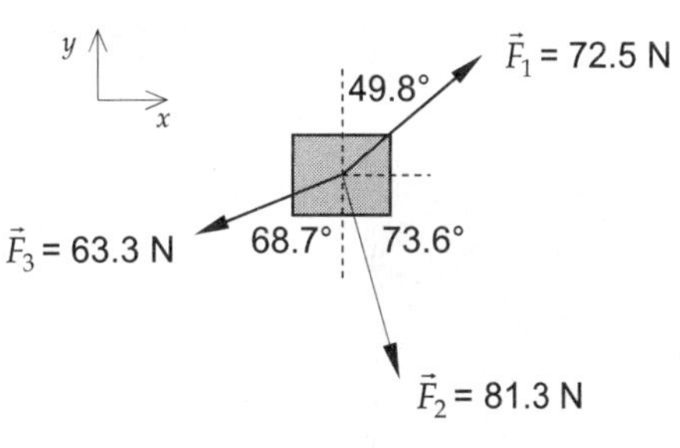

Exercise 1-8

Find the sum of the three force vectors in Exercise 1-7.

Exercise 1-9

An object is pulled by a string, which makes an angle of 28.5° with the ground. The tension force, $\vec{F}_T$, on the object is 103 N. Resolve the tension force using the coordinate system given.

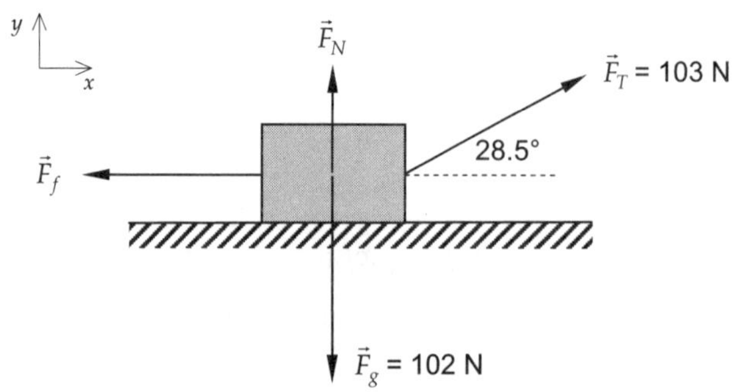

Exercise 1-10

The forces acting on the object in Exercise 1-9 are balanced. That is, the resultant of the forces is zero. Given that the object's weight is 102 N, downward, find the normal force, $\vec{F}_N$, and the friction force, $\vec{F}_f$.

Exercise 1-11

Three identical objects are moving as shown. Resolve the momentum vectors of objects A and B.

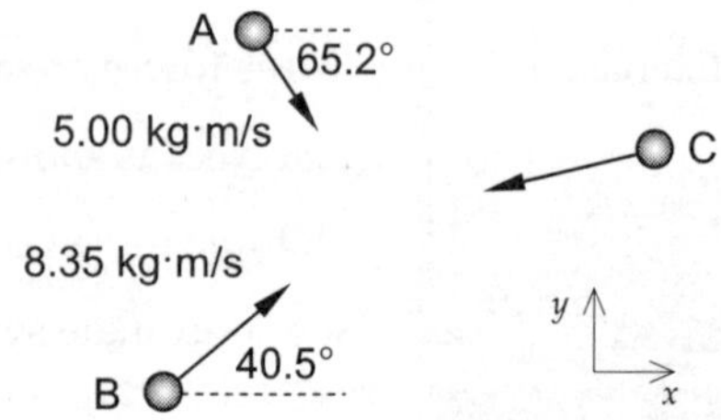

Exercise 1-12

If the vector sum of the three momentum vectors in Exercise 1-11 is zero, find the momentum of object C. (Find the magnitude and direction.)

Exercise 1-13

Given that the three force vectors add up to zero, find the magnitudes of the two unknown forces, $\vec{F}_1$ and $\vec{F}_2$. (Draw a vector diagram using the tip-to-tail method, and the force vectors should form a triangle. Use sine law to find the magnitudes.)

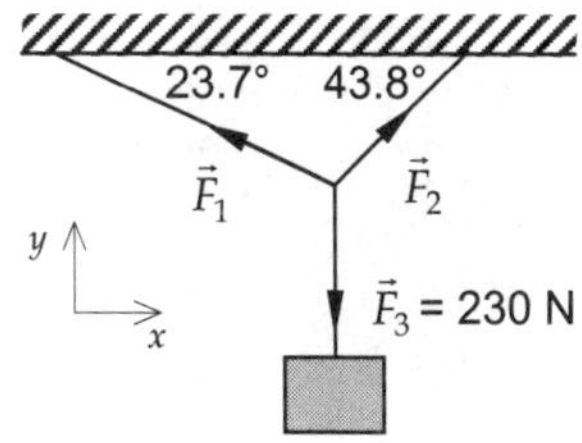

Exercise 1-14

Resolve $\vec{F}_1$ and $\vec{F}_2$ you found in Exercise 1-13 into components using the coordinate system given. Add the three force vectors by components and confirm that their sum is indeed zero.

Exercise 1-15

A long, slender rod is supported by a cable as shown. Resolve the force of gravity, $\vec{F}_g$, and the tension force, $\vec{F}_T$, using the coordinate system given. (Note that the x-axis is parallel to the rod, and the y-axis is perpendicular to it.)

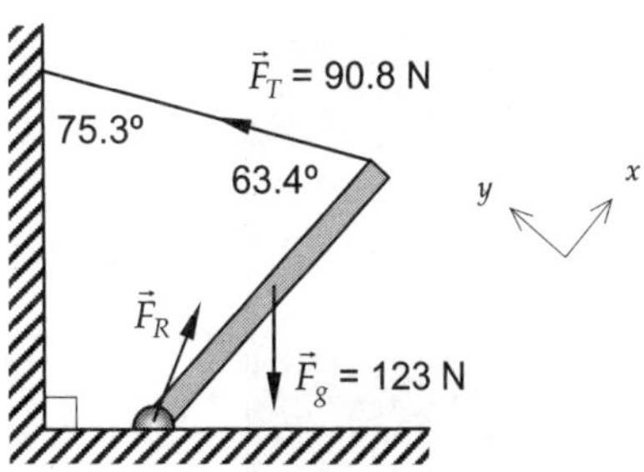

Exercise 1-16

If the vector sum of the forces in Exercise 1-15 is zero, what is the force at the hinge, $\vec{F}_R$? Find its magnitude and direction.

For Exercises 1-17 to 1-21, use the following velocity vectors:

$\vec{A}$ = 2.7 m/s, 27° below $+x$-axis

$\vec{B}$ = 8.1 m/s, 81° above $-x$-axis

$\vec{C}$ = 3.6 m/s, 36° above $+x$-axis

$\vec{D}$ = 5.5 m/s, 42° below $-x$-axis

$\vec{E}$ = 1.7 m/s, along $+y$-axis

Exercise 1-17

Draw the five vectors and resolve each into x- and y-components.

Exercise 1-18

Find the vector sum of $\vec{A}$, $\vec{C}$, and $\vec{E}$.

Exercise 1-19

Perform the following vector calculation: $2.0\vec{B}-\vec{D}+\vec{C}$

Exercise 1-20

Perform the following vector calculation: $\vec{A}-\vec{B}+3.0\vec{C}$

Exercise 1-21

If $a\vec{A}+b\vec{D}=\vec{E}$, find the constants a and b.

PROBLEMS ON CHAPTER 1: VECTORS

(Answer Key on p. 365)

1 - 5. Fill in the blanks

1. A __________ is a quantity with magnitude only.
2. A vector is a quantity possessing both __________ and __________.
3. When two or more vectors are added, the result is another vector called a __________ vector.
4. Two dimensional vectors can be represented as a vector sum of two orthogonal __________.
5. Multiplying a vector by a positive scalar changes its __________ only.

6 - 15. True/False questions

6. _____ The speedometer of a car measures a vector quantity.
7. _____ The magnitude of a vector cannot be negative.
8. _____ The magnitude of a vector is a scalar.
9. _____ If a quantity has a magnitude, it is a scalar.
10. _____ If a vector is added to a scalar, the sum is a vector.
11. _____ If two vectors possess the same magnitude, they are identical vectors.
12. _____ When a vector is multiplied by -1, its direction is reversed with no change in its magnitude.
13. _____ The magnitude of a vector may be smaller than those of its components.
14. _____ If a vector points in the direction of negative x-axis, its y-component is zero.
15. _____ Specifying the direction of a vector as '30° south of east' is the same as specifying it as '60° north of east'.

16 - 22. Multiple choice questions

16. Which of the following pairs contains only vectors?

    a) mass, time
    b) distance, velocity
    c) force, acceleration
    d) energy, power

17. Which of the following statements may be <u>false</u>?

    a) If two vectors point in the same direction, the magnitude of their vector sum is greater than that of either vector.

    b) If two vectors are perpendicular to each other, the magnitude of their vector sum is greater than that of either vector.

    c) If two vectors are in opposite directions, the magnitude of their vector sum is smaller than that of either vector.

    d) If two vectors point in opposite directions, the direction of their vector sum is the same as one of the vectors.

18. For the vector shown below, which of the following diagrams shows a correct way to resolve it into components? (Components are drawn in dotted lines.)

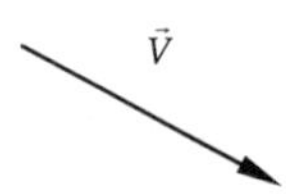

a)
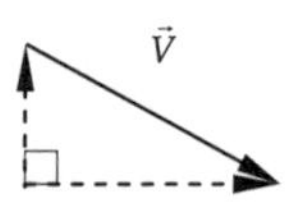

b)
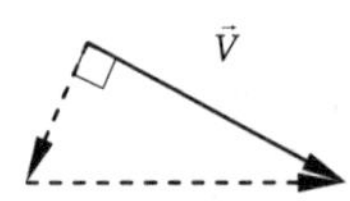

c)
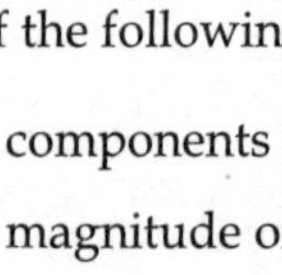

d)
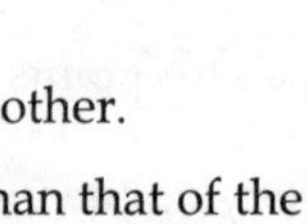

19. A vector is resolved into two components, and neither component is zero. Which of the following statements is <u>false</u>?

    a) The components are perpendicular to each other.

    b) The magnitude of each component is less than that of the vector.

    c) The sum of the components' magnitudes is the magnitude of the vector.

    d) Doubling the components' magnitudes also doubles the vector's magnitude.

20. There are ten vectors of unequal magnitude. Which statement must be true?

    a) There is no pair of vectors whose vector sum is zero.

    b) There is no trio of vectors whose vector sum is zero.

    c) The sum of any pair of vectors is zero.

    d) The sum of any trio of vectors is zero.

21. For the two vectors shown below, which of the following diagrams shows the correct method to find $\vec{A}-\vec{B}$?

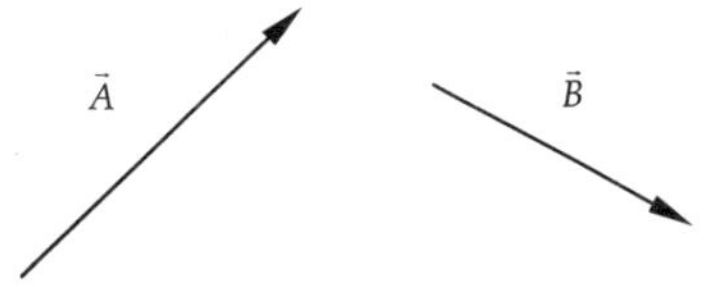

a)

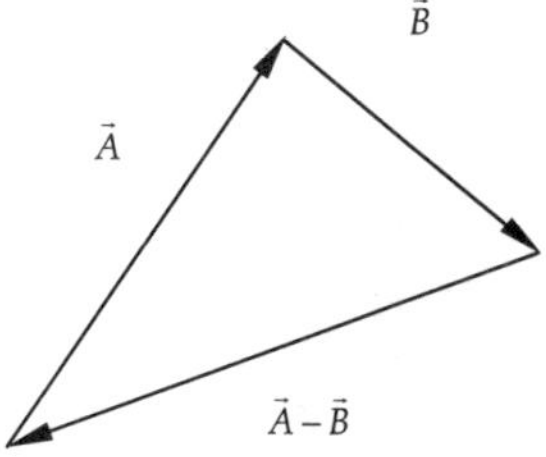

b)

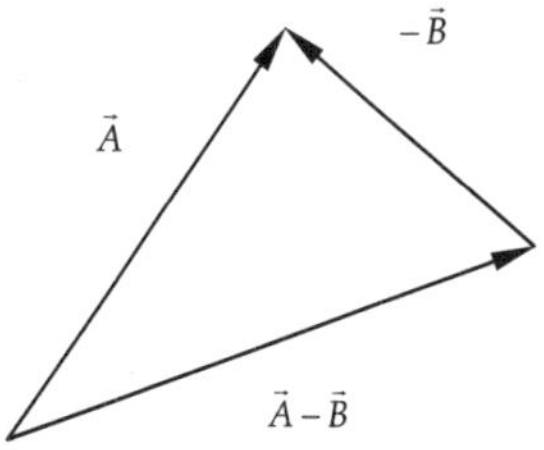

c)

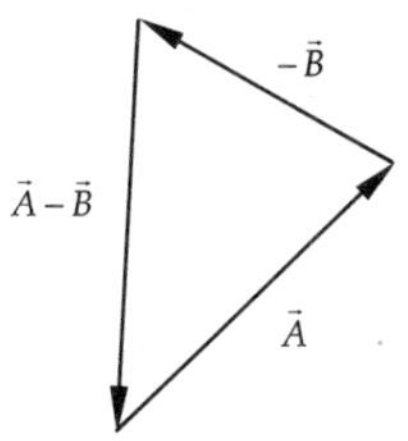

d)

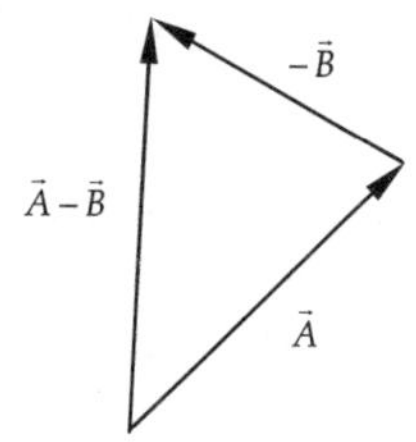

22. The vector sum of three vectors is zero. If the vectors' magnitudes are 3.0 m, 4.0 m, and 5.0 m, respectively, what can we conclude about their directions?

   a) At least two of the vectors are in the same direction.
   b) Exactly two of the vectors are in opposite directions.
   c) None of the vectors are in the same direction.
   d) None of the above

23 – 35. Written-response questions

23. A train is moving at a constant speed of 18 m/s up a hill which is 6.8° to the horizontal. What is the vertical component of the train's velocity?

24. A vector whose magnitude is 9.2 m makes a 130° angle with the +*x*-axis and a 140° angle with the +*y*-axis. Find the vector's *x*- and *y*-components.

25. A vector is given by its components, $A_x = 3.5\,\text{N}$ and $A_y = 5.7\,\text{N}$. What is the magnitude of the vector and the angle it makes with the +*x*-axis?

26. An airplane is climbing at an angle of 18° to the horizontal, with the sun directly overhead. The shadow of the plane is moving across the ground at 200 km/h.

    a) What is the actual speed of the plane?

    b) How long will it take the plane to reach an altitude of 10 km?

27. Resolve the two force vectors shown below into their x- and y-components.

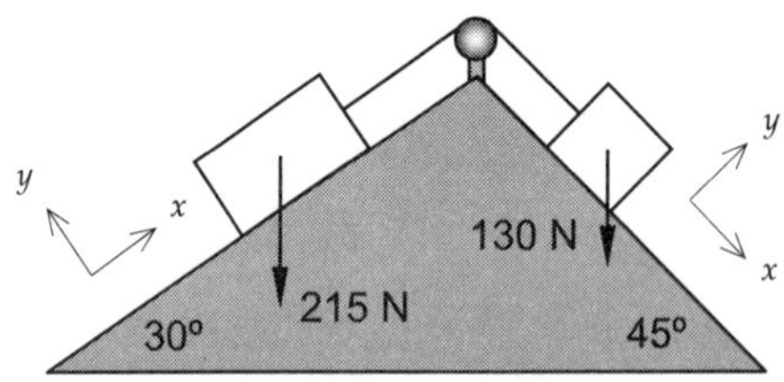

28. Redo Exercise 1-15 and 1-16 using the coordinate system in Exercise 1-9.

29. Find the sum of the following vectors:

    a) $\vec{A}$ = 7.5 m, south; $\vec{B}$ = 2.5 m, west

    b) $\vec{A}$ = 5.0 N, 30° north of east; $\vec{B}$ = 10 N, 45° south of east

    c) $\vec{A}$ = 13.5 m/s, 35.0° above the $+x$-axis; $\vec{B}$ = 23.0 m/s, 15.5° below the $+x$-axis

30. You are on a treasure hunt, and your map says, "Walk due west for 100 paces, then walk 40.0° N of W for 82 paces, and then walk due north for 60 paces." How far north and how far west of your starting point is the treasure buried?

31. Given three vectors $\vec{A}$, $\vec{B}$, and $\vec{C}$, find $-\vec{A}-2.3\vec{B}+3.5\vec{C}$.

    a) $\vec{A}$ = 2.70 m, 25° E of S; $\vec{B}$ = 8.15 m, 85° N of E; $\vec{C}$ = 5.12 m, 30° S of W

    b) $\vec{A}$ = 71 N, 15° above the $+x$-axis; $\vec{B}$ = 33 N, 20° below the $-x$-axis; $\vec{C}$ = 29 N, 48° below the $+x$-axis

32. Three vectors $\vec{A}$, $\vec{B}$, and $\vec{C}$ add together so that the resultant vector is zero. $\vec{A}$ points 75.0° N of E, $\vec{B}$ points due west, and $\vec{C}$ points south and has a magnitude of 185 m/s. Find the magnitudes of $\vec{A}$ and $\vec{B}$. (Draw a diagram.)

33. A sailor leaves a port and travels for one hour and forty minutes at 3.0 km/h, 65° south of west, and then for another two hours at 1.4 km, 75° west of south, to arrive at an island. After resting in the island for a day, the sailor leaves the island and heads back to the port. If he wants to travel straight, in what direction and how far does he need to go?

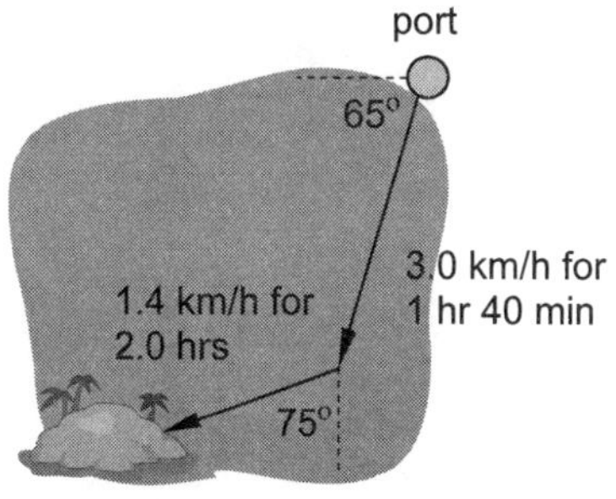

34. What are the x- and y-components of the vector that must be added to the following three vectors so that the sum of the four vectors is zero?

$\vec{A}$ = 115 km, 50° south of west

$\vec{B}$ = 200 km, 40° south of east

$\vec{C}$ = 175 km, 20° north of west

35. A free-body diagram of a long rod is shown below. Given that the sum of the force vectors is zero, find the unknown force, $\vec{F}_R$. (Find both magnitude and direction.)

2

CHAPTER 2

KINEMATICS

Movement and change are two cornerstones of Physics. In Physics 11 you learned how to quantify motion in one dimension, the kinematic equations of motion, and when they can be used. In Physics 12 we look at motion in two dimensions, particularly projectile motion and relative velocity. If you have ever played tennis, golf, or basketball, you certainly know about projectiles and the shape of the curve they make as they fly through the air. If you have ever tried to swim across a river in which a current is flowing, you will have experienced relative motion.

Now we can apply the kinematic equations and vector analysis to investigating these kinds of motion. It is true that we live in a three-dimensional world, but once you learn how to work in two dimensions, it is easy to extend it to three, particularly as what happens vertically is independent of what is happening horizontally. This is an important principle that you will meet in this chapter and throughout this course.

Indeed, the next three chapters build on the work you did in Physics 11 and lay down the most important concepts you need to understand in order to progress in this subject. These concepts include forces and Newton's laws of motion; and the conservation of momentum and energy.

In this chapter, you will:

✓ review kinematics from Physics 11

✓ test your knowledge of kinematics from Physics 11

✓ learn the characteristics of projectile motion and how to analyze it

✓ learn relative motion

✓ solve exercise problems on projectile motion and relative motion

✓ test your knowledge of kinematics from Physics 12

Review of Physics 11: Kinematics

1. Position ($\vec{x}$)

- **Position** is a vector quantity that refers to an object's location. It indicates <u>how far</u> and <u>in what direction</u> an object is located with respect to a reference point.
- An object's position is relative to the location of the observer, as shown in Fig. 2-1.

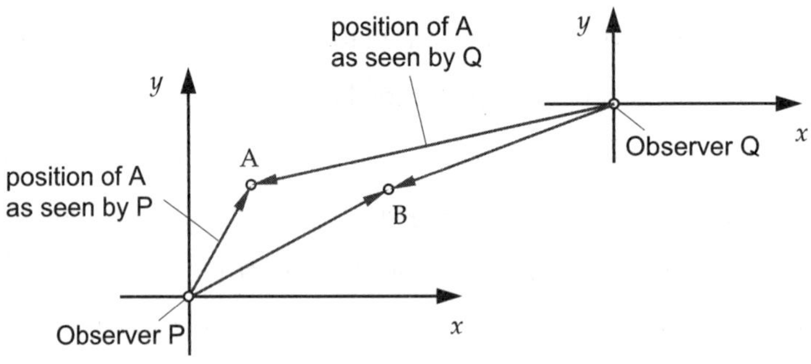

Figure 2-1: Position vectors

2. Displacement ($\vec{d}$) and Distance (D)

- **Displacement** is a vector quantity that refers to a <u>change in</u> position.
- **Distance** is the length of the actual path of travel. It is a scalar quantity.
- We can represent a displacement graphically by drawing a straight arrow from an initial position to a final position. An example is shown in Fig. 2-2, from which we find the following relationship between the position vectors and the displacement vector:

$$\vec{x}_i + \vec{d} = \vec{x}_f \quad [\text{m}] \qquad \text{... (2.1)}$$

We can re-write Eq. (2.1) to express displacement in terms of the initial and final position vectors:

$$\vec{d} = \vec{x}_f - \vec{x}_i = \Delta\vec{x} \quad [\text{m}] \qquad \text{... (2.2)}$$

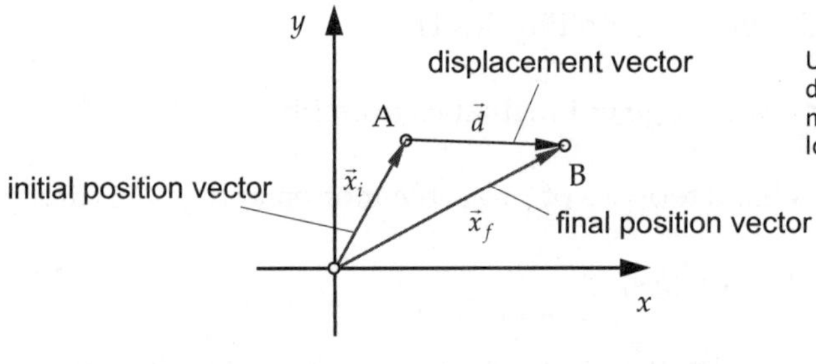

Unlike position vectors, displacement vectors do not depend on the location of the observer.

Figure 2-2: Drawing a displacement vector

- Displacement depends only on the initial and final positions. It does not depend on the actual path of travel, but distance does.

3. Velocity ($\vec{v}$) and Speed (v)

- **Velocity** is a vector quantity that indicates the rate at which an object is changing its position. There are two kinds of velocity: average velocity and instantaneous velocity.
- To find average velocity, measure the displacement (or change in position) of an object and divide it by the time of travel. Mathematically it is expressed as

$$\vec{v}_{\text{avg}} = \frac{\Delta\vec{x}}{\Delta t} = \frac{\vec{d}}{\Delta t} \text{ [m/s]} \qquad \text{... (2.3)}$$

- Instantaneous velocity is the velocity at an instant in time. Its direction is always the same as the direction in which an object is moving at that instant.

  Example: A basketball bouncing on the floor has the instantaneous velocity vector pointing upward while rising and downward while falling.

- **Speed** is a scalar quantity that indicates the rate at which an object is covering the ground. An object's average speed over a period of time is calculated by dividing the distance traveled by the time taken.

> average velocity = displacement / time
> average speed = distance / time

$$v_{\text{avg}} = \frac{D}{\Delta t} \text{ [m/s]} \qquad \text{... (2.4)}$$

- An object's instantaneous speed is the magnitude of its instantaneous velocity.

  Example: If an object's instantaneous velocity is 80 km/h, east, then its instantaneous speed is 80 km/h.

4. Acceleration ($\vec{a}$)

- **Acceleration** is a vector quantity that indicates the rate at which an object is changing its velocity. Mathematically it is written as

> Acceleration is NOT velocity divided by time! It is a change in velocity divided by time.

$$\vec{a} = \frac{\Delta\vec{v}}{\Delta t} = \frac{\vec{v}_f - \vec{v}_i}{\Delta t} \text{ [m/s}^2\text{]} \qquad \text{... (2.5)}$$

($\vec{v}_i$ and $\vec{v}_f$ are the instantaneous velocities at the initial and final times, respectively.)

Example: If an object moving at 25 m/s, north, changes its velocity to 35 m/s, north, in 2.0 seconds, its average acceleration will be

$$\vec{a} = \frac{35\,\text{m/s} - 25\,\text{m/s}}{2.0\,\text{s}} = 5.0 \text{ m/s}^2\text{, north}$$

- The direction of an acceleration vector is <u>independent</u> of the direction in which an object is moving.

  Example: You might think that an object thrown vertically upward experiences an upward acceleration while it is rising. However, the object always experiences a downward acceleration while in the air, whether it is rising or falling.

5. Graphical Analysis of One-Dimensional Motion

- In a position vs. time graph, velocity is the <u>slope</u> of the graph. The sign of the slope (that is, + or –) indicates the <u>direction</u> of the velocity vector, and the steepness of the slope represents the <u>magnitude</u> of the velocity, or speed.

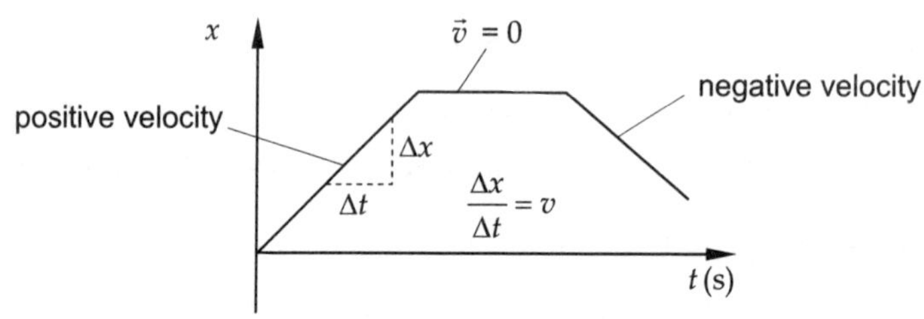

Figure 2-3: Velocity as the slope of a position vs. time graph

- In a velocity vs. time graph, acceleration is the <u>slope</u> of the graph. The sign of the slope (that is, + or –) indicates the <u>direction</u> of the acceleration vector, and the steepness of the slope represents the <u>magnitude</u>.

- In a velocity vs. time graph, the <u>area</u> between the graph and the horizontal axis represents displacement. The area above the axis is considered positive, and the one beneath the axis is considered negative.

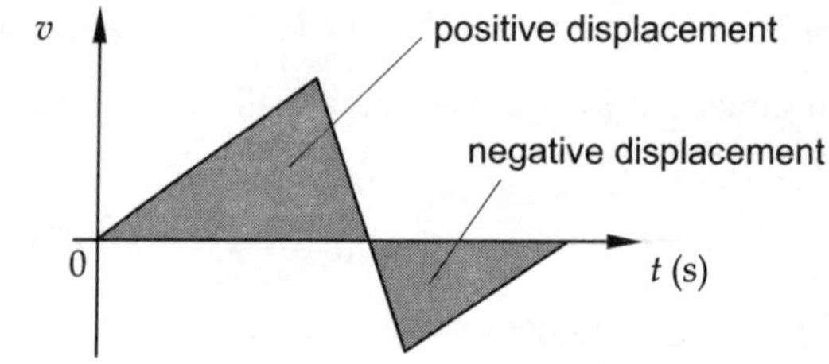

Figure 2-4: Displacement as the area under a velocity vs. time graph

6. Free-Falling Objects

- A **free-falling object** is an object that is under the influence of the gravitational force alone.

All objects traveling through the air are free-falling objects.

- All free-falling objects experience a constant acceleration of 9.8 m/s² towards the center of the Earth.
- Objects moving in mid-air lose speed while rising and gain speed while falling. They have the minimum speed at the highest elevation of the flight.

7. Kinematic Equations

- When acceleration is constant, the velocity vs. time graph is a straight line, and we can derive a set of equations to analyze such a motion. These equations are called the **kinematic equations**:

$$\vec{v}_f = \vec{v}_i + \vec{a}t \qquad [\text{m/s}] \qquad \ldots (2.6)$$

$$\vec{d} = \vec{v}_i t + \frac{1}{2}\vec{a}t^2 \qquad [\text{m}] \qquad \ldots (2.7)$$

$$v_f^2 = v_i^2 + 2\vec{a}\cdot\vec{d} \qquad [\text{m}^2/\text{s}^2] \qquad \ldots (2.8)$$

$$\vec{d} = \vec{v}_{avg}\cdot t = \left(\frac{\vec{v}_i + \vec{v}_f}{2}\right)\cdot t \qquad [\text{m}] \qquad \ldots (2.9)$$

- You may apply kinematic equations to any object experiencing a constant acceleration (whether it is positive, negative, or zero).

8. How to Solve Kinematic Problems

Watch out for the directions of vector quantities. If you can attach proper signs to displacement, velocity, and acceleration, most kinematic problems are fairly straightforward.

1. Read the problem statement carefully and make sure it is about uniformly accelerated motion. Draw a simple diagram that describes the motion, if one is not given.
2. Set the positive direction of motion and determine the values of displacement, velocity, and acceleration based on that direction.
3. Choose an appropriate kinematic equation from Eq. (2.6) to Eq. (2.9). The equation of choice depends on which variable you need to solve for and what data are given to you.
4. Substitute the data into the chosen equation and make sure you have only one unknown. Solve for the unknown. Think about whether your answer makes sense.

Test Your Knowledge of Physics 11: Kinematics

(Answer Key on p. 366)

1 - 10. True/False questions

1. _____ The position of an object depends on the location of the observer.
2. _____ Displacement of a body depends on the actual path of travel.
3. _____ If the displacement of an object is zero, it means the object has not moved.
4. _____ When the displacement of a body is zero, its average velocity is also zero.
5. _____ The velocity of an object may vary even though its speed does not change.
6. _____ The direction of an object's acceleration is always the same as that of its velocity.
7. _____ The slope of a velocity vs. time graph represents displacement.
8. _____ A free-falling body always experiences a constant acceleration of 9.8 m/s^2 towards the ground.
9. _____ The kinematic equations can be applied to analyze the motion of objects moving with constant acceleration.
10. _____ When two objects with different masses are dropped simultaneously from the same height, the heavier object hits the ground first.

11 - 20. Multiple choice questions

11. Which vector quantity is measured in units of m/s^2?

a) position	b) displacement
c) velocity	d) acceleration

12. An object is moving at constant speed along a curved path. Which of the following statements is true?
    a) Its velocity remains constant.
    b) Its velocity constantly changes.
    c) Its velocity changes once in the middle of the curve.
    d) We cannot determine whether its velocity changes unless the exact path of travel is known.

13. Which of the following is true for an object accelerating at 12 m/s^2, north?

    a) The direction of its displacement is north.
    b) The direction of its velocity is north.
    c) Its speed is increasing.
    d) Its instantaneous velocity is changing.

14. Which of the following is true for an object thrown straight up in the air?

    a) The magnitude of its acceleration is 9.8 m/s^2 and the direction is up.
    b) The magnitude of its acceleration is 9.8 m/s^2 and the direction is down.
    c) There is no acceleration.
    d) The acceleration cannot be determined unless we know its initial speed.

15. An object is thrown vertically upward at 23 m/s from the ground and falls back to the ground a short time later. Which of the following statements is false?

    a) The displacement of the object is zero.
    b) The final speed of the object at the ground is equal to its initial speed.
    c) The final velocity of the object at the ground is equal to its initial velocity.
    d) The final acceleration of the object at the ground is equal to its initial acceleration.

For Problems 16 to 18, refer to the graphs shown below:

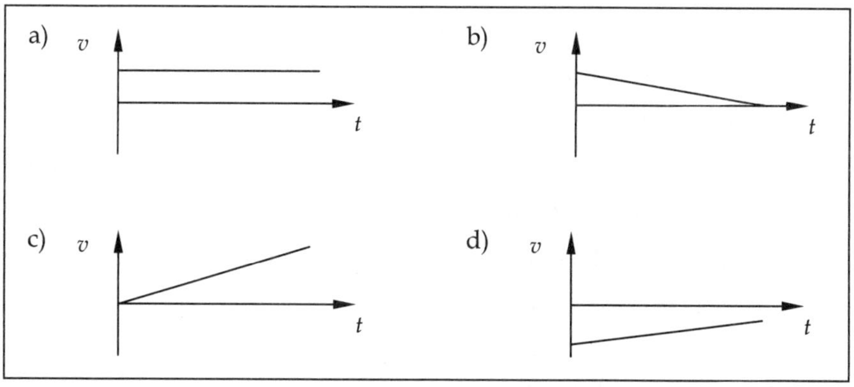

16. Which graph shows a displacement in the negative direction?

17. Which graph shows an object's speed increasing?

18. Which graph shows an acceleration in the negative direction?

19. Which of the following position vs. time graphs represents the motion of an object thrown vertically upward?

a) 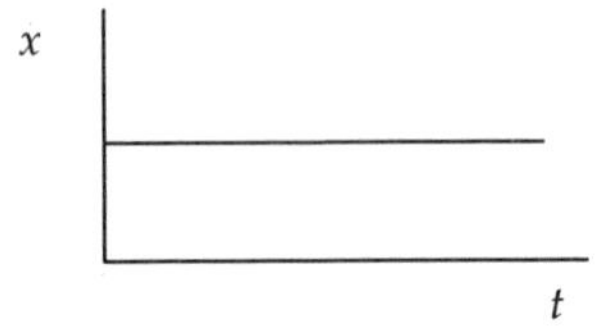

b) 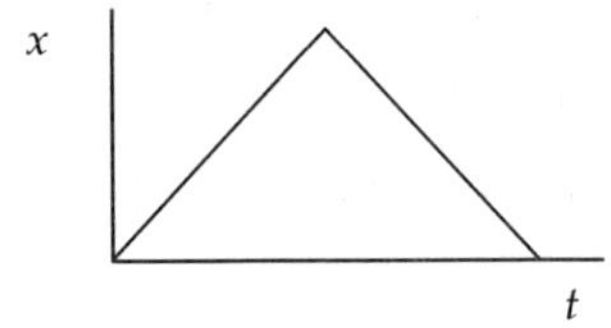

c) 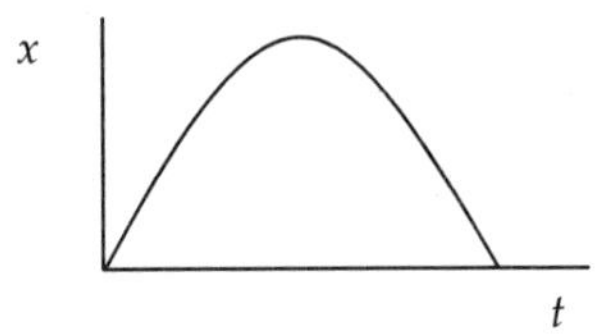

d) 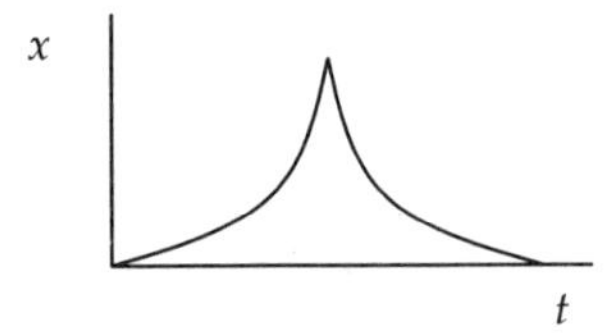

20. Which of the following describes the motion shown in the position vs. time graph shown?

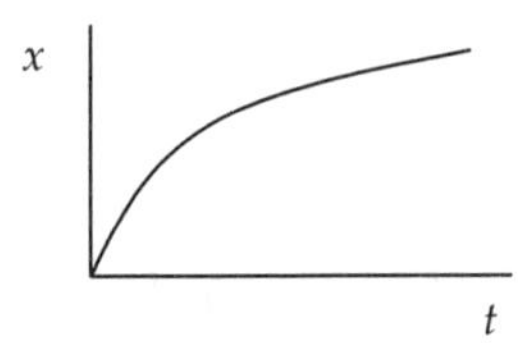

a) The displacement is constant.
b) The velocity is constant.
c) The velocity is changing.
d) The acceleration is zero.

21 – 35. Written-response questions

21. Find the displacement of each object:
   a) A basketball, dropped from 1.5 m above the floor, bounces a few times and comes to rest on the floor.
   b) A coin, tossed vertically upwards from 0.4 m above the ground, is caught on its way down at 0.9 m above the ground.
   c) Tommy walks 1.6 km to school in the morning and comes back home after school via the same route.

22. A Ferris wheel with a radius of 55 m is rotating at a rate of 0.25 revolutions per minute. If Becky gets on the Ferris wheel at its lowest point, calculate (a) her displacement and distance of travel after half a revolution and (b) her average velocity and average speed after half a revolution.

23. Jim is driving west at 60 km/h when he remembers that he has left his physics homework at home. Within 30 seconds he is driving east at 40 km/h. Find the average acceleration of Jim's car. (Be careful of the units!)

24. A car moves east at 20 m/s for 10 seconds, then accelerates at 1.2 m/s^2 east for 12 seconds, and finally accelerates at 4.8 m/s^2 west for 15 seconds. Draw a velocity vs. time graph to illustrate this motion. What is the final velocity of this car?

25. The velocity vs. time graphs of two objects in one-dimensional motion are shown below.

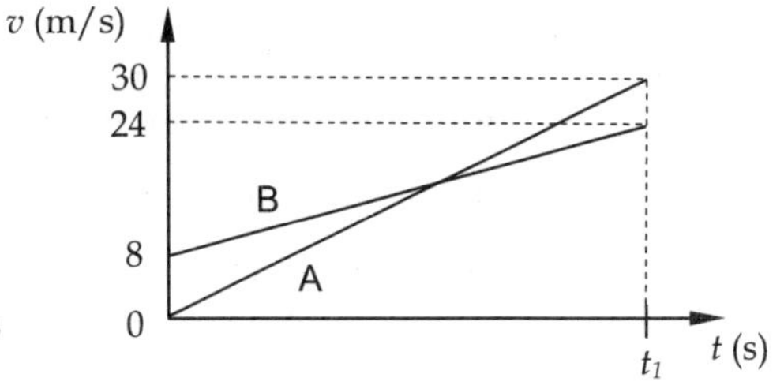

a) Compare the displacements of the objects between $t = 0$ and $t = t_1$.

b) Compare the average velocities of the objects between $t = 0$ and $t = t_1$.

c) Compare the instantaneous velocities of the objects at $t = t_1$.

d) Compare the average accelerations of the objects between $t = 0$ and $t = t_1$.

26. An object initially moving to the left at 22 m/s accelerates uniformly for 8.0 seconds, by which time it is traveling at 45 m/s to the right.

a) What was the average acceleration of the object?

b) How long did it take for the object to come to rest momentarily?

c) What was the total displacement of the object during the 8.0 second period? What about its distance of travel?

27. A ball is thrown vertically upwards at 25 m/s. Sketch a velocity vs. time graph describing the motion of this ball for the next 4.0 seconds. What is the ball's velocity after 1.5 s and 3.5 s?

28. A car traveling at 60 km/h strikes a tree. The front end compresses, and the driver comes to rest after traveling 0.70 m. Calculate the average acceleration of the driver during this collision. Express your answer in m/s^2 and in g's, where $g = 9.8\ m/s^2$.

29. A truck leaves skid marks for 32 m on a highway as it comes to a halt. Assuming an average deceleration of 10 m/s^2, estimate the speed of the truck before braking.

30. A helicopter is descending vertically with a speed of 6.00 m/s. At a height of 120 m above the ground, a package is dropped to a man stranded on an island directly below. How long will it take to reach him?

31. A train travels 68 km at an average velocity of 88 km/h east, waits for 10.0 minutes at a station, and then runs for half an hour at an average velocity of 72 km/h east. Calculate (a) the total displacement of the train and (b) the average velocity for the whole trip.

32. A subway train starts from rest at a station and accelerates uniformly at 1.8 m/s^2 for 15 seconds. Then it runs at a constant speed for the next 35 seconds before decelerating at 2.0 m/s^2 until it comes to rest at the next station. Find the distance between the stations and the average speed of the train.

33. An object accelerates uniformly from rest at 2.5 m/s^2 for 2.2 seconds. If a second object is to cover the same distance within 1.7 seconds with the initial velocity of 3.2 m/s, what should be its minimum acceleration?

34. An object is dropped from the observation deck of the CN Tower, located 342 m above the ground. Ignoring air resistance, find (a) the displacement after 3.5 s; (b) the time it takes the object to travel half the distance to the ground; and (c) the average velocity for the entire flight.

35.* A coin is tossed vertically upward at 21 m/s. Find the distance the coin travels <u>during the third second</u> of its flight.

What's New in Physics 12?

2-1. Projectile Motion

A **projectile** is an object that is thrust into the air. It experiences the force of gravity and no other force.† In short, it is a free-falling body. The path a projectile takes during its flight is called its **trajectory**, and its shape is part of a parabola that opens downward, as shown in Fig. 2-5. This curved path is the result of two independent motions, a horizontal motion with a uniform velocity (because of no horizontal force) and a vertical motion with a constant acceleration due to the gravitational force.

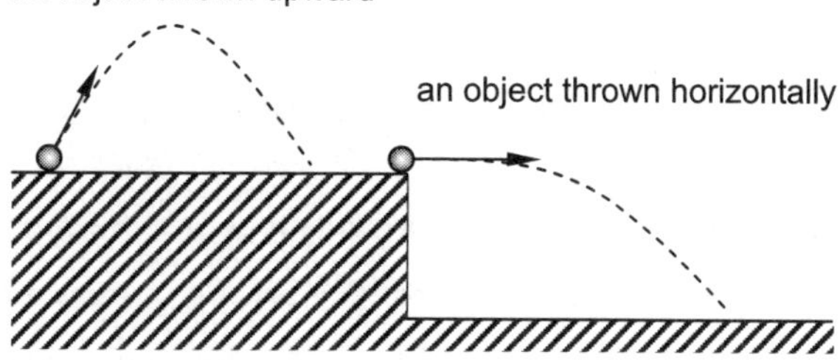

Figure 2-5: Examples of two-dimensional projectile motion

Since a projectile's acceleration is constant, we may apply kinematic equations to analyze its motion. Consider the trajectory in Fig. 2-6, which shows a projectile's initial velocity ($\vec{v}_i$), its velocity after some time ($\vec{v}_f$), the displacement ($\vec{d}$), and the acceleration ($\vec{g}$). (Note that a projectile always accelerates vertically downward at 9.8 m/s^2.) Resolving each vector into components and putting them into Eqs. (2.6) to (2.9) (on p. 31), we can find some important characteristics of projectile motion.

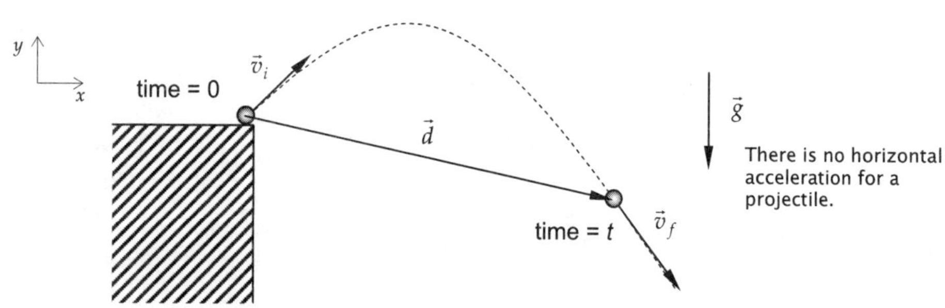

Figure 2-6: Analysis of projectile motion

† The force of air resistance is ignored in Physics 12.

For instance, Fig. 2-7, which represents Eq. (2.6) graphically with the above vectors, shows that the horizontal component of the velocity remains constant. In fact, this is true for the entire flight. That is, at any point in its trajectory, the horizontal component of a projectile's velocity is the same as that of its initial velocity. On the other hand, the vertical component changes as a result of downward acceleration. The way it changes is identical to the way the velocity changes for an object thrown up vertically. Namely, its magnitude decreases at the rate of 9.8 m/s per second while rising and increases by 9.8 m/s in each second while falling.

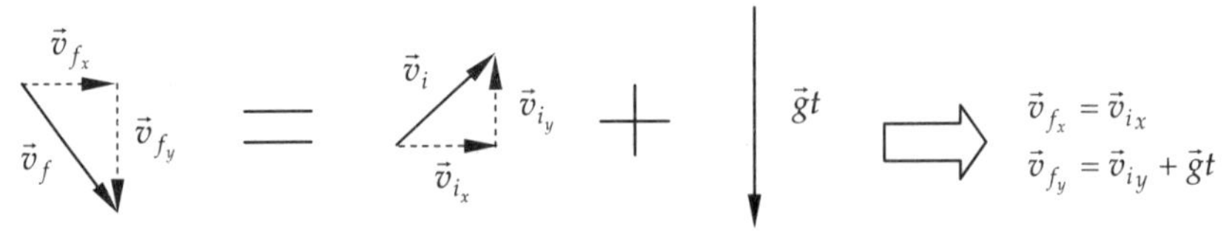

Figure 2-7: Vector diagram for $\vec{v}_f = \vec{v}_i + \vec{a}t$

Considering both velocity components together, we can see that a projectile loses speed while ascending and gains speed while descending (see Fig. 2-8). Indeed, the speed of a projectile depends only on the elevation, and the higher the elevation of a projectile, the lower its speed. Consequently, the speed becomes a minimum at the highest altitude, where the vertical component of the velocity disappears and only the horizontal component remains. For an object thrown either horizontally or downward, its speed during the entire flight will always be greater than the initial speed.

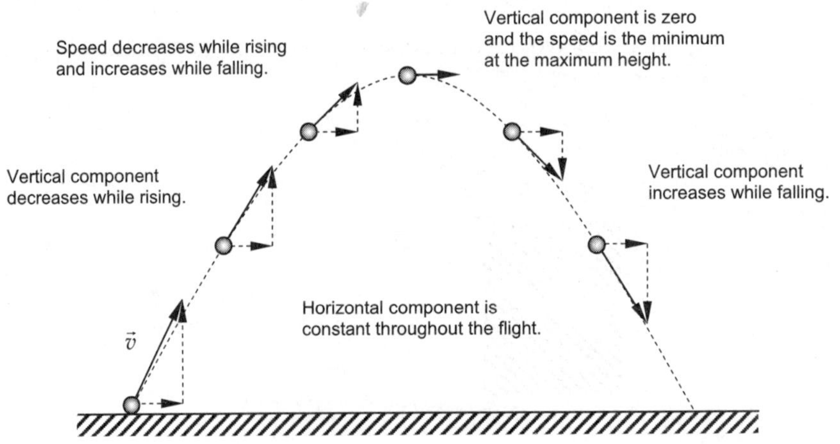

Figure 2-8: Characteristics of projectile motion

When solving for a projectile's velocity as it hits the ground, be careful not to assume it is zero. You must find the velocity just before the impact with the ground, not afterwards. Once a projectile touches the ground, it experiences a force exerted by the ground and is no longer a projectile.

2-2. Kinematic Equations for Projectile Motion

As you saw in the previous section, the horizontal motion of a projectile is independent of its vertical motion, and vice versa. That is, what happens vertically (because of the gravitational pull) has no effect on the projectile's horizontal velocity. Therefore, we may derive two sets of equations, one for the horizontal motion and the other for the vertical motion.

In the previous section, you learned that the horizontal component of a projectile's velocity is constant throughout the flight. Thus,

$$\vec{v}_{f_x} = \vec{v}_{i_x} \qquad \text{... (2.10)}$$

Then the horizontal component of a projectile's displacement, or the projectile's **range**, is simply the product of the horizontal component of its velocity and the time of flight:

$$\vec{d}_x = \vec{v}_{i_x} t \qquad \text{... (2.11)}$$

We can confirm Eq. (2.11) in Fig. 2-9, which represents Eq. (2.7) with the vectors in Fig. 2-6. The horizontal component of the left-hand side ($\vec{d}_x$) is equal to the horizontal component of the vector sum of the right-hand side, which is $\vec{v}_{i_x} t$.

The horizontal motion of a projectile is described by Eqs. (2.10) and (2.11). They are simple because of the absence of any horizontal acceleration.

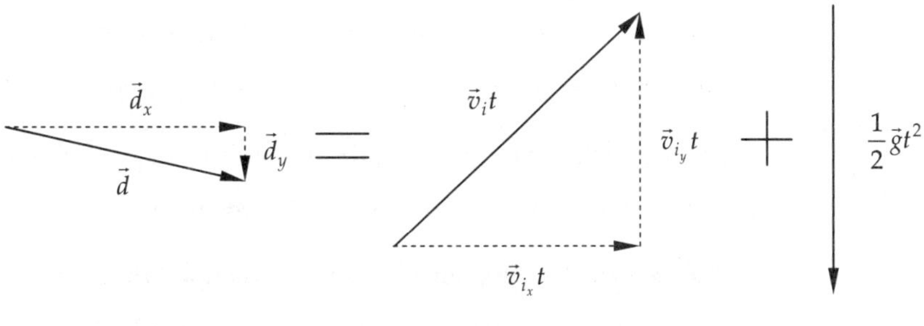

Figure 2-9: Vector diagram for $\vec{d} = \vec{v}_i t + \frac{1}{2}\vec{a}t^2$

As for the vertical motion, the kinematic equations are somewhat more complicated. In Figs. 2-7 and 2-9, the sums of the vertical components yield the following equations:

$$\vec{v}_{f_y} = \vec{v}_{i_y} + \vec{g}\,t \qquad \text{... (2.12)}$$

$$\vec{d}_y = \vec{v}_{i_y} t + \frac{1}{2}\vec{g}\,t^2 \qquad \text{... (2.13)}$$

When we solve Eq. (2.12) for t and substitute the resulting expression into Eq. (2.13), we arrive at the following equation:

$$v_{f_y}^2 = v_{i_y}^2 + 2\vec{g}\cdot\vec{d}_y \qquad \text{... (2.14)}$$

Finally, solving Eq. (2.12) for $\vec{g}$ and substituting the resulting expression into Eq. (2.13), we find an equation without the gravitational acceleration:

$$\vec{d}_y = \left(\frac{\vec{v}_{i_y} + \vec{v}_{f_y}}{2}\right)\cdot t = \vec{v}_{y_{\text{avg}}}\,t \qquad \text{... (2.15)}$$

2-3. How to Solve Projectile Motion Problems

When solving projectile motion problems, you will be using the kinematic equations derived in the previous section. In general, the problems ask for the time of flight (t), the range of the projectile (d_x), the velocity some time after it was launched ($\vec{v}$), and so on. Because there are a number of equations to choose from, many students feel that choosing the right equation(s) is the most important part of solving the problem. Although this is true, many students do not recognize another crucial factor: attaching the proper signs to the vector quantities in the kinematic equations. Never forget that kinematic equations are vector equations. Therefore, you need to pay close attention to the direction of these vectors. As long as you do that, you should be able to handle most projectile motion problems with ease. The following six-step solution procedure is suggested:

1. Read the problem carefully and draw a simple diagram that describes the motion of the projectile, if one is not already given.
2. Set the +x- and +y-directions. It is recommended that you always set the +y-direction to be upward. (So, the value of the gravitational acceleration ($\vec{g}$) becomes -9.8 m/s^2.)

3. Resolve all the vectors given in the problem into components using the coordinate system you set in Step 2.

For resolving a vector into components, see Section 1-1, p. 10.

IMPORTANT! 4. Write down everything you know from the problem statement, making sure the vector components have proper signs.

5. Select from Eqs. (2.10) to (2.15) an equation that contains the quantity you need to solve for. For example, if you need to find the time of flight (t), choose one from Eqs. (2.11), (2.12), and (2.13). Which equation you actually choose depends on the given data. If you are looking for the range (d_x), choose Eq. (2.11). If you are asked to find the velocity ($\vec{v}$), first get its components separately. Then the magnitude and direction of the velocity can be obtained from

$$v = \sqrt{v_x^2 + v_y^2} \qquad \ldots (2.16)$$

$$\theta = \tan^{-1}\left(\frac{v_y}{v_x}\right) \qquad \ldots (2.17)$$

6. Remind yourself of the characteristics of projectile motion, covered in Section 2-1 (p. 37), and think about whether your answer makes sense. For instance, the time of flight or the speed can <u>never</u> be negative, and, if the answer indicates that the projectile is rising when you think it should be falling, you should suspect a mistake in the calculation.

Let's apply the above solution procedure to an example problem. Suppose in Fig. 2-6 (p. 37), the initial velocity ($\vec{v}_i$) is 47 m/s, 44° above the horizontal, and the ball is traveling at 55 m/s at time t. Here is how to obtain the time t:

1. The picture that describes the ball's motion is given in Fig. 2-6.
2. Use the coordinate system shown in Fig. 2-6. That is, $+x$-direction is to the right, and $+y$-direction is upward.
3. Resolve the initial velocity vector as shown below:

$$\vec{v}_{i_x} = 47\,\text{m/s} \cdot \cos 44° = 33.8\,\text{m/s}$$

$$\vec{v}_{i_y} = 47\,\text{m/s} \cdot \sin 44° = 32.6\,\text{m/s}$$

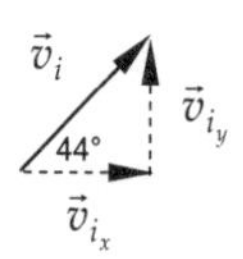

4. The horizontal component of the ball's velocity stays constant, so $\vec{v}_{f_x} = \vec{v}_{i_x} = 33.8\,\text{m/s}$. The ball's speed at time t, v_f, is 55 m/s. Lastly, the

acceleration of the ball is -9.8 m/s². It is negative because the vector is directed downward. We list all known quantities in the following table:

$\vec{v}_{i_x}$	$\vec{v}_{i_y}$	$\vec{v}_{f_x}$	$v_f = \sqrt{v_{f_x}^2 + v_{f_y}^2}$	$\vec{g}$	t
33.8 m/s	32.6 m/s	33.8 m/s	55 m/s	-9.8 m/s²	?

$$v_{f_y} = \sqrt{v_f^2 - v_{f_x}^2} = \sqrt{(55\,\text{m/s})^2 - (33.8\,\text{m/s})^2} = 43.4\ \text{m/s}$$

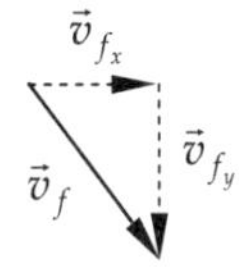

With v_{f_x} and v_f, we can calculate v_{f_y} as shown above.

Then $\vec{v}_{f_y} = -43.4$ m/s (negative because it points downward).

(Although the problem did not state that the ball was falling, you can figure out that it was falling because its speed would be less than 47 m/s if it were still rising.)

5. Since we do not know $\vec{d}_x$ or $\vec{d}_y$, the only equation we can use to solve for t is Eq. (2.12). Substitute the numbers into the equation and solve for t:

$$-43.4\,\text{m/s} = 32.6\,\text{m/s} + \left(-9.8\,\text{m/s}^2\right)\cdot t \implies t = 7.76\ \text{s} \cong 7.8\,\text{s}$$

Hence, the ball spends 7.8 s in the air before its speed reaches 55 m/s. Now, let's find out the displacement of the ball during that 7.8 s. To obtain the horizontal component of the displacement, use Eq. (2.11):

$$\vec{d}_x = 33.8\,\text{m/s}\cdot(7.76\,\text{s}) = 262\ \text{m}$$

As for the vertical component, we may use either Eq. (2.13) or (2.14). Choose Eq. (2.14) and solve for $\vec{d}_y$:

$$(-43.4\,\text{m/s})^2 = (32.6\,\text{m/s})^2 + 2\left(-9.8\,\text{m/s}^2\right)\vec{d}_y \implies \vec{d}_y = -41.8\ \text{m} \cong -42\,\text{m}$$

The negative sign in the solution indicates that the ball has fallen 42 m below the initial altitude. The magnitude and direction of the displacement is then

$$d = \sqrt{d_x^2 + d_y^2} = \sqrt{(262\,\text{m})^2 + (42\,\text{m})^2} = 265\ \text{m} \cong 270\ \text{m}$$

$$\theta = \tan^{-1}\left(\frac{d_y}{d_x}\right) = \tan^{-1}\left(\frac{42\,\text{m}}{262\,\text{m}}\right) = 9.05° \cong 9.1°$$

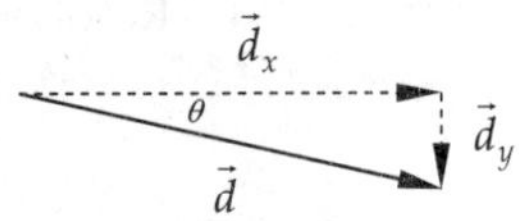

Thus, the displacement of the ball is 270 m, 9.1° below the horizontal.

2-4. Relative Velocity

The reported velocity of an object depends on the motion of the observer. For instance, if you are driving a car at 50 km/h, you will perceive that the road in front of you is moving toward you at 50 km/h. However, someone standing on the road will state that you are the one that is moving. Who do you think is right? Before answering, think about another example. You might think that you are not moving when you are sitting on a chair reading this book, but an astronaut in outer space will argue that you are indeed moving—and moving fast!—as the Earth rotates. In these examples, everyone is right. They are describing the motion that they see. Therefore, we need to specify which observer or object the motion is relative to (for example, an airplane flying at 950 km/h relative to the ground or a boat cruising at 20 km/h relative to the river current).

Let's study this concept of relative velocity further using vector analysis. First, we denote the velocity of a as seen by b as ${}_a\vec{v}_b$ (thus, b is observing the motion of a). Then from the discussion in the previous paragraph, we can derive the following equation:

$$ {}_a\vec{v}_b = -{}_b\vec{v}_a \quad \text{... (2.18)} $$

In other words, the speed of a as seen by b is the same as the speed of b as seen by a, but the directions of motion are exactly opposite.

With that in mind, all the problems about relative velocity can be solved easily by applying the following relationship between relative velocities:

$$ {}_a\vec{v}_b + {}_b\vec{v}_c = {}_a\vec{v}_c \quad \text{... (2.19)} $$

Note that the order of subscripts is important. The second subscript in the first vector must match the first subscript in the second vector, and then disappears in the resultant vector. (In Eq. (2.19) the subscript b vanishes in the resultant vector.) This is called the **'chain-link' method** of combining vectors to eliminate one of the subscripts.

For one-dimensional motions, you can find the relative velocities easily without Eq. (2.18) or (2.19). Consider the following example. Cathy is standing on the ground and Ken is riding a tricycle at 4 m/s relative to the ground, as shown in Fig. 2-10. Then, relative to Ken, Cathy is approaching him at 4 m/s while Cathy sees Ken approaching her at the same speed. Using the vector convention introduced above, this can be written as

$${}_C\vec{v}_K = -{}_K\vec{v}_C \qquad \text{(}C\text{ for Cathy, }K\text{ for Ken)}$$

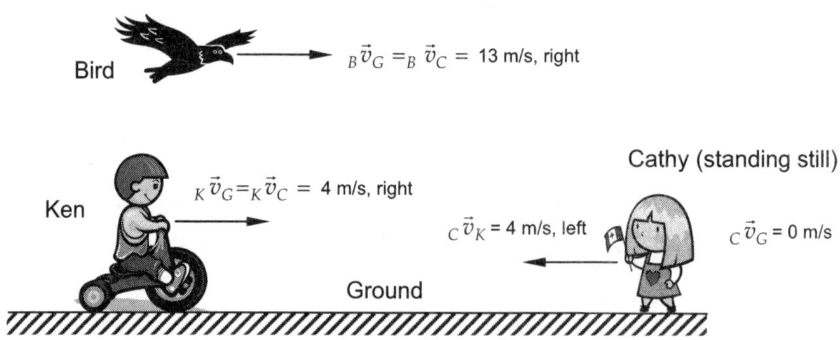

Figure 2-10: Relative velocity for one-dimensional motion

Now, if Cathy sees a bird flying at 13 m/s in the same direction as Ken is moving, how would Ken describe the bird's motion? To him, the bird appears to be flying at 9 m/s, away from him. Most students can find this answer without using any of the above vector equations, but let's confirm that we will obtain the same answer by using them. What we need to find is the velocity of the bird (*B*) relative to Ken (*K*), which, according to Eq. (2.19), can be written as

$${}_B\vec{v}_K = {}_B\vec{v}_G + {}_G\vec{v}_K \qquad \text{(}G\text{ for Ground)}$$

These subscripts should match!

Because ${}_G\vec{v}_K = -{}_K\vec{v}_G$, we can re-write the above equation as

$${}_B\vec{v}_K = {}_B\vec{v}_G - {}_K\vec{v}_G$$

Since ${}_B\vec{v}_G$ is 13 m/s, right, and $-{}_K\vec{v}_G$ is 4 m/s, left, the resultant vector, ${}_B\vec{v}_K$, is 9 m/s, right. So we do get the same answer, but the former approach of relying on your intuition took a lot less work and time. Then you might wonder why we should consider using these equations in the first place. Well, consider the following example of two-dimensional motion.

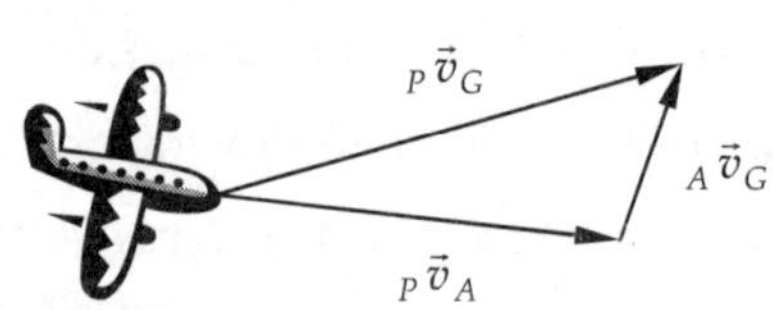

Figure 2-11: Relative velocity for two-dimensional motion

Suppose the airplane in Fig. 2-11 is flying at 240 km/h, 6.0° S of E, through the air. (The plane's speed with respect to the air is called its airspeed.) However, when seen from the ground, the plane is moving at 295 km/h, 15° N of E, because of the wind. (The plane's speed with respect to the ground is called its ground speed.) Then what is the velocity of the wind (or the air) relative to the ground? Although you can guess the general direction of the wind, it is not easy to calculate the magnitude and direction of its velocity in your head. We use Eq. (2.19) instead to set up the following equation:

$${}_P\vec{v}_A + {}_A\vec{v}_G = {}_P\vec{v}_G \qquad (A \text{ for air, } G \text{ for ground, } P \text{ for plane})$$

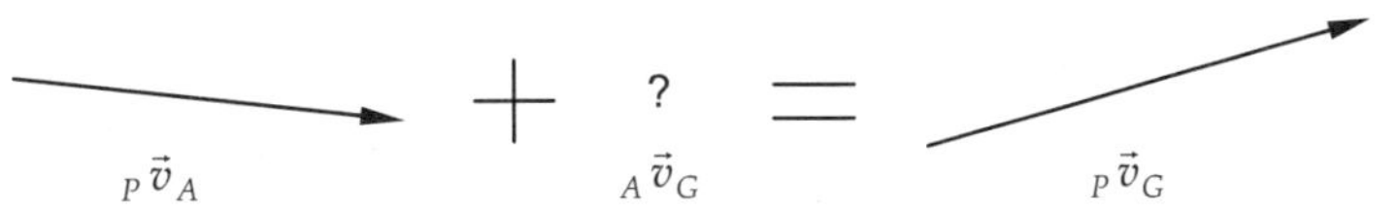

Thus,

$${}_A\vec{v}_G = {}_P\vec{v}_G - {}_P\vec{v}_A$$

Set the +x- and +y-directions to be east and north, respectively, and resolve each vector into their components:

	angle with x-axis	x-component (km/h)	y-component (km/h)
$_P\vec{v}_G$	15°	$290 \cdot \cos 15° = 280$	$290 \cdot \sin 15° = 75.1$
$-_P\vec{v}_A$	6.0°	$-240 \cdot \cos 6.0° = -239$	$240 \cdot \sin 6.0° = 25.1$
$_A\vec{v}_G$	θ	$280 + (-239) = 41$	$75.1 + 25.1 = 100.2$

Using the components of $_A\vec{v}_G$, we can find its magnitude and direction as

$${}_Av_G = \sqrt{(41\,\text{km/h})^2 + (100.2\,\text{km/h})^2} = 108\,\text{km/h} \cong 110\,\text{km/h}$$

$$\theta = \tan^{-1}\left(\frac{100.2}{41}\right) = 67.7° \cong 68°$$

north of east because both components are positive

Therefore, the wind's velocity is 110 km/h, 68° N of E, relative to the ground.

As you can see from this example, once you set up an equation by expressing an unknown relative velocity in terms of given relative velocities, the rest of the process is straightforward; it is just adding or subtracting two-dimensional vectors, which we covered in Chapter 1.

PHYSICS 12 ANALYZED: KINEMATICS

2-1. Projectile Motion

- A projectile is an object under the influence of the gravitational force only.
- The horizontal component of a projectile's velocity is constant during the entire flight.
- A projectile gains speed while falling and loses speed while rising.
- A projectile has a minimum speed at the highest altitude of its trajectory.

2-2. Kinematic Equations for Projectile Motion

- Kinematic equations for horizontal motion:

$$\vec{v}_{f_x} = \vec{v}_{i_x} \qquad \vec{d}_x = \vec{v}_{i_x} t$$

- Kinematic equations for vertical motion:

$$\vec{v}_{f_y} = \vec{v}_{iy} + \vec{g}\,t \qquad \vec{d}_y = \vec{v}_{i_y} t + \frac{1}{2}\vec{g}\,t^2$$

$$v_{f_y}^2 = v_{i_y}^2 + 2\vec{g}\cdot\vec{d}_y \qquad \vec{d}_y = \left(\frac{\vec{v}_{iy} + \vec{v}_{f_y}}{2}\right)\cdot t$$

2-3. How to Solve Projectile Motion Problems

1. Read the problem carefully and draw a simple diagram if one is not given.
2. Set the +*x*- and +*y*-directions.
3. Resolve all the vectors given in the problem into components. (Make sure you assign proper signs!)
4. Write down everything you know from the problem statement.
5. Select an appropriate kinematic equation and solve for the unknown.

2-4. Relative Velocity

- The velocity of an object relative to each observer is different if the observers themselves are moving relative to one another.
- Apply the following two equations: ${}_a\vec{v}_b = -{}_b\vec{v}_a$ and ${}_a\vec{v}_b + {}_b\vec{v}_c = {}_a\vec{v}_c$

EXERCISES ON CHPATER 2: KINEMATICS

Exercise 2-1

An object is thrown 42.5° above the horizontal at 11.3 m/s as shown right.

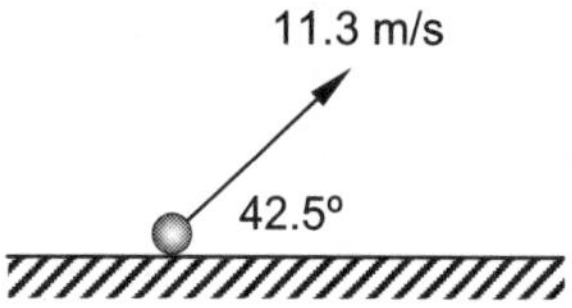

a) What is the maximum height reached by this projectile?

b) What is the speed of the projectile at the maximum height?

Exercise 2-2

A missile is launched with a speed of 175 m/s at 35.2° to the horizon as shown. How high above the ground is the missile 9.00 s after launch?

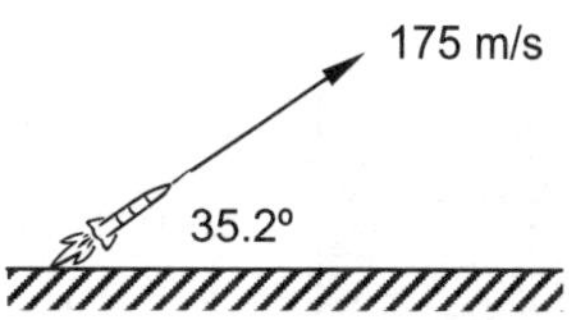

Exercise 2-3

A projectile is launched over level ground with a speed of 53 m/s at 62° above the horizontal.

a) What is the velocity of this projectile 3.1 s after launch? (Find both magnitude and direction.)

b) What is the time of flight of this projectile?

Exercise 2-4

A goalkeeper kicks a soccer ball 83 cm above the ground at 16 m/s, 38° above the horizontal.

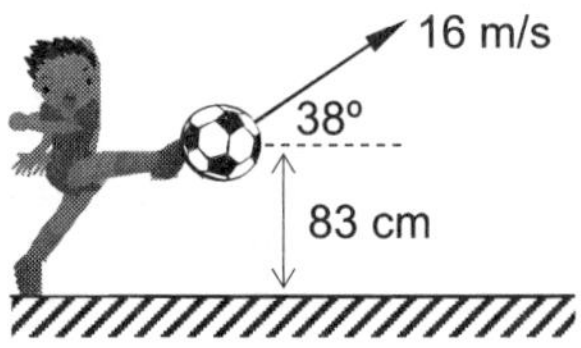

a) How long does the ball fly before hitting the ground?

b) What is the range of the ball? (That is, how far does it fly?)

Exercise 2-5

A ball rolls off the edge of a vertical cliff. After 0.25 s, the ball is 7.8 m away from the side of the cliff, as shown. How far from the base of the cliff is the ball when it lands on the ground, 65 m below the top of the cliff?

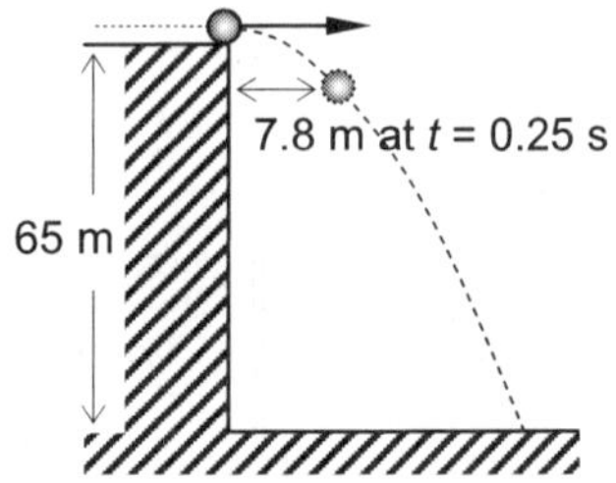

Exercise 2-6

A diver jumps off a platform with a speed of 4.8 m/s at 72° above the horizontal, and travels 3.5 m horizontally before entering water, as shown.

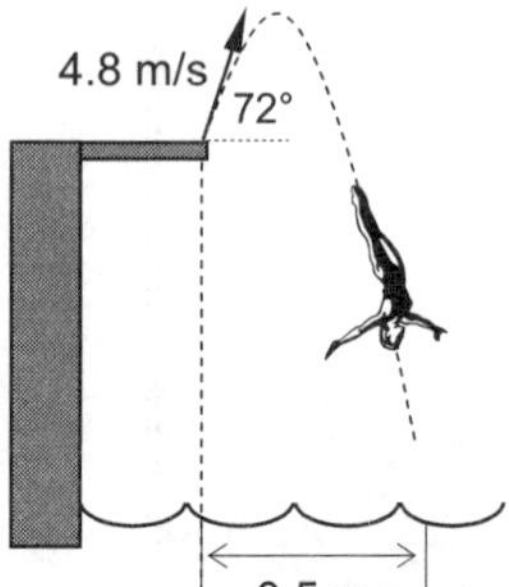

a) How long does it take for the diver to reach her maximum height?

b) How high above the water is the platform?

Exercise 2-7

A missile is fired with an initial speed of 75 m/s at 40.0° above the horizontal and hits a target on the other side of the mountain as shown. The target is located 390 m away from where the missile was launched.

a) How much time does it take for the missile to hit the target?

b) With what velocity does the missile hit the target? (Find both magnitude and direction.)

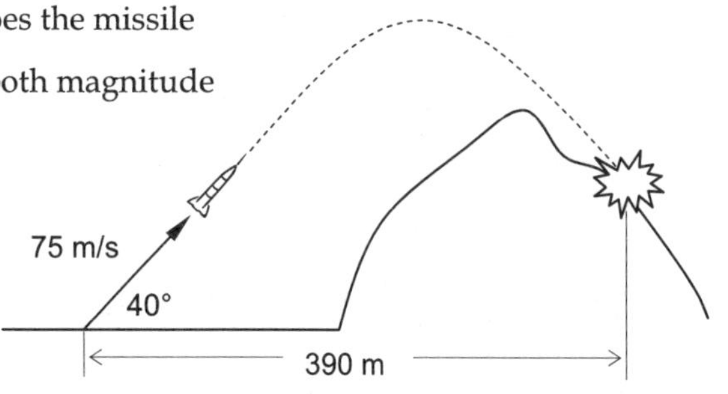

Exercise 2-8

A projectile is launched at 15 m/s, 25° above the horizontal, from the top of a 120-m high cliff. Will it ever reach the speed of 40 m/s? If so, how high above the ground is the ball when its speed is 40 m/s?

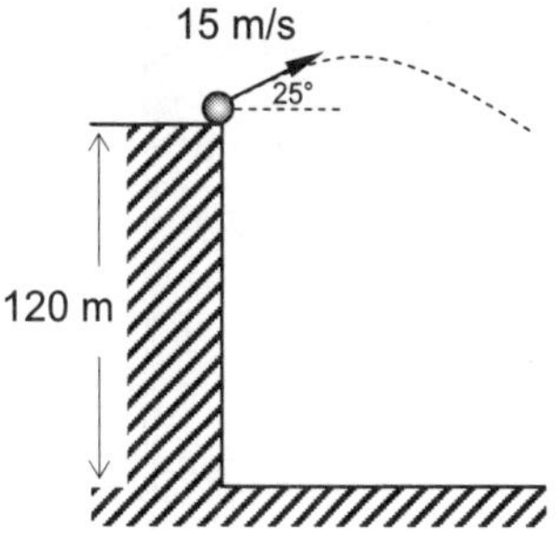

Exercise 2-9

A rock is thrown horizontally at 32 m/s from the top of a 75-m high cliff. What is the speed of the rock when it is 35 m above the ground?

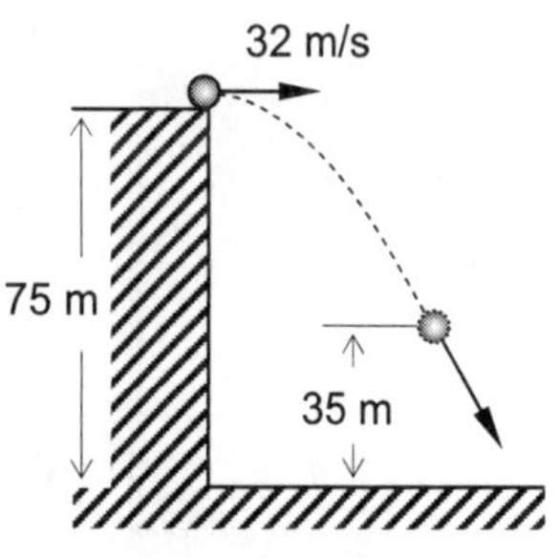

Exercise 2-10

An expert skier jumps off a ski slope at 15 m/s, 29° below the horizontal. He then falls 12 m to the ground below. What is his 'airtime'? (That is, how long is he in the air before landing?)

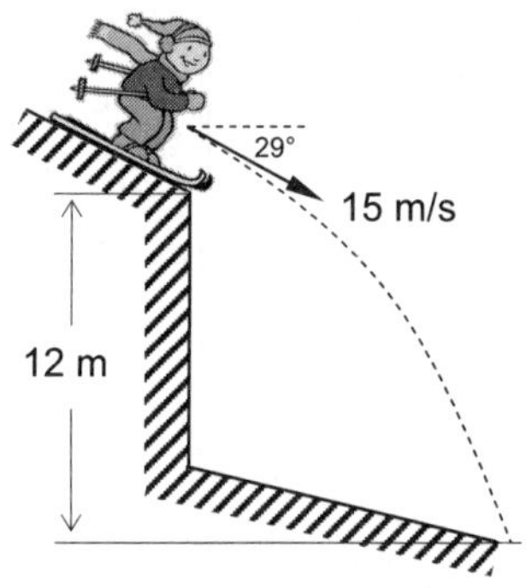

For Exercises 2-11 to 2-13, consider the projectiles shown below. All projectiles have the same initial speed of 15 m/s, but are thrown at different angles.

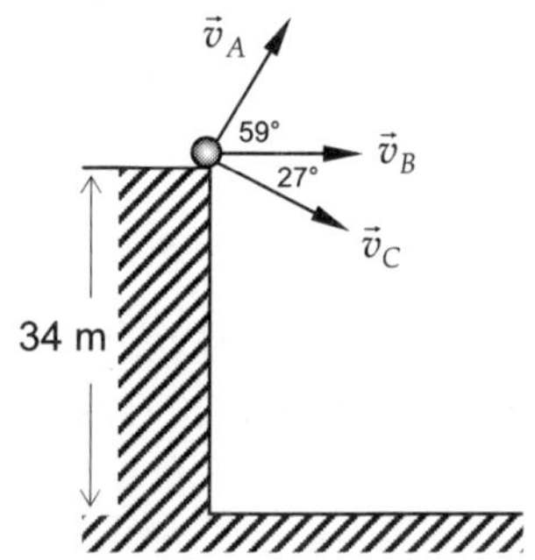

Exercise 2-11

Find the time of flight of each projectile and determine which object hits the ground first.

Exercise 2-12

Find the impact speed of each projectile and determine which object has the greatest speed on impact.

Exercise 2-13

Find the range of each projectile and determine which object travels the farthest before landing.

Exercise 2-14

A boat whose speed in still water is 1.2 m/s heads due east across a river. Because of the river current, the boat's path relative to the shore is 18° south of east.

a) Find the speed of the current.

b) How long does it take the boat to cross the river?
(The river is 90 m wide.)

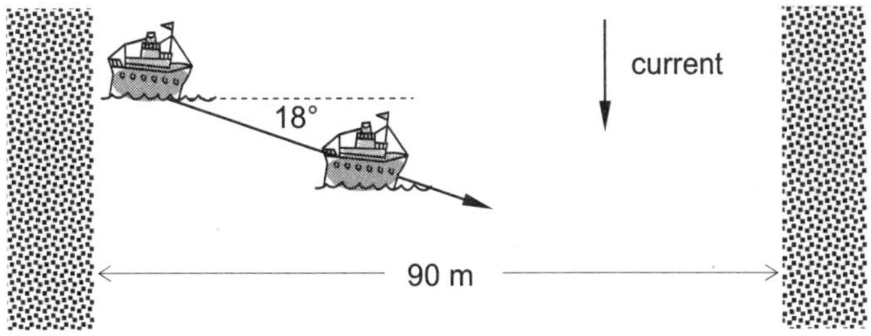

Exercise 2-15

An airplane heading due south is traveling 23° west of south relative to the ground because of the wind from the east. If the ground speed of the plane is 130 m/s, what is the wind speed?

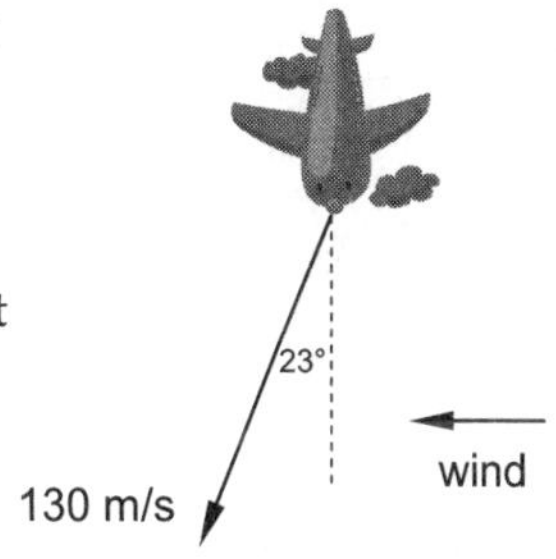

Exercise 2-16

A shark swims due east at 8.3 m/s relative to the ocean current, which itself is flowing north at 1.3 m/s relative to the ocean floor. What is the shark's velocity relative to the ocean floor?

Exercise 2-17

A boat whose speed in still water is 7.5 m/s needs to reach the destination directly east. If the current flows from the south at 2.4 m/s, in what direction should the boat head?

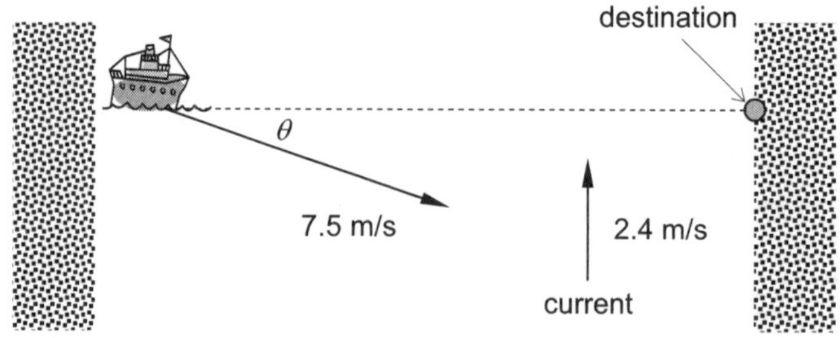

Exercise 2-18

An airplane is heading due west with an airspeed of 77 m/s. If the wind's velocity relative to the ground is 13 m/s, 49° west of north, what is the plane's velocity relative to the ground? (Find both magnitude and direction.)

Exercise 2-19

A pilot leaves Vancouver at noon and heads to Toronto, which is 3350 km, 7.27° south of east from Vancouver. The wind's average velocity relative to the ground is 67.2 m/s, 5.89° north of east. If the plane is to arrive in Toronto four hours later, what should be the plane's velocity relative to the wind?

PROBLEMS ON CHAPTER 2: KINEMATICS

(Answer Key on p. 366)

1 – 8. Fill in the blanks

1. A __________ vector refers to an object's location.
2. The __________ of an object is a change in its position.
3. The __________ of travel is the length of an actual path taken by an object.
4. The __________ of an object indicates the rate at which it is changing its position.
5. The __________ of an object is a scalar that indicates how fast it is traveling.
6. An __________ vector indicates the rate at which an object is changing its velocity.
7. When an object moves with constant acceleration, its motion can be analyzed using a set of equations known as the __________ equations.
8. A __________ is an object that is thrust into the air. The path it takes while traveling in the air is called its __________.

9 – 19. True/False questions

9. _____ A hockey puck sliding on ice is a projectile.
10. _____ A projectile experiences constant acceleration.
11. _____ A projectile experiences no horizontal force.
12. _____ The vertical component of a projectile's velocity is constant.
13. _____ The speed of a projectile is zero at the highest altitude.
14. _____ The acceleration of a projectile is zero at the highest altitude.
15. _____ The speed of a projectile increases while it is losing altitude.
16. _____ One may use the kinematic equations to analyze projectile motion.
17. _____ The direction of a projectile's acceleration is upward while it is rising.
18. _____ The motion of an object described by two different observers will be identical if the observers are not moving relative to each other.
19. _____ The fact that golfers aim slightly to the right of the hole when wind blows from right to left can be explained by the concept of relative velocity.

20 – 25. Multiple choice questions

20. A missile is fired as shown. What is the direction of its acceleration at p?

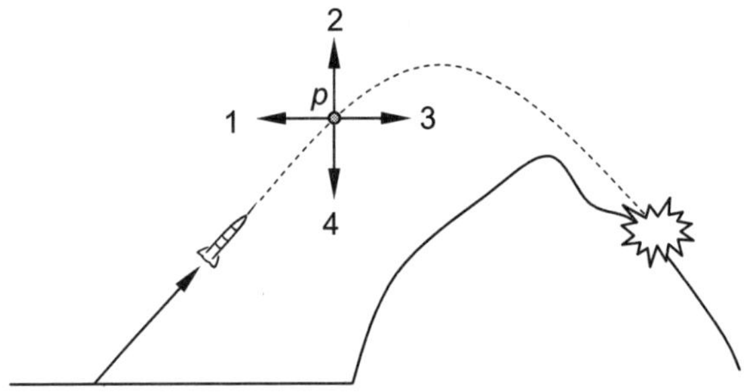

a) 1 b) 2 c) 3 d) 4

21. Which of the following graphs best represents the <u>speed</u> vs. time graph for the missile in Problem 20?

a) 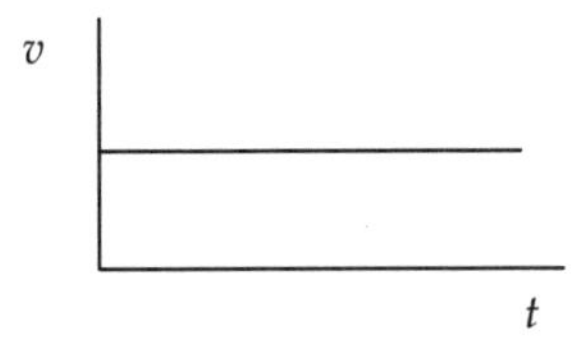

b) 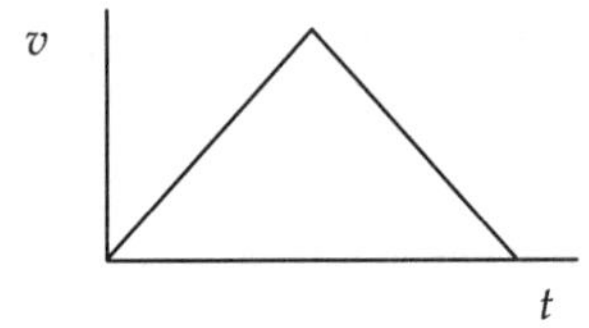

c) 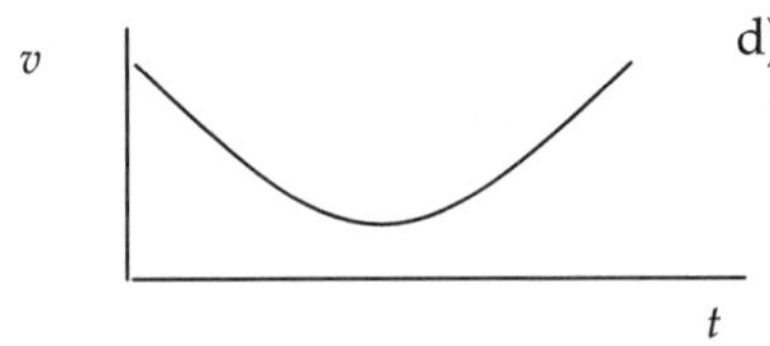

d) 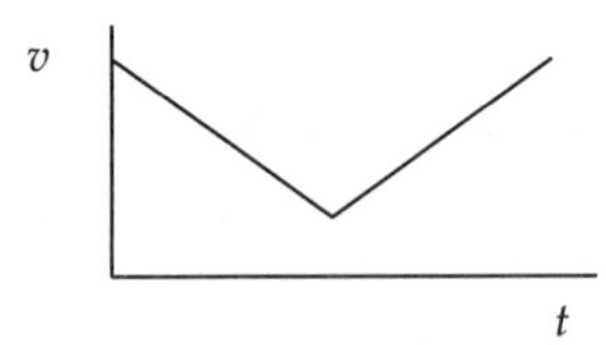

22. Which of the following graphs best represents the <u>range</u> vs. time graph for the missile in Problem 20?

a) 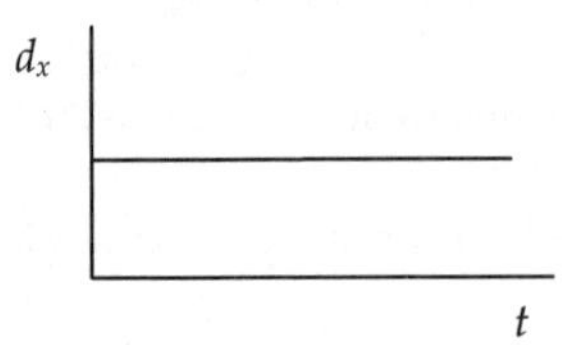

b) 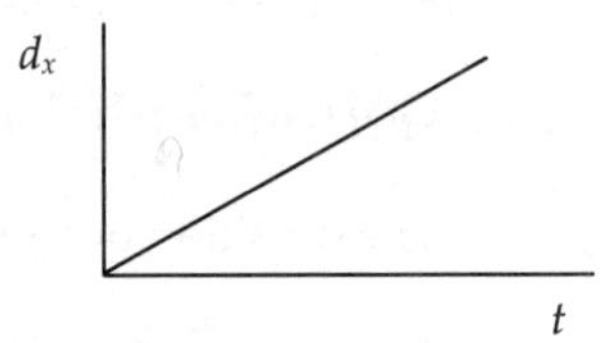

c) 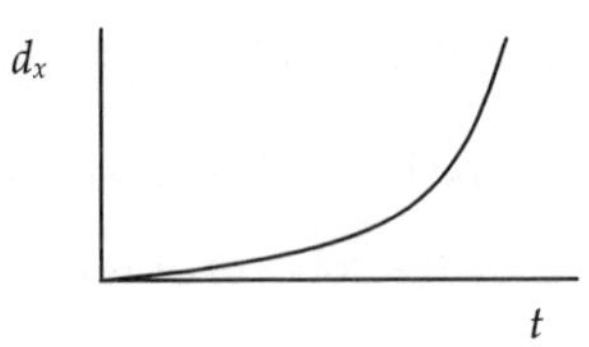

d) 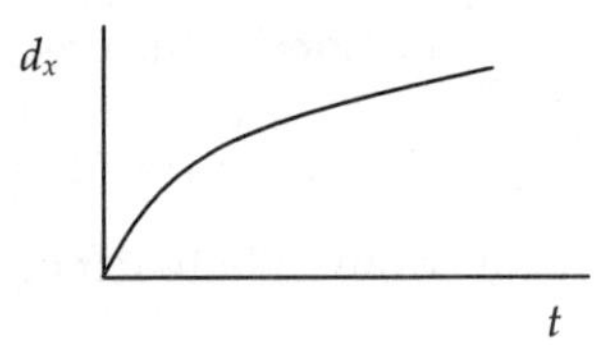

23. From the roof of a school building, Jane throws a pencil horizontally and at the same time drops a pen to the ground below. Ignoring the effect of wind or air resistance, which of the objects will hit the ground first?

a) the pencil
b) the pen
c) Both will hit the ground simultaneously.
d) We cannot determine which object hits the ground first unless we know their masses.

24. Ken, Cathy, and a bird are moving relative to the ground as shown. Which of the following statements is <u>false</u>?

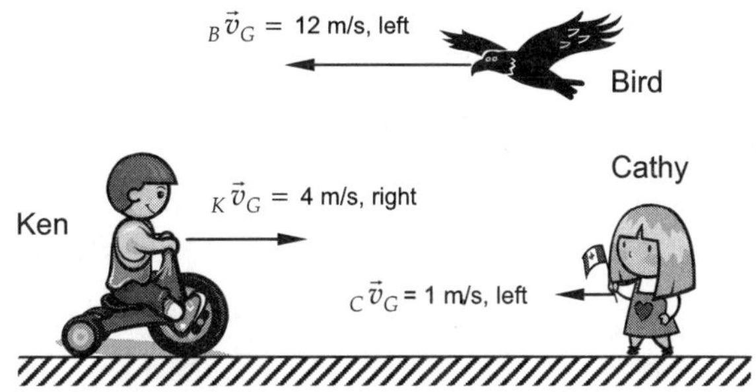

a) Relative to Cathy, Ken is moving at 5 m/s to the right.
b) Relative to Ken, the bird is flying at 16 m/s to the left.
c) Relative to the bird, Cathy is walking at 11 m/s to the right.
d) Relative to Cathy, the ground is not moving.

25. On a rainy day a car (*C*) is traveling east (relative to the ground (*G*)). Assuming raindrops (*R*) are falling vertically (relative to the ground), which of the following vector diagrams correctly shows the velocity of raindrops as seen by the driver in the car, $_R\vec{v}_C$?

a)

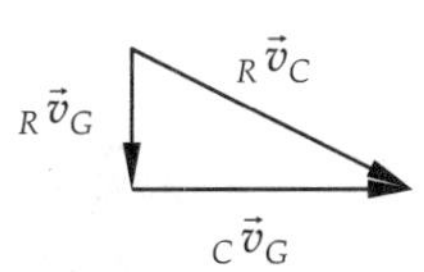

b)

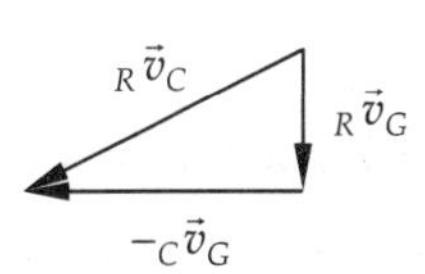

c)

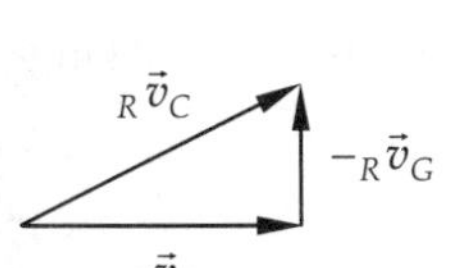

d)

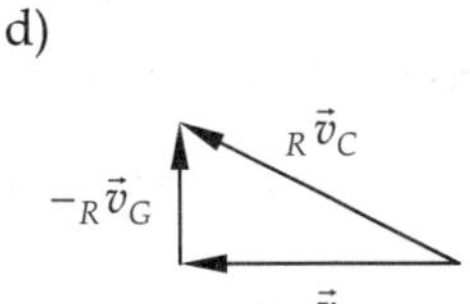

26 – 40. Written-response questions

26. A 15-kg object is thrown at 8.3 m/s, 71° above the ground. Sketch the horizontal speed vs. time graph for this object while it is in the air. Show the value of the horizontal speed at t = 0 s and at t = 2.5 s, and calculate the slope of the graph.

27. For the object in Problem 26, draw the vertical speed vs. time graph. Show the value of the vertical speed at t = 0 s and at t = 2.5 s, and calculate the slope of the graph. Indicate the point on the graph where the object reaches the maximum height.

28. A ball, initially 1.5 m above the ground, is thrown into the air at 15 m/s, 15° above the horizontal, as shown. Find (a) the maximum height the ball reaches; (b) the ball's velocity after 0.83 seconds; and (c) the total time of flight.

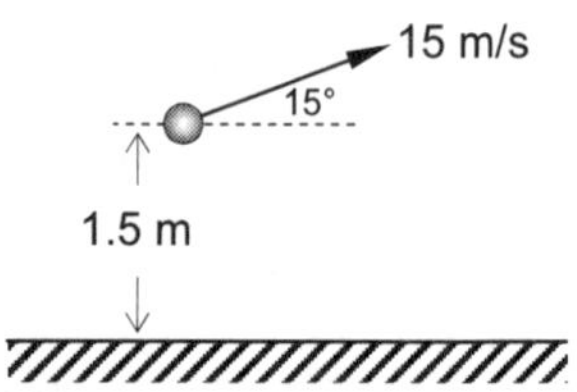

29. A balloon is floating 5.3 m above the ground and 7.7 m away from you. If you are to hit the balloon with a baseball thrown 0.72 m above the ground at 60° above the horizontal, what should be the initial speed of the ball?

30. In Problem 29, if the baseball is thrown at 13 m/s, 67° above the horizontal, will it fly over or under the balloon? Show your calculations.

31. An object is thrown horizontally from an unknown height. As it falls, it passes a 2.4-m tall window in 0.15 seconds as shown. Determine the height from which the object was thrown relative to the top of the window.

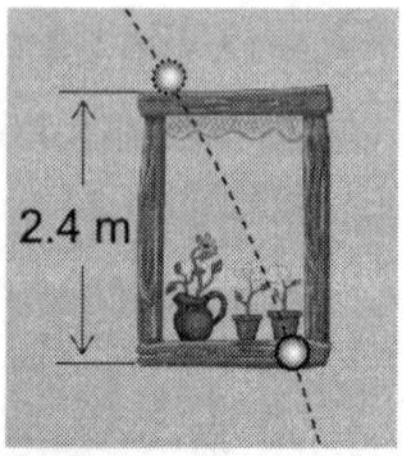

32. James is running away from his enemy but finds himself near the edge of a 28-m cliff. He estimates that water is 12 m from the base of the cliff, as shown. What's the minimum speed he needs to jump off in order to reach the water below?

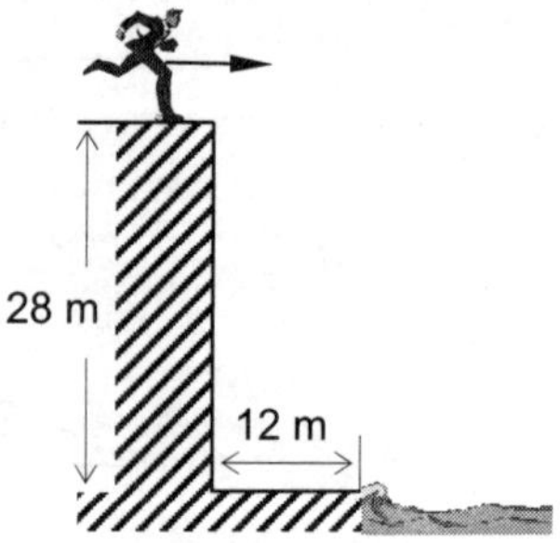

33. A bullet traveling at 315 m/s strikes a wooden block at rest. Upon impact, the bullet is embedded within the block, which starts to slide on a horizontal surface at 17 m/s. Because of friction, the block slows down for 2.2 seconds before falling to the ground below as shown. If the block travels 1.9 m horizontally before hitting the ground, find (a) the speed with which the block left the horizontal surface; (b) the deceleration of the block while it was moving on the horizontal surface; and (c) the distance it traveled on the horizontal surface. (Assume the block does not move until the bullet is fully embedded and comes to rest within it.)

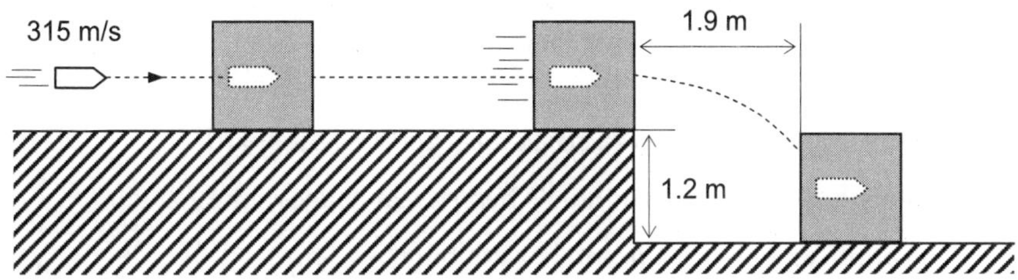

34. A ball is thrown from the top of a 65-m cliff at 13 m/s, 27° above the horizontal. At the same time, Eric starts running from the base of the cliff as shown. If Eric wants to receive the ball before it hits the ground, what should be his minimum speed? How does the answer change if the height of the cliff was 55 m?

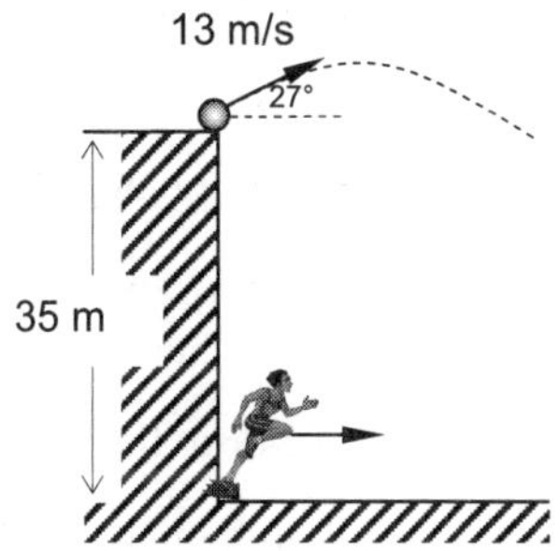

35. An object, initially traveling at 25 m/s, starts decelerating at 2.5 m/s², 32 m before the edge of a 27-m cliff. Determine whether the object ever falls off the cliff. If it does, find (a) the distance from the base of the cliff to the landing point and (b) the object's speed at impact. If it doesn't fall, find (a) how far it travels before coming to a stop and (b) the maximum acceleration needed to ensure the object falls off the cliff.

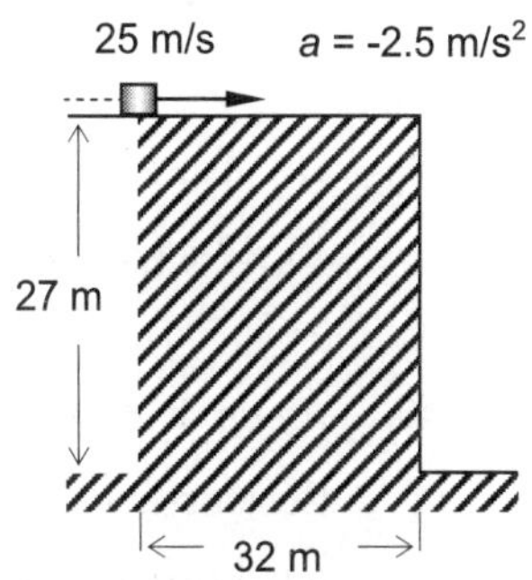

36.* A cannon is aimed at a target located 72 m away from the cannon and 20 m above the ground. The cannon ball, when fired, will have the initial speed of 32 m/s, but it has to clear a 13-m wall between the cannon and the target, as shown. At what angle should the cannon be fired?

[Hint: Use the following identity: $\sin^2\theta + \cos^2\theta = 1$]

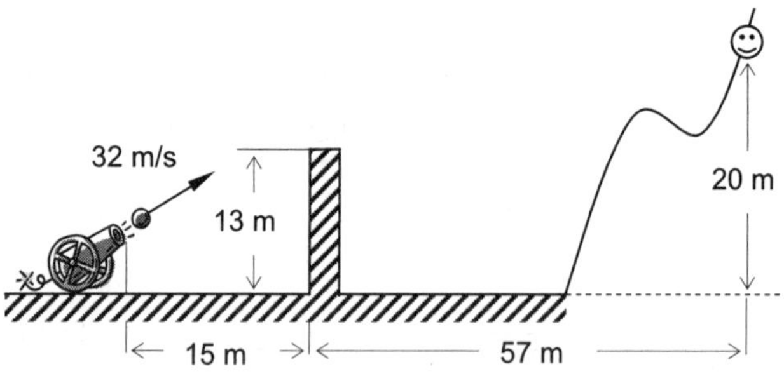

37.* A missile, initially at rest, accelerates vertically upward at 35 m/s² from the ground. When it is 25 m above the ground, an object is thrown horizontally at 12 m/s relative to the missile. Find (a) the velocity of the object relative to the ground; (b) the time it takes the object to reach the ground; (c) the speed of the missile when the object lands on the ground; and (d) the impact velocity of the object relative to the missile.

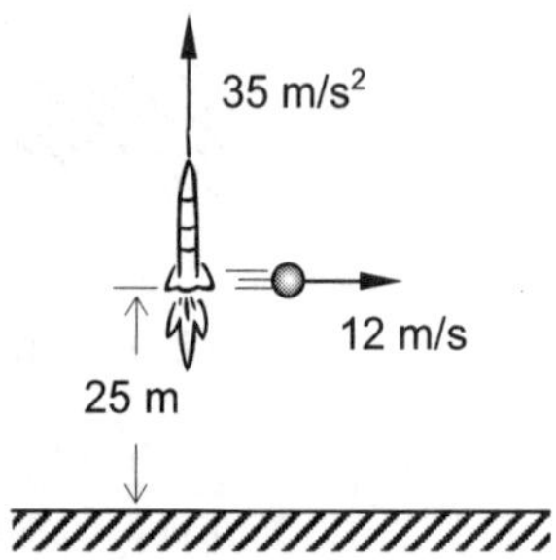

38. Suppose the car in Problem 25 is traveling at 70 km/h. From the inside of the car, it appears that the rain is coming down nearly horizontally, at 10° below the horizontal. Find the speed with which the raindrop is falling relative to the car and that relative to the ground.

39. An object is thrown at 22 m/s, 35° N of E. If the wind blows from the east at 13 m/s, what is the displacement of the object after 8.5 s?

40. A ship is moving due south at a speed of 32 km/h relative to the ocean current. If the ship is seen to be moving at a speed of 25 km/h, 17° E of S, relative to the land, what is the velocity of the ocean current relative to the land?

3

CHAPTER 3

DYNAMICS

There is really nothing new that needs to be learned in this chapter, as you have already covered the core concepts in Physics 11. Here you will simply extend your understanding and learn to solve two-dimensional problems. Perhaps the most important skill you need to solve those problems is the ability to resolve force vectors into components (covered in Chapter 1). Of course, that skill will not help very much if you have forgotten the fundamentals of dynamics. Hence, take your time to review various types of forces, Newton's laws of motion, how to construct a free-body diagram, and how to solve dynamic problems in general. In this chapter, drawing a free-body diagram will be given special emphasis because of its importance in solving dynamic problems.

In this chapter, you will:

- ✓ review dynamics from Physics 11
- ✓ test your knowledge of dynamics from Physics 11
- ✓ learn more about drawing a free-body diagram
- ✓ learn how to solve two-dimensional dynamics problems
- ✓ solve exercise problems on two-dimensional dynamics
- ✓ test your knowledge of dynamics from Physics 12

Review of Physics 11: Dynamics

1. Force ($\vec{F}$)

- **Force** is an interaction between two objects producing an attraction (a pull) or a repulsion (a push).
- Force is a vector and is measured in Newtons [N].

2. Gravitational Force ($\vec{F}_g$)

- **Gravitational force** is an attractive force between two pieces of matter.
- When one object is the Earth and the other object is on or near the surface of the Earth, the gravitational force of the Earth acting on that object is called its **weight**, whose magnitude is obtained from

$$F_g = m\,g \qquad \text{... (3.1)}$$

  where g is the gravitational acceleration and is equal to 9.8 m/s^2 on or near the surface of the Earth.

  On the surface of the Moon, the gravitational acceleration is 1.6 m/s^2.

- The direction of the gravitational force is always towards the center of the Earth.

  Example: The weight of a 5.2-kg object is $5.2\,\text{kg} \times 9.8\,\text{m/s}^2 \cong 51$ N, towards the center of the Earth.

  The unit of weight is the Newton [N], not the kilogram [kg].

3. Normal Force ($\vec{F}_N$)

- **Normal force** is the force that a surface exerts on an object lying upon it.
- The direction of the normal force is always perpendicular to and away from the surface, as shown below.

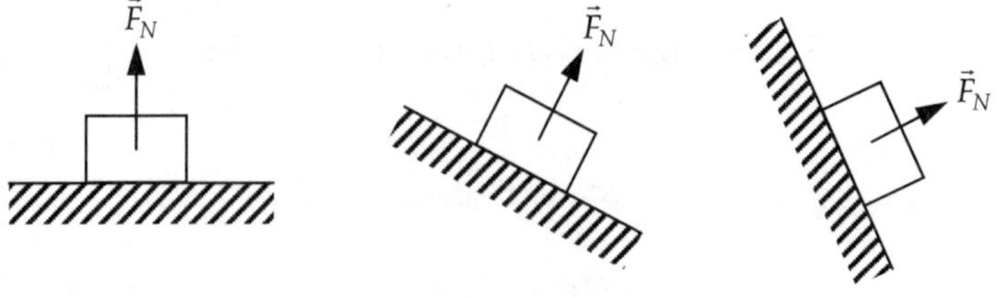

Figure 3-1: Directions of normal forces

- The normal force is an accommodating force as its magnitude changes depending on the situation.

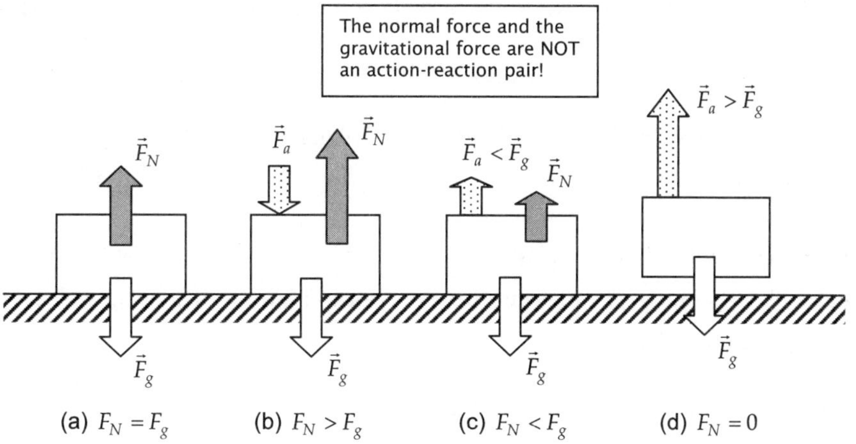

Figure 3-2: Normal forces in various situations

4. Tension Force ($\vec{F}_T$)

- **Tension force** is the force transmitted through a string when it is taut.
- The direction of the tension force on an object is always along the string and away from the object.
- The magnitude of the tension force depends on the situation. It is thus an accommodating force.

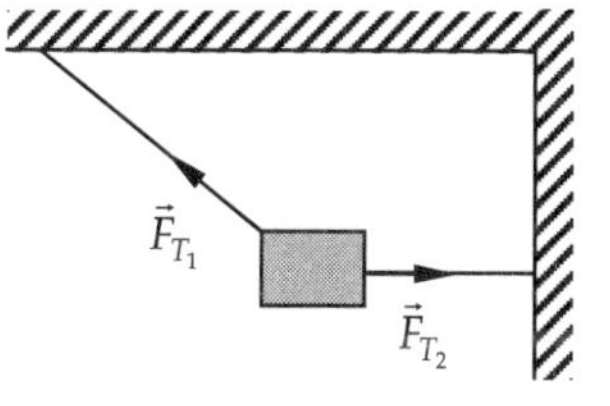

Figure 3-3: Tension forces

- Objects connected by a string always move at the same speed and accelerate at the same rate.

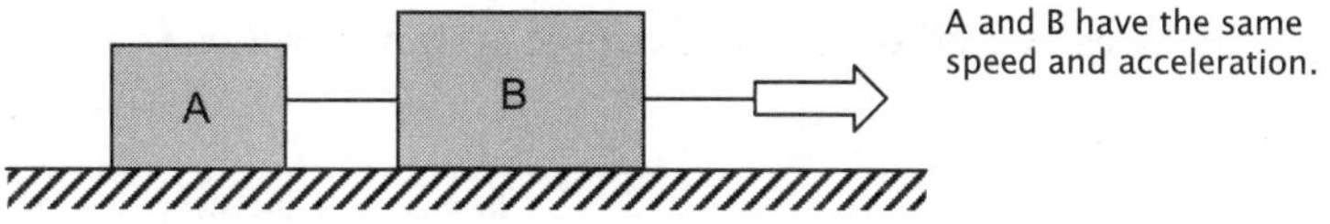

Figure 3-4: Speed and acceleration of moving objects joined by a string

5. Friction Force ($\vec{F}_f$)

- **Friction force** is the force exerted by a surface as an object slides or tries to slide over it.
- The friction force always acts parallel to the surface and to oppose the motion.
- The magnitude of the friction force on a moving object is given by

$$F_f = \mu F_N \qquad \text{... (3.2)}$$

where μ is the **coefficient of friction**, which depends only on the nature of the materials in contact. The magnitude of the normal force indicates how closely those materials are in contact.

- The friction force on a stationary object is an accommodating force; its magnitude varies depending on the situation.

Example: Find the friction force on the object shown right if (a) the object is stationary and (b) the object is accelerating. (μ = 0.33)

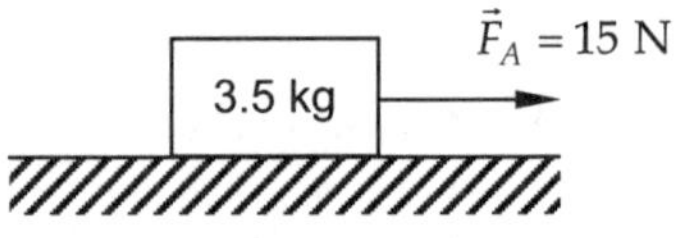

(a) When the object is stationary, the forces on it must be balanced. Thus,

$$F_f = 15 \text{ N, left}$$

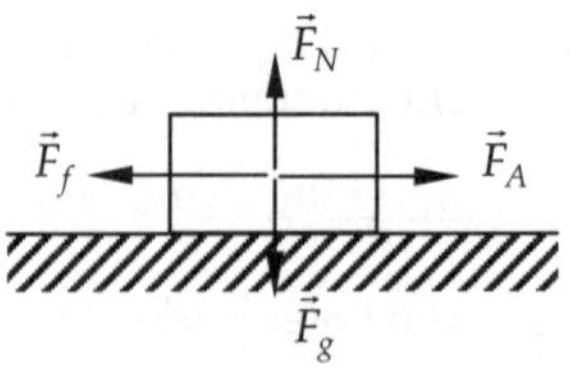

(b) Since the object is moving, apply Eq. (3.2). The magnitude of the normal force is equal to that of the gravitational force, mg.

$$F_f = 0.33(3.5\,\text{kg})(9.8\,\text{m/s}^2) \cong 11 \text{ N, left}$$

6. Free-Body Diagram

- A **free-body diagram** shows an object of interest and all the forces acting on it (but not the ones exerted by it). An example is shown below.

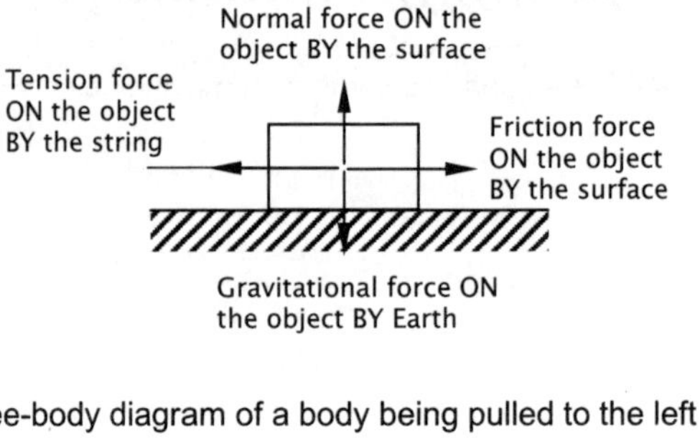

Figure 3-5: Free-body diagram of a body being pulled to the left by a string

7. Newton's First Law of Motion

- The First Law states that, if the forces acting on a body are balanced, the motion of the body does not change. That is, a moving object continues to move in a straight line at constant speed, and a motionless object stays at rest.
- The following three statements have identical meaning:
  a. The velocity of an object is constant.
  b. An object is not accelerating.
  c. Forces on an object are balanced.
- The tendency of a body to maintain its state of motion is called **inertia**. The greater the inertia of a body, the harder it is to change its state of motion. Newton's First Law is also called the **law of inertia**.

8. Newton's Second Law of Motion

- The Second Law states that, if the forces acting on a body are <u>not</u> balanced, the body will accelerate in the direction of the <u>net</u> force. **Net force** is the vector sum of all the forces acting <u>on</u> a body, and the acceleration it produces is directly proportional to the net force and inversely proportional to the body's mass. That is,

> 1 N is defined as the amount of net force required to accelerate a mass of 1 kg by 1 m/s².

$$\vec{a} = \frac{\vec{F}_{net}}{m} \quad [\text{N/kg} = \text{m/s}^2] \qquad \ldots (3.3)$$

Example: Three forces are acting on a 12-kg object as shown. To find its acceleration, we calculate the net force on the object, which is

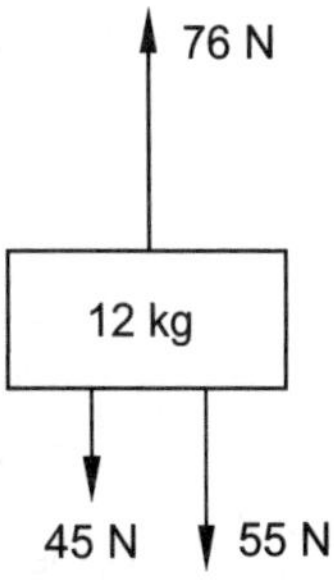

$$\vec{F}_{net} = -45\,\text{N} + (-55\,\text{N}) + 76\,\text{N} = -24\,\text{N}, \text{ or } 24\text{ N downward}$$

Dividing the net force by the object's mass gives the acceleration:

$$\vec{a} = \frac{-24\,\text{N}}{12\,\text{kg}} = -2.0\ \text{m/s}^2 \text{ or } 2.0\ \text{m/s}^2 \text{ downward}$$

9. Newton's Third Law of Motion

in magnitude — in direction

- The Third Law states that, for every action force, there is an equal and opposite reaction force. This law is known as the **action-reaction law** and explains the interaction between two objects or forces, when one influences the other.

$$\vec{F}_{\text{on A by B}} = -\vec{F}_{\text{on B by A}} \qquad \dots (3.4)$$

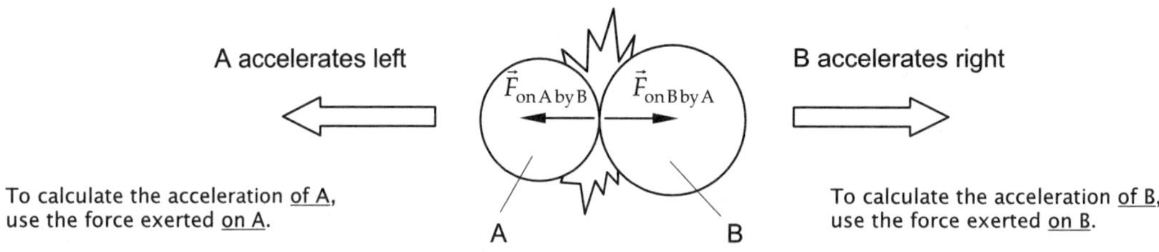

Figure 3-6: Action and reaction forces

- This Law explains most kinds of propulsion such as a rocket moving away from the Earth's surface, a swimmer moving through the water, etc.

10. How to Solve Dynamic Problems

1. Read the problem statement carefully and look for phrases such as "constant velocity" or "smooth surface." The former means that the forces on the object are balanced, and the latter allows you to ignore friction.
2. Draw a free-body diagram for each mass.
3. Set the direction in which the object is moving as positive, and set up an equation of motion for each mass using Newton's second law:

$$\underbrace{\vec{F}_{\text{net}}}_{\text{dynamic side}} = \underbrace{m\vec{a}}_{\text{kinematic side}} \qquad \dots (3.5)$$

Use free-body diagram! → dynamic side

kinematic side ← Use kinematic equations!

To find the net force, go to the free-body diagram of the object. Add the forces opposing the motion and subtract it from the forces encouraging the motion.

4. Check that there is only one unknown in the equation and solve for it. For a system of two masses joined by a string, there should be two equations.

Test Your Knowledge of Physics 11: Dynamics

(Answer Key on p. 367)

1 – 10. True/False questions

1. _____ Newton per kilogram [N/kg] is equal to meter per second squared [m/s²].
2. _____ The larger the mass of a body, the greater its inertia.
3. _____ If an object is accelerating, the forces on it are balanced.
4. _____ Newton's Third Law states that the acceleration of a body is directly proportional to net force.
5. _____ The unit of weight is the kilogram [kg].
6. _____ The magnitude of the normal force is always equal to that of the weight.
7. _____ Normal and tension forces are accommodating forces.
8. _____ If an object is stationary, it does not experience any friction force.
9. _____ The free-body diagram of an object shows the forces exerted by the object.
10. _____ For an object moving on a smooth surface, the friction force on it is zero, and so is the normal force on it because $F_f = \mu F_N$.

11 – 20. Multiple choice questions

11. Which of the following quantities has the same direction as the net force vector?

    a) position vector b) displacement vector
    c) velocity vector d) acceleration vector

12. When aboard an elevator starting to descend, you feel lighter than usual for a brief moment. Which Newton's law explains this phenomenon?

    a) Newton's First Law b) Newton's Second Law
    c) Newton's Third Law d) Newton's laws cannot explain it.

13. When you kick a wall, your foot hurts. The harder you kick, the more your foot hurts. Which Newton's law explains this phenomenon?

    a) Newton's First Law b) Newton's Second Law
    c) Newton's Third Law d) Newton's laws cannot explain it.

14. Which of the following objects is accelerating?

a) an object at rest
b) an object moving at constant speed on a level ground
c) an object thrown vertically upward
d) an object acted on by two forces: 20 N, east and 20 N, west

15. Which of the following statements must be true for an object whose velocity is changing?

a) Its speed is changing.
b) Its acceleration is zero.
c) It is moving along a straight line.
d) Forces acting on the object are not balanced.

16. The free-body diagram of an object being pulled by a string on level ground is shown right. Which of the following forces is the reaction force to the normal force, $\vec{F}_N$?

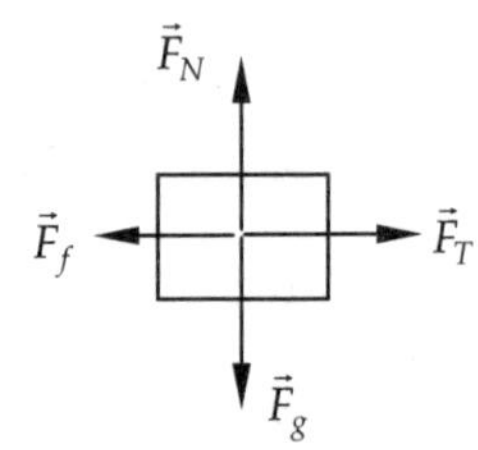

a) the tension force, $\vec{F}_T$ b) the gravitational force, $\vec{F}_g$

c) the friction force, $\vec{F}_f$ d) The reaction force is not shown in this diagram.

17. Object A weighs more than object B when they are weighed on an unknown planet. Which of the following statements must be true?

a) A has the greater mass than B.
b) B has the greater inertia than A.
c) A has the greater density than B.
d) B has the greater volume than A.

18. Besides the coefficient of friction, the magnitude of the friction force on a moving object depends on

a) the object's weight b) the normal force on the object
c) the object's speed d) the net force on the object

19. An object is moving down an inclined surface at constant velocity. Which of the following statements must be true?

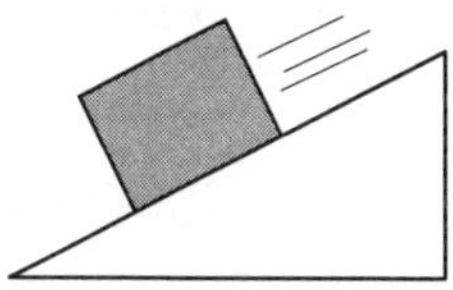

   a) There is a normal force acting upward.
   b) There is an applied force pushing the object down the slope.
   c) There is a friction force acting parallel to the surface.
   d) There are only two forces acting on the object.

20. An object is being pulled up a smooth inclined surface by a rope. Which of the following forces should <u>not</u> be included in the free-body diagram?

   a) normal force   b) friction force
   c) tension force   d) gravitational force

21 – 35. Written-response questions

21. A 93-kg man, training for the World's Strongest Man competition, lifts 245 kg of weights off the ground, as shown. What is the magnitude of the normal force on his body? Does the answer change when he drops the weights to the ground? If so, explain how the answer changes.

22. An 1100-kg automobile traveling at 16 m/s comes to a complete stop over a distance of 93 m. What is the magnitude of the net force on the vehicle during its deceleration?

23. Force F, when applied to mass m_1, causes it to accelerate at 2.4 m/s^2. When the same force is applied to mass m_2, it accelerates at 3.9 m/s^2. If the same force is applied to accelerate both objects together, what is their acceleration?

24. Liz and James are initially skating at constant velocity when James suddenly pushes Liz from behind with a 150-N force. If Liz and James are 46 kg and 64 kg, respectively, find their accelerations while the force is being applied.

25. A string that can withstand the maximum tension force of 240 N is used to lift a 14-kg object. What is the maximum acceleration the object can attain without snapping the string?

26. Two masses connected by a string are hanging from a ceiling by another string as shown. Find the tension force in each string.

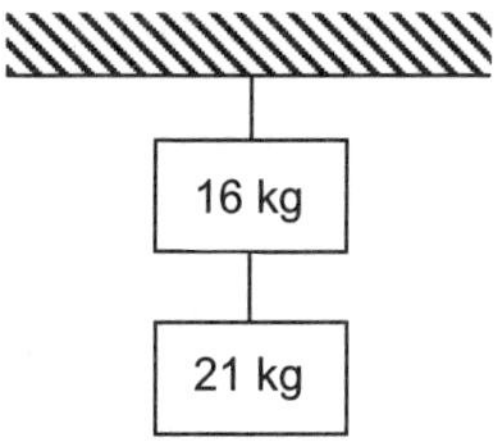

27. A 59-N force is applied horizontally on a 9.3-kg object on level ground. If the object remains motionless despite the applied force, what is the magnitude of the friction force on the object? (The coefficient of friction between the ground and the object is 0.65.)

28. A 5.1-kg object is pressed against the wall by a horizontal force, as shown. Draw the free-body diagram of the object.

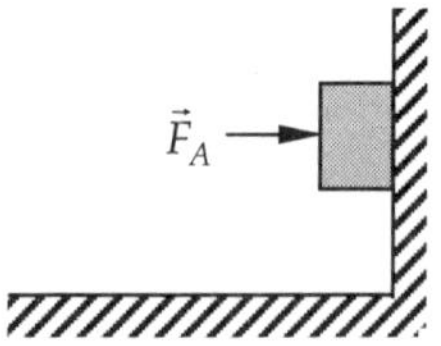

29. In Problem 28, the object is held stationary by the horizontal applied force. If the coefficient of friction between the object and the wall is 0.23, find the magnitude of the minimum applied force required.

30. A 25-kg block is pushed by a 500-N applied force along a level surface. If the resulting acceleration of the block is 13 m/s², find the coefficient of friction between the block and the surface.

31. A 3.7-kg object, initially at rest 3.2 m above the ground, is lowered vertically by a string. If the object reaches the ground in 2.1 s, what is the magnitude of the tension force in the string?

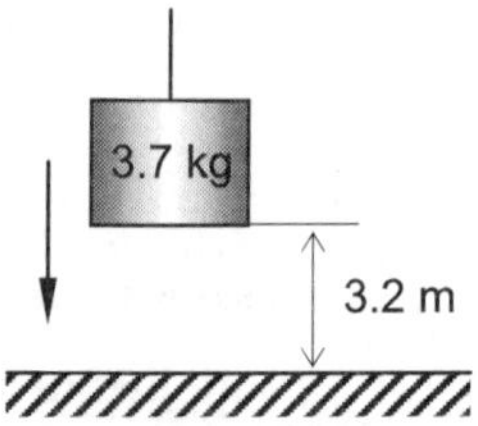

32. A 72-kg man stands on a scale while accelerating in an elevator. If the scale reads 888 N, what is the acceleration of the elevator? (Find both magnitude and direction.)

33. A car of mass 1100 kg is traveling at a constant velocity of 60 km/h, east. The coefficient of friction between the tires and the road is 0.45.

    a) Draw a free-body diagram showing the magnitude and direction of all the forces acting on the car.

    b) The driver of the car sees the traffic light in front turning red and steps on the brake, only to find that the brakes are not working. If the car is 35 m before the intersection when this happens, would it be able to come to a stop before entering the intersection? Show your calculations.

34. A system of masses shown below is pulled by 99-N force to the right. If the coefficient of friction is 0.29, find the acceleration of the system and the tension force between the masses.

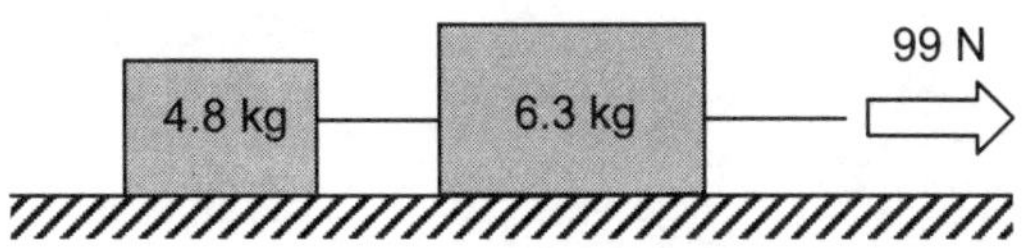

35. A 66-N force is applied horizontally to accelerate the two objects shown. If the coefficient of friction between the floor and the boxes is 0.36, what is the magnitude of the force exerted by the 4.5-kg object on the 3.5-kg object?

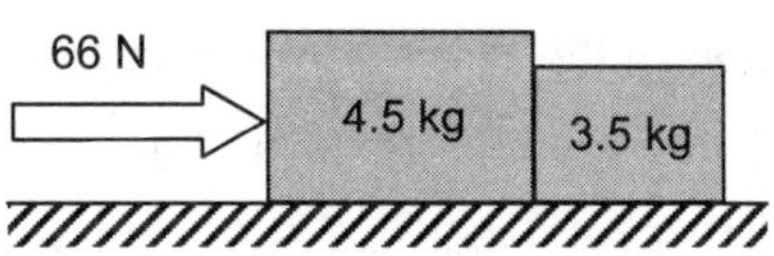

What's New in Physics 12?

3-1. More on Drawing a Free-Body Diagram

It is worth repeating how to construct a correct free-body diagram as it is such an important step in solving any dynamic problem. But before reviewing the process of drawing a free-body diagram, let's classify various types of forces into two groups: **contact forces** and **non-contact forces**. As the names suggest, the former requires contact between two bodies, and the latter does not. Non-contact forces include the gravitational, electric, and magnetic forces. (Until you are introduced to electromagnetism later in the course, the only non-contact force you will come across is the gravitational force.) All other forces such as the normal, friction, tension, and applied forces are contact forces, as shown in Table 3-1.

Table 3-1: Contact and non-contact forces

Force	Contact force	Non-contact force
Gravitational force ($\vec{F}_g$)		✓
Normal force ($\vec{F}_N$)	✓	
Tension force ($\vec{F}_T$)	✓	
Friction force ($\vec{F}_f$)	✓	
Applied force ($\vec{F}_A$)	✓	
Electric force ($\vec{F}_E$)		✓
Magnetic force ($\vec{F}_M$)		✓

Don't be concerned about these forces for now. (Electric force, Magnetic force)

Thus, when drawing a free-body diagram of an object, you should draw the gravitational force vector first and then look for the things the object is in contact with. For example, if the object is in contact with a surface, draw the normal force perpendicular to the surface and the friction force parallel to the surface. (If the surface is smooth, friction does not exist.) If it is in contact with a string, draw the tension force along the string. Finally, if it is in contact with another body, draw the force applied by that body. Usually, the direction of the applied force is obvious from the description of the situation.

Consider the projectile shown with its trajectory in Fig. 3-7. To construct the free-body diagram of the projectile, first draw the force of gravity downward and then see if it is touching anything. Since nothing is in contact with it, there is no additional force to be included in the free-body diagram.

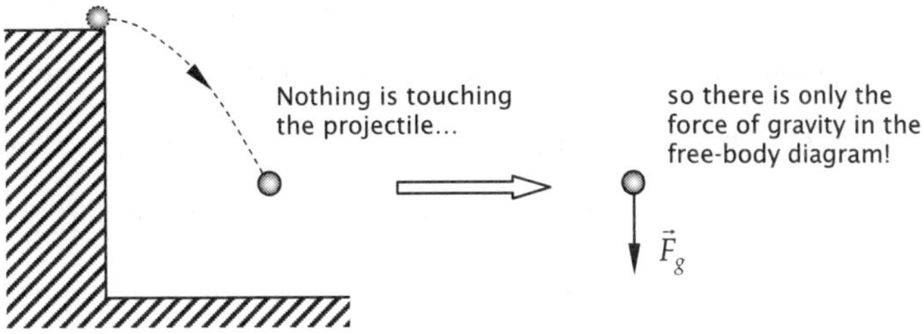

Figure 3-7: Free-body diagram of a projectile

Now consider an object at rest on an inclined surface, as shown in Fig. 3-8. We see that the object is in contact with the surface, which can exert normal and friction forces. In this case, both forces are present. We know the friction force exists because the net force cannot be zero otherwise. (Since the object is at rest, the forces on it must be balanced.) The friction force acts up the slope since, without it, the object would slide down the slope. As the object is in contact with nothing else, there is no additional force to be included in the free-body diagram.

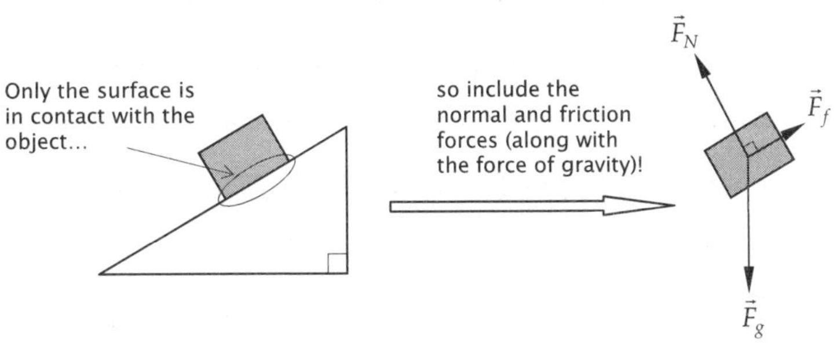

Figure 3-8: Free-body diagram of an object at rest on an inclined surface

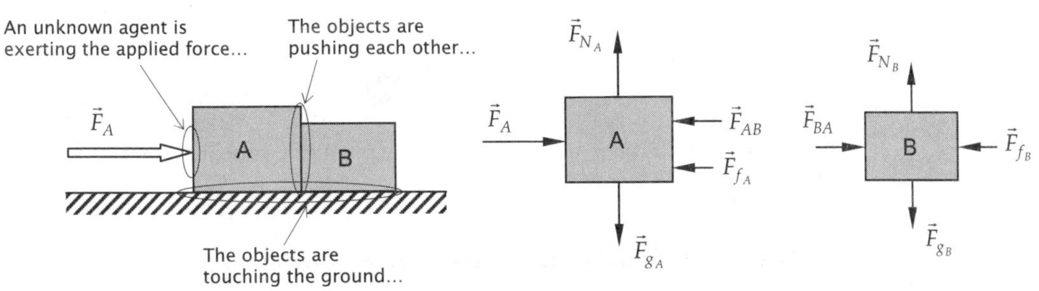

Figure 3-9: The free-body diagrams of objects on a horizontal smooth surface

Finally, consider a system of two objects being pushed to the right at constant speed on level ground, as shown in Fig. 3-9. The bigger object (A), in contact with the surface and the other object (B), experiences the normal, friction, and the applied force by B ($\vec{F}_{AB}$), in addition to the gravitational force and the applied force to the right exerted by an unknown agent ($\vec{F}_A$). ($\vec{F}_{AB}$ is to the left because it acts to resist the motion of A.) As for B, its free-body diagram includes the gravitational, normal, friction, and the applied force by A ($\vec{F}_{BA}$). Note that $\vec{F}_{AB}$ and $\vec{F}_{BA}$ form an action-reaction pair; they are the same in magnitude and opposite in direction.

The net force on an object is the sum of all the forces that appear in its free-body diagram and is not an applied force. Therefore, the net force vector should never appear in a free-body diagram.

3-2. How to Solve Two-dimensional Dynamic Problems

When solving a dynamic problem, always start by drawing a free-body diagram. If a problem involves a system of objects, you probably need to draw a free-body diagram for each object (as shown in Fig. 3-9). After this initial step, the rest of the solution procedure is relatively straightforward and can be summarized as follows:

1. Set up the coordinate system. The x-axis is usually chosen to be parallel to the surface, and the y-axis perpendicular to it.
2. Resolve each force vector in the free-body diagram into the x- and y-components.
3. Set up the following two equations using Newton's Second Law of motion, one for the x-components and the other for the y-components.

Use the free-body diagram for the left-hand side! →

$$\vec{F}_{\text{net}_x} = m\,\vec{a}_x \quad \dots (3.6)$$

$$\vec{F}_{\text{net}_y} = m\,\vec{a}_y \quad \dots (3.7)$$

4. If there are two or more objects in the problem, repeat steps 1 to 3 for each object.
5. Solve for the unknowns in the equations. After you obtain an answer, think about whether it is reasonable. Be aware that the direction of the acceleration vector is independent of the direction of the motion.

Let's apply the above solution procedure to a couple of examples. The first example is a 6.0-kg object sliding down a 30° slope at constant speed, as shown in Fig. 3-10. We are asked to find the coefficient of friction between the object and the surface.

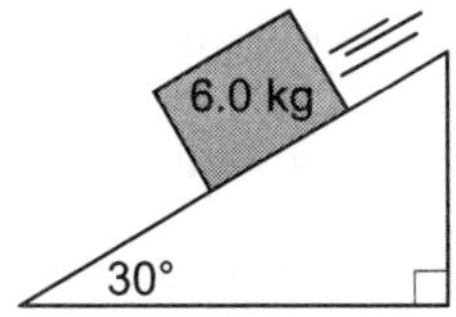

Figure 3-10: An object sliding down an inclined surface

The first step is to draw the free-body diagram, shown in Fig. 3-11 (a). Then set the x-axis parallel to the surface and the y-axis perpendicular to it, as shown in Fig. 3-11 (b). With this coordinate system, the only vector in the free-body diagram that needs to be resolved is the gravitational force. Its x- and y-components are shown in Fig. 3-11 (c). The gravitational force on the object is

$$F_g = m\,g = 6.0\,\text{kg}\left(9.8\,\text{m/s}^2\right) = 58.8\,\text{N}$$

According to Fig. 3-11 (c), the x- and y-components of the gravitational force are

$$F_{g_x} = m\,g\sin\theta = 58.8\,\text{N}\cdot\sin 30° = 29.4\,\text{N}$$

$$F_{g_y} = m\,g\cos\theta = 58.8\,\text{N}\cdot\cos 30° = 50.9\,\text{N}$$

Note that we use the sine ratio to find the x-component in this case.

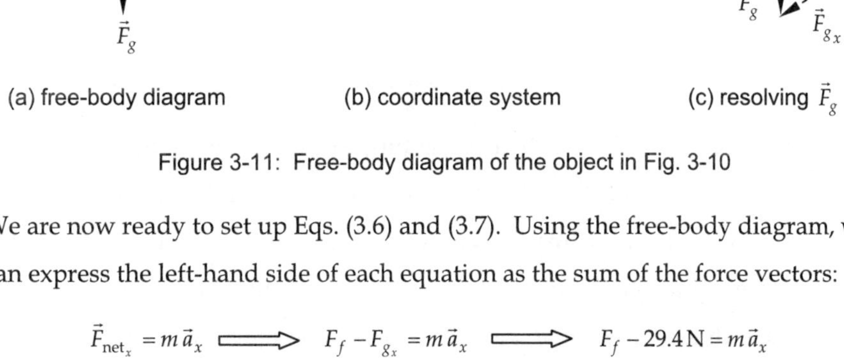

Figure 3-11: Free-body diagram of the object in Fig. 3-10

We are now ready to set up Eqs. (3.6) and (3.7). Using the free-body diagram, we can express the left-hand side of each equation as the sum of the force vectors:

$$\vec{F}_{\text{net}_x} = m\,\vec{a}_x \Longrightarrow F_f - F_{g_x} = m\,\vec{a}_x \Longrightarrow F_f - 29.4\,\text{N} = m\,\vec{a}_x$$

$$\vec{F}_{\text{net}_y} = m\,\vec{a}_y \Longrightarrow F_N - F_{g_y} = m\,\vec{a}_y \Longrightarrow F_N - 50.9\,\text{N} = m\,\vec{a}_y$$

Since the object is moving at constant velocity, its acceleration is zero. Hence, $\vec{a}_x = \vec{a}_y = 0$. The above equations can then be solved for the normal and friction forces:

$$F_f - 29.4\,\text{N} = 0 \implies F_f = 29.4\,\text{N}$$

$$F_N - 50.9\,\text{N} = 0 \implies F_N = 50.9\,\text{N}$$

Applying Eq. (3.2) (on p. 62), we finally obtain the coefficient of friction:

$$29.4\,\text{N} = \mu(50.9\,\text{N}) \implies \mu = 0.577 \cong 0.58$$

The next example is the system of two masses shown in Fig. 3-12 (a). The 25-kg mass is on a smooth inclined surface, and we are to find the magnitude of the tension force and the acceleration of the system when the masses are released from the given position. As always, start by drawing free-body diagrams. Each mass requires one free-body diagram. Besides the gravitational force, the 25-kg object experiences the normal and tension forces as it is in contact with the surface and the string. There is no friction force on it because the surface is smooth. As for the 15-kg mass, only two forces act on it, the gravitational and tension forces.

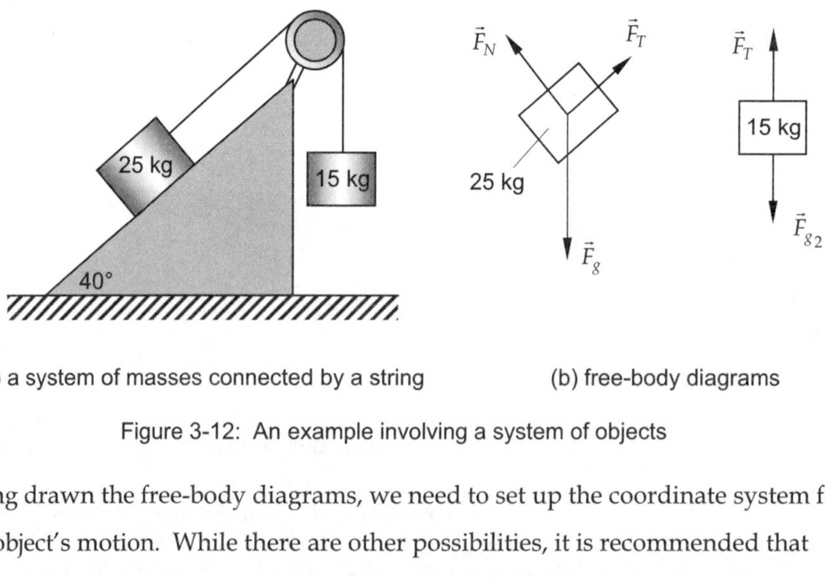

(a) a system of masses connected by a string (b) free-body diagrams

Figure 3-12: An example involving a system of objects

Having drawn the free-body diagrams, we need to set up the coordinate system for each object's motion. While there are other possibilities, it is recommended that you set up the coordinate systems as shown in Fig. 3-13 (a). Notice that the x-axis for the 15-kg is directed downward while that for the 25-kg mass points up the incline. This combination of coordinate systems allows both objects to move in the same direction in their respective coordinate system. In other words, if the 15-kg mass falls and accelerates in its $+x$-direction (which is downward), the 25-kg mass also accelerates in its $+x$-direction (which is up the incline).

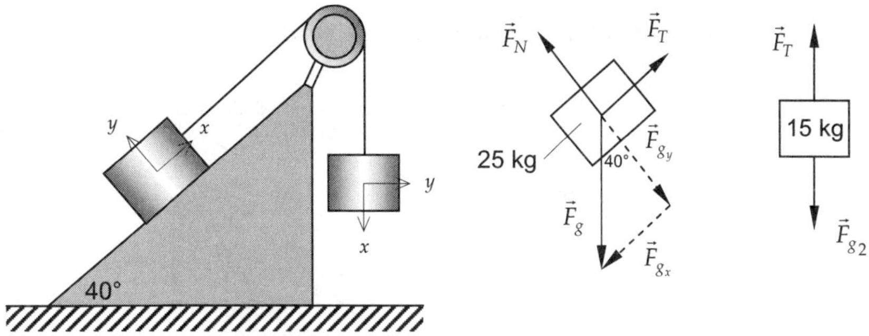

(a) setting up coordinate systems (b) free-body diagrams

Figure 3-13: Resolving force vectors for the system of objects in Fig. 3-12

Now that coordinate systems have been defined, we resolve the force vectors. The only vector that needs to be resolved is the gravitational force on the 25-kg mass, shown in Fig. 3-13 (b) with its components, $\vec{F}_{g_x}$ and $\vec{F}_{g_y}$. They are

$$F_{g_x} = mg\sin\theta = 25\,\text{kg}\cdot(9.8\,\text{m/s}^2)\sin 40° = 157.5\,\text{N}$$

$$F_{g_y} = mg\cos\theta = 25\,\text{kg}\cdot(9.8\,\text{m/s}^2)\cos 40° = 187.7\,\text{N}$$

Then the equations of motion for the 25-kg mass become

$$\vec{F}_{\text{net}_x} = m\vec{a}_x \Longrightarrow F_T - F_{g_x} = m\vec{a}_x \Longrightarrow F_T - 157.5\,\text{N} = 25\,\text{kg}\cdot\vec{a}_x$$

$$\vec{F}_{\text{net}_y} = m\vec{a}_y \Longrightarrow F_N - F_{g_y} = m\vec{a}_y \Longrightarrow F_N - 187.7\,\text{N} = 25\,\text{kg}\cdot\vec{a}_y$$

Since the 25-kg mass moves only along the inclined surface, $\vec{a}_y = 0$. Thus,

$$F_N - 187.7\,\text{N} = 0 \Longrightarrow F_N = 187.7\,\text{N} \cong 190\,\text{N}$$

2 equations & 2 unknowns!

As for the 15-kg mass, we only need to set up the equation in the x-direction:

$$\vec{F}_{\text{net}_x} = m_2\,\vec{a}_x \Longrightarrow F_{g_2} - F_T = m_2\,\vec{a}_x \Longrightarrow 15\,\text{kg}\cdot(9.8\,\text{m/s}^2) - F_T = 15\,\text{kg}\cdot\vec{a}_x$$

We now have a system of two equations and two unknowns (F_T and $\vec{a}_x$). Adding the equations together produces another equation in terms of $\vec{a}_x$:

$$\begin{array}{r} F_T - 157.5\,\text{N} = 25\vec{a}_x \\ +\,15\,\text{kg}\cdot(9.8\,\text{m/s}^2) - F_T = 15\,\text{kg}\cdot\vec{a}_x \\ \hline 15\,\text{kg}\cdot(9.8\,\text{m/s}^2) - 157.5\,\text{N} = 40\,\text{kg}\cdot\vec{a}_x \end{array} \Longrightarrow \vec{a}_x = -0.26\,\text{m/s}^2$$

The negative sign in the answer indicates that the masses move in the $-x$-direction. That is, the 25-kg mass slides down the slope while the 15-kg mass is pulled up by the string. Substituting -0.26 m/s² for $\vec{a}_x$ in one of the two equations, we find that the magnitude of the tension force is 150 N.

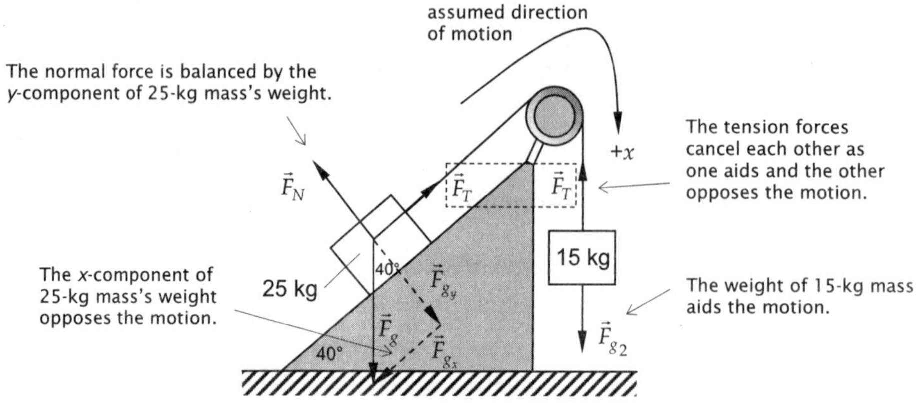

Figure 3-14: Treating a system of masses as one

In case you are interested in only the acceleration of the masses, there is an alternative, simpler way to find it. We treat the system of masses as one object and assume its direction of motion as shown in Fig. 3-14. Set the assumed direction of motion as positive. Then the forces aiding the motion are positive, and those opposing the motion are negative. The equation of motion is written as

$$\vec{F}_{\text{net}} = F_{\text{aid}} - F_{\text{oppose}} = F_{g_2} - \cancel{F_T} + \cancel{F_T} - F_{g_x} = (m_1 + m_2)\vec{a}$$

Substitute the numbers into the equation and solve for the acceleration of the system:

$$15\,\text{kg}\cdot(9.8\,\text{m/s}^2) - 25\,\text{kg}\cdot(9.8\,\text{m/s}^2)\sin 40^\circ = (15\,\text{kg} + 25\,\text{kg})\vec{a}$$

$$\vec{a} = -0.26\,\text{m/s}^2$$

To find the magnitude of the tension force, you would need to set up an equation of motion for either the 15-kg mass or the 25-kg mass and substitute the value of the acceleration into that equation.

PHYSICS 12 ANALYZED: DYNAMICS

3-1. More on Drawing a Free-Body Diagram

➢ Use the following procedure to draw a free-body diagram:

1. Draw the object of interest.
2. Draw the force of gravity.
3. Look for things in contact with the object.
   - If in contact with a surface, draw $\vec{F}_N$ and $\vec{F}_f$. (If the surface is smooth, there is no friction.)
   - If in contact with a string, draw $\vec{F}_T$.
   - If in contact with another object, draw $\vec{F}_A$.

3-2. How to Solve Two-dimensional Dynamic Problems

1. Draw the free-body diagram.
2. Set up the coordinate system. The x-axis is usually chosen to be parallel to the surface, and the y-axis perpendicular to it.
3. Resolve each force vector in the free-body diagram into its components.
4. Set up the following equations of motion:

$$\vec{F}_{\text{net}_x} = m\vec{a}_x \text{ and } \vec{F}_{\text{net}_y} = m\vec{a}_y$$

5. If there are two or more objects, repeat steps 1 to 4 for each object.
6. Solve for the unknown(s) in the equations.

EXERCISES ON CHAPTER 3: DYNAMICS

Exercise 3-1

A 0.26-kg pendulum bob is held motionless by a horizontal force, $\vec{F}_A$, at the position shown.

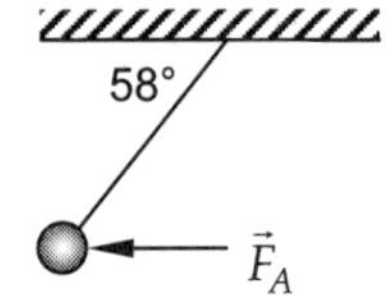

a) Draw the free-body diagram of the pendulum bob.

b) Find the magnitude of the applied force on the pendulum bob.

Exercise 3-2

A force of 25 N, 25° north of east, and a force of 10 N, 10° east of south, are acting on a 5.0-kg mass as shown. Find the acceleration of the mass.

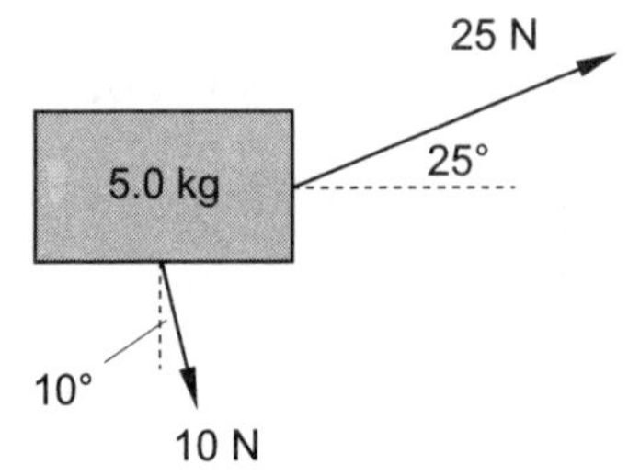

Exercise 3-3

An object is sliding down a frictionless slope as shown.

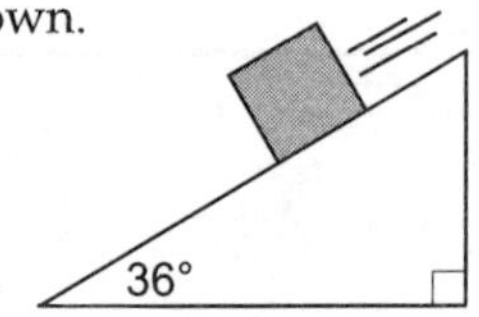

a) Draw the free-body diagram of the object.

b) Find the acceleration of the object.

Exercise 3-4

Mrs. Strong is pulling a 15-kg luggage with a rope at constant speed along the floor, as shown below. She is exerting a 50-N force on the rope.

a) Draw the free-body diagram of the luggage.

b) Calculate the magnitudes of the normal and friction forces on the luggage.

c) Find the coefficient of friction between the luggage and the floor.

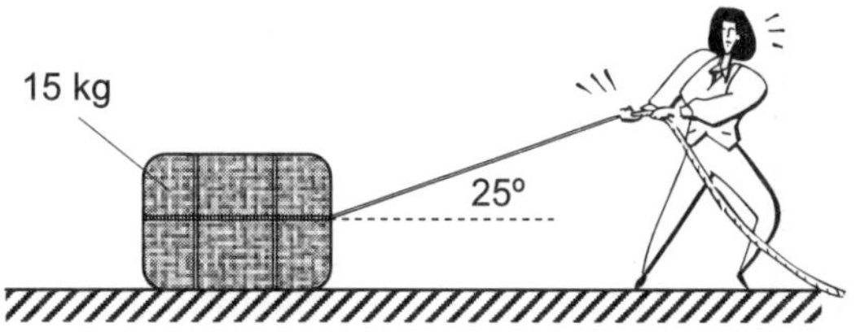

Exercise 3-5

A helicopter accelerates horizontally while carrying a 220-kg water balloon by a rope as shown.

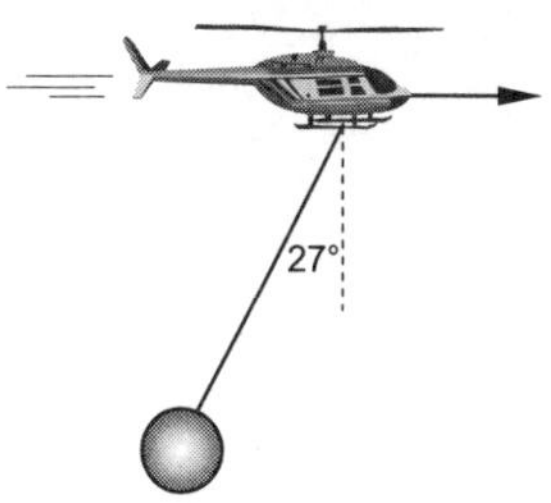

a) Draw the free-body diagram of the water balloon. (Ignore air resistance.)

b) Find the magnitude and direction of the helicopter's acceleration.

Exercise 3-6

A 23-kg block is at rest on a 23° slope as shown. Find the magnitudes of the friction and normal forces on the block.

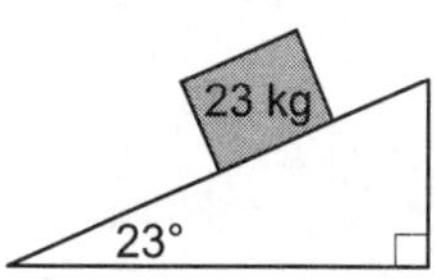

Exercise 3-7

A 170-N force is applied to a 68-kg man in his 23-kg wheelchair as shown.

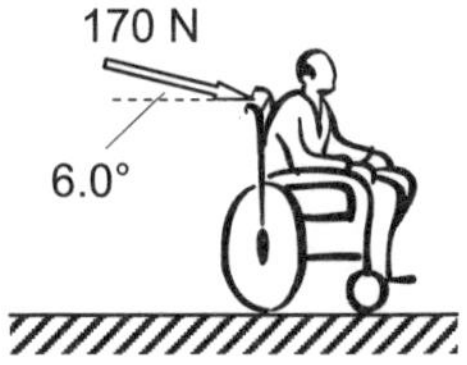

a) Draw the free-body diagram of the man and his wheelchair.

b) Find the magnitude of the normal force.

c) If the coefficient of friction is 0.18, find the acceleration of the man and his wheelchair.

Exercise 3-8

A 75-kg skier accelerates down a 28° slope at 1.8 m/s^2.

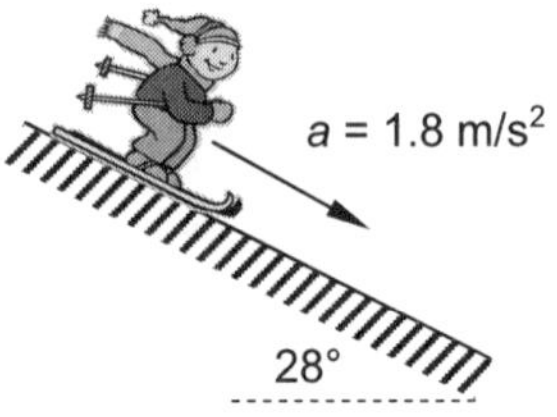

a) Draw the free-body diagram of the skier.

b) Find the magnitude of the friction force on the skier and the coefficient of friction between the skier and the slope.

Exercise 3-9

The system of objects in the diagram below is pulled to the right by an applied force and accelerates at 0.72 m/s^2.

a) Ignoring friction, find the tension force in each of the three strings.

b) Assuming the coefficient of friction is 0.37, find the magnitude of the applied force.

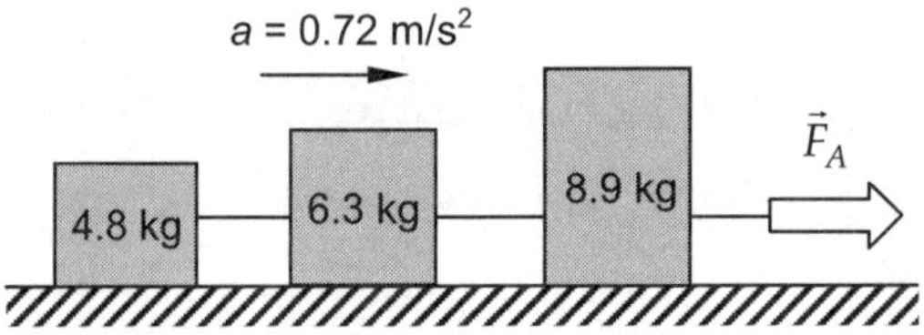

Exercise 3-10

A 1.5-kg box is pulled up a smooth incline by a string as shown and accelerates up the incline at 0.77 m/s².

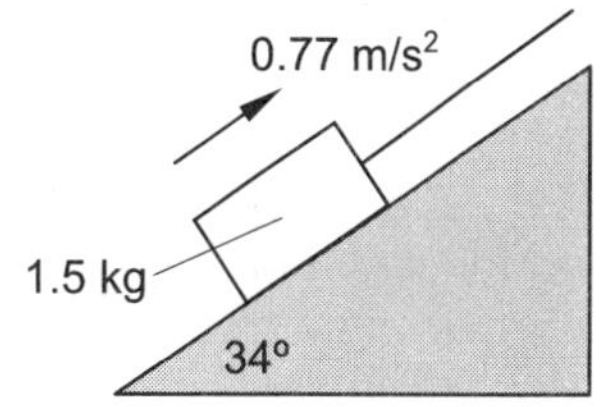

a) Draw the free-body diagram of the box.

b) Find the magnitude and direction of each and every force acting on the box.

Exercise 3-11

Two objects connected by a string are released from the position shown.

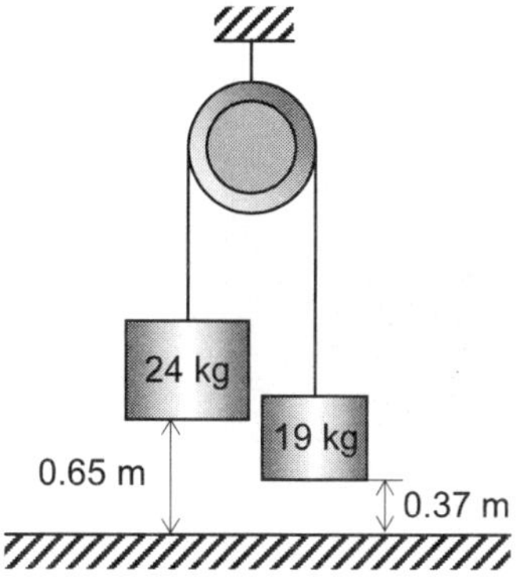

a) Find the acceleration of the objects and the tension in the string.

b) Determine time taken for the 24-kg object to reach the ground.

c) Find the velocity of the 19-kg object when the 24-kg object hits the ground.

Exercise 3-12

A 13-kg object is pushed up an incline by a horizontal applied force as shown in the diagram. The coefficient of friction between the object and the surface is 0.40. Find the magnitude of the applied force required to move the object at constant speed.

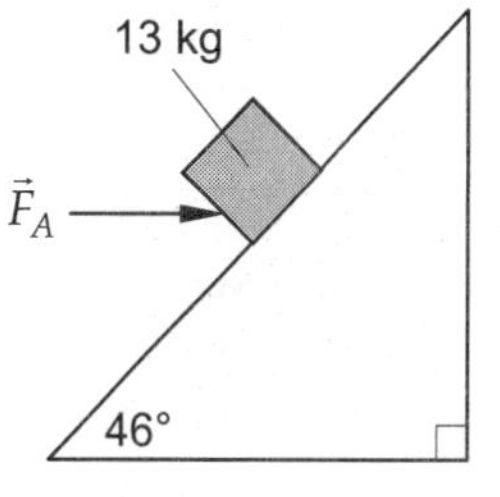

Exercise 3-13

The system of masses shown right accelerates at 2.2 m/s². Find the coefficient of friction between the 5.0-kg mass and the surface.

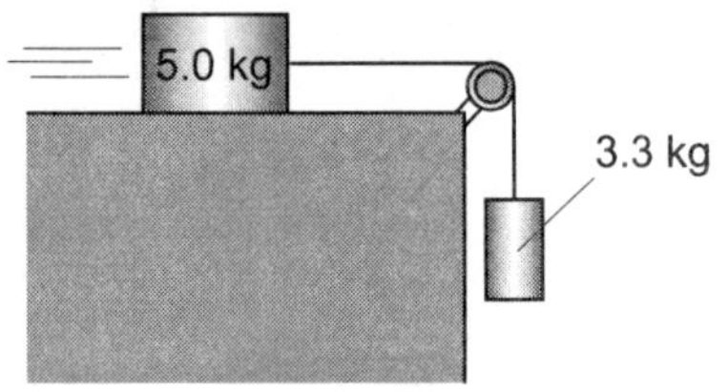

Exercise 3-14

A skier starts from rest and travels 33 m down a ski slope in 8.3 s. Find the angle the slope makes with the horizontal. (Ignore friction.)

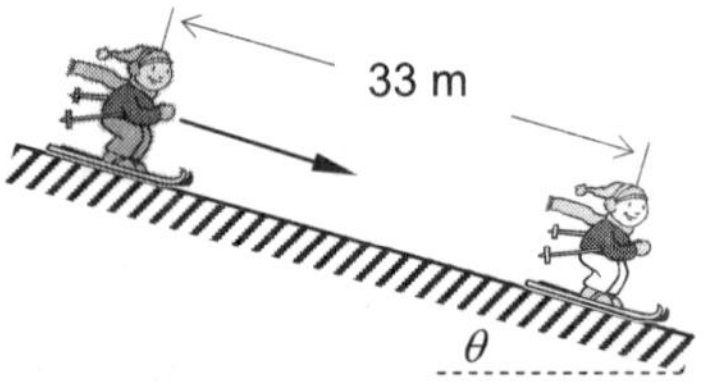

Exercise 3-15

Two cylinders connected by a string are released from the position shown. Find the magnitude of the tension force and the 3.2-kg cylinder's acceleration. (Ignore friction.)

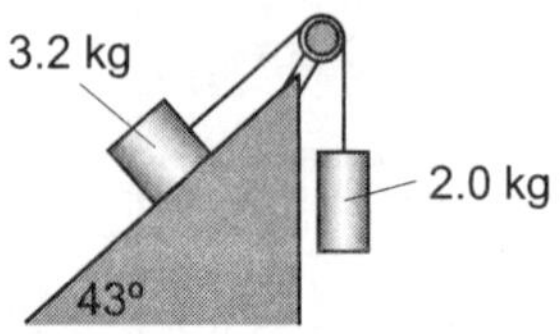

Exercise 3-16

A hockey puck is moving at 13 m/s when it starts climbing an incline. It travels 7.1 m from the base of the incline before sliding back down the slope, as shown.

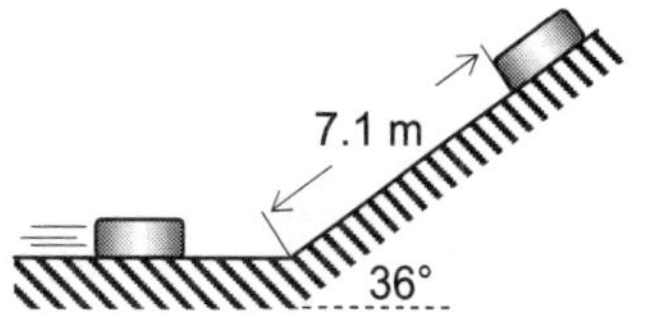

a) Draw the free-body diagram of the puck as it moves up the incline.

b) Find the coefficient of friction between the puck and the incline.

Exercise 3-17

Two objects are connected by a string as shown. The 7.5-kg object is on a frictionless slope, and the 9.0-kg object is on a horizontal surface where the coefficient of friction is 0.31. Find the magnitude of the tension force and the magnitude of the objects' acceleration.

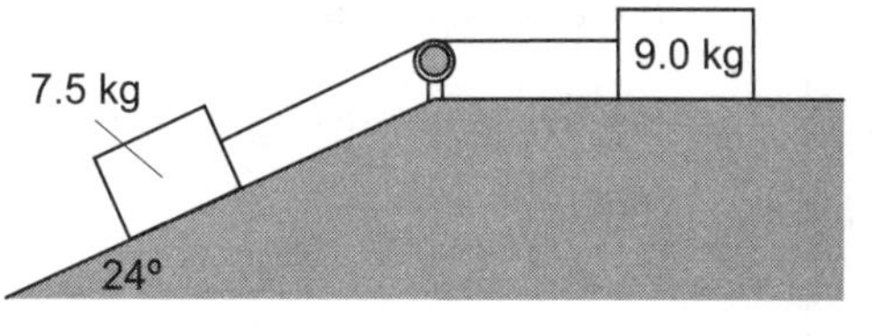

Exercise 3-18

A 2.4-kg object is sliding down a 30° slope at 2.2 m/s at the position shown right. If the coefficient of friction between the object and the slope is 0.63, how far does the block travel before coming to rest?

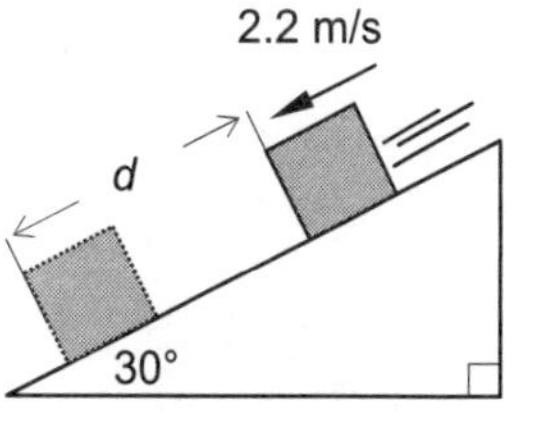

Exercise 3-19

For the system of masses shown, find the acceleration of the 8.0-kg mass when the system is released assuming (a) no friction and (b) $\mu = 0.14$.

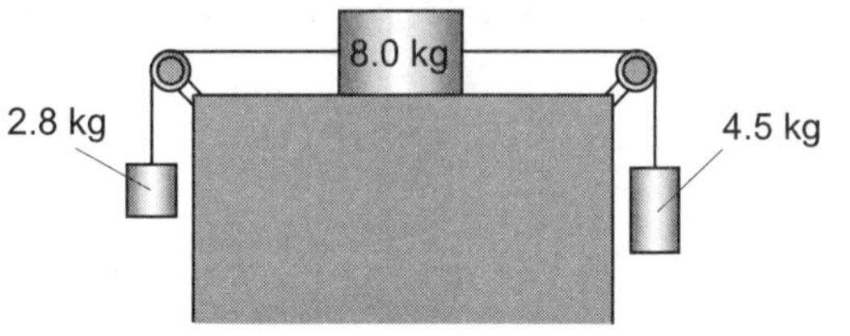

Exercise 3-20

A hockey puck weighing 0.23 kg is sliding up an incline with an initial speed of 12 m/s, as shown. The coefficient of friction between the puck and the surface is 0.35.

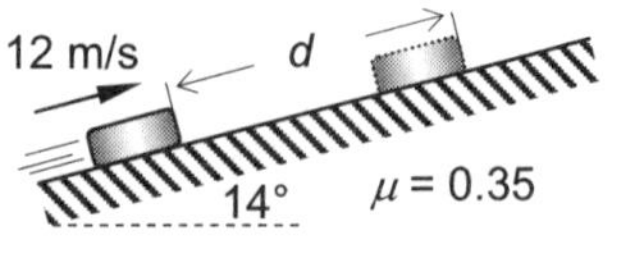

a) Find the magnitude of the friction force on the puck.

b) Find the magnitude of the puck's acceleration.

c) Determine the distance the puck travels before sliding back down the incline.

Exercise 3-21

Two masses connected by a string are released from the position shown below on a frictionless surface. Find the acceleration of the masses and the magnitude of the tension force.

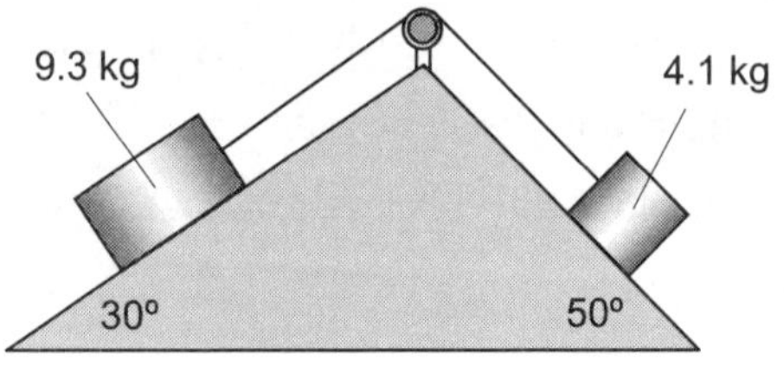

PROBLEMS ON CHAPTER 3: DYNAMICS

(Answer Key on p. 367)

1 - 10. Fill in the blanks

1. __________ is an interaction between two objects producing an attraction or a repulsion.
2. The __________ force is an attractive force between two pieces of matter.
3. The __________ force, exerted by a surface, is perpendicular to the surface.
4. The __________ force, also exerted by a surface, acts to oppose an object's motion.
5. The __________ force is the force transmitted through a string when it is taut.
6. The magnitude of the friction force on a moving object is the product of the __________ of __________ and the magnitude of the normal force on the object.
7. A __________ __________ shows an object of interest and all the forces acting on it.
8. The tendency of a body to maintain its state of motion is called __________.
9. __________ __________ law states that the acceleration of a body is proportional to the net force acting on it.
10. __________ __________ law states that, for every action, there is an equal and opposite reaction force.

11 - 19. True/False questions

11. _____ The gravitational force exists only between an object and the Earth.
12. _____ Tension forces always point along the string, away from the object.
13. _____ If an object near the surface of the Earth is in contact with nothing, no force appears in the free-body diagram of the object.
14. _____ If the normal force acting on an object is zero, there is no friction force acting on it, either.
15. _____ If the friction force acting on an object is zero, there is no normal force acting on it, either.
16. _____ For an object moving on an inclined surface, the direction of the friction force is always parallel to the surface, up the incline.

17. _____ For an object moving on an inclined surface, the magnitude of the normal force acting on it is greater than that of its weight.

18. _____ The gravitational and normal forces always appear together in a free-body diagram.

19. _____ The net force vector is usually included in a free-body diagram.

20 – 30. Multiple choice questions

20. Which of the following is true about an object set in motion?

a) It wants to come to a stop.
b) It wants to maintain its motion.
c) It wants to accelerate.
d) None of the above because we do not know the net force acting on the object.

21. Which of the following is supported by Newton's Second Law?

a) $\vec{a} = 0$
b) $\vec{a} = \dfrac{\Delta \vec{v}}{\Delta t}$
c) $\vec{a} = \dfrac{\vec{F}_{net}}{m}$
d) $\vec{F}_{\text{on A by B}} = -\vec{F}_{\text{on B by A}}$

22. A stationary object on an incline remains motionless when the angle of incline, θ, is gradually increased. As the angle θ increases,

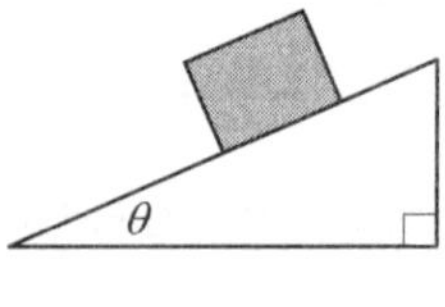

a) the magnitude of the object's weight increases.
b) the magnitude of the normal force on the object increases.
c) the magnitude of the friction force on the object increases.
d) the magnitude of the net force on the object increases.

23. When a system of masses is released from the position shown, its acceleration and the tension in the string are a and F_T, respectively. How do they compare with those obtained on a frictionless surface, a' and F_T'?

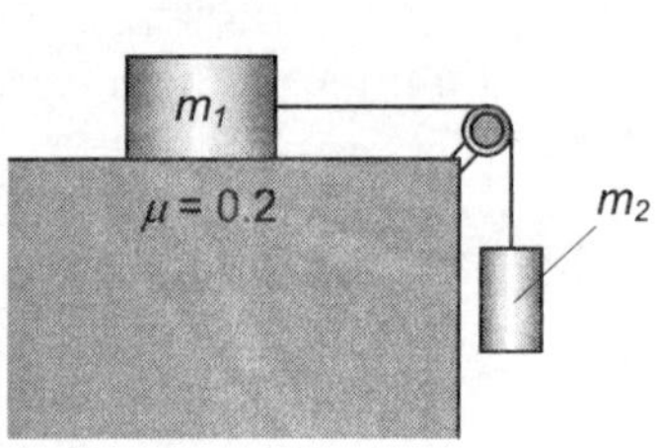

a) $a > a'$ and $F_T > F_T'$
b) $a > a'$ and $F_T < F_T'$
c) $a < a'$ and $F_T > F_T'$
d) $a < a'$ and $F_T < F_T'$

For Problems 24 to 26, refer to the graphs shown below:

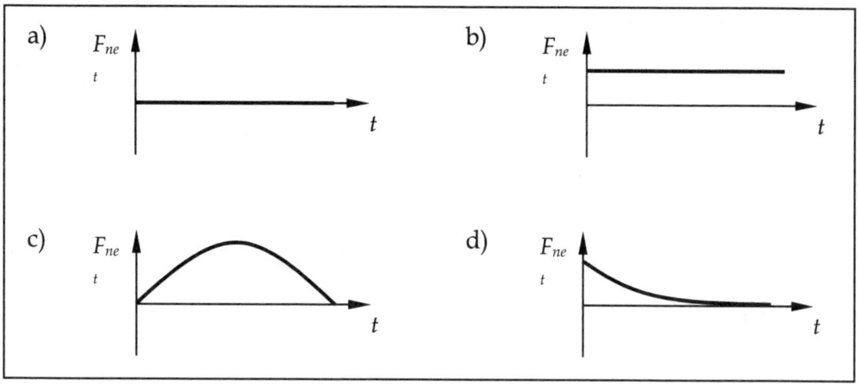

24. An object is thrown in the air as shown. Which of the above graphs best represents the magnitude of the net force on the object vs. time?

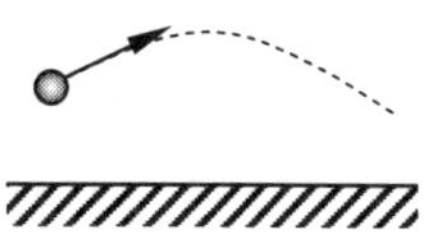

25. An object is sliding down a 45° slope, where the coefficient of friction is 1.0. Which of the above graphs best represents the magnitude of the net force on the object vs. time?

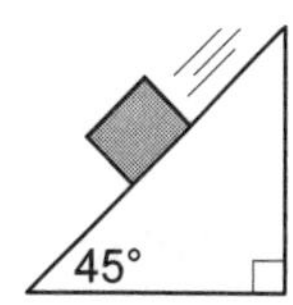

26. An object is released at the top of a smooth, curved slope as shown. Which of the above graphs best represents the magnitude of the net force on the object vs. time?

27. A pendulum is attached to the ceiling of a cart. When the cart accelerates to the right on a horizontal surface, the pendulum swings to the left as shown. Which of the following is the correct free-body diagram of the pendulum bob?

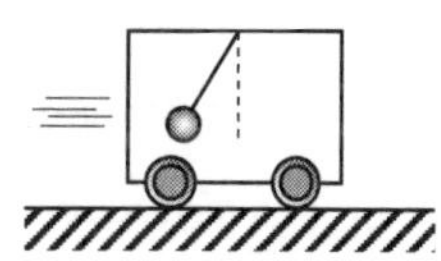

a)  b)  c)  d) 

28. Which of the following is the direction of the net force on the pendulum bob in Problem 27?

a) b) c)  d) 

29. What happens to the pendulum in Problem 27 when the cart starts to slow down while continuing to move to the right?

    a) The pendulum swings farther to the left.
    b) The pendulum swings to the right and stops at the vertical line.
    c) The pendulum swings to the right and passes the vertical line.
    d) There is no movement of the pendulum relative to the cart.

30. When the system of masses shown below is released, it is seen to move to the left. If the friction between the masses and the surface is negligible, it must be true that

    a) $m_1 > m_2$
    b) $\theta_1 > \theta_2$
    c) $m_1 \sin\theta_1 > m_2 \sin\theta_2$
    d) $m_1 \cos\theta_1 > m_2 \cos\theta_2$

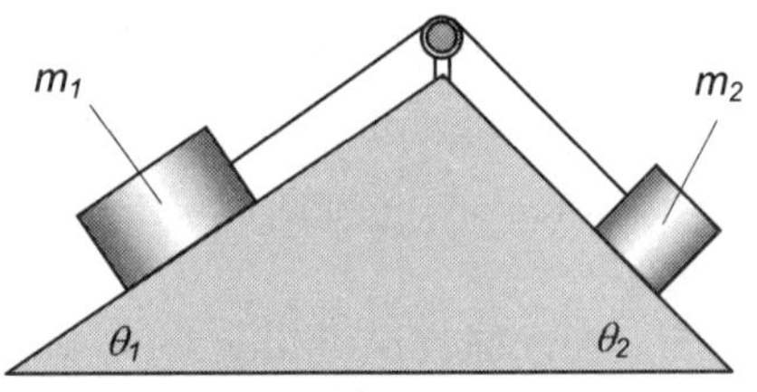

31 – 40. Written-response questions

31. A 35-g bullet moving at high speed strikes a wooden block at rest on an incline as shown. Upon impact, the bullet is embedded within the block, and they start moving at 15 m/s. If the coefficient of friction between the block and the ground is 0.53, how far does the block travel before coming to a stop? Show the free-body diagram of the block while it is moving. (Assume the block does not move until the bullet is fully embedded and comes to rest within it.)

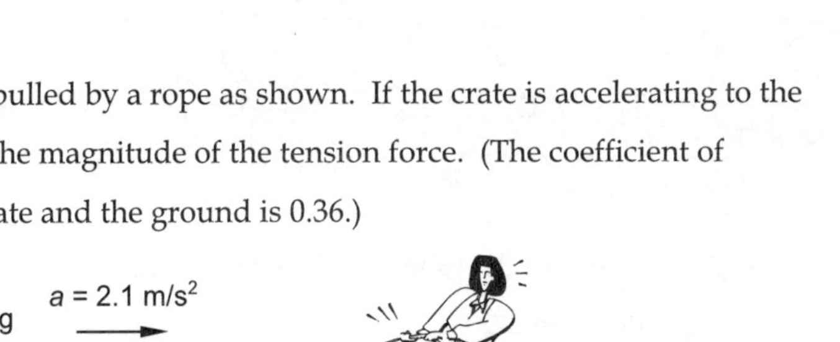

32. A 23-kg crate is being pulled by a rope as shown. If the crate is accelerating to the right at 2.1 m/s², find the magnitude of the tension force. (The coefficient of friction between the crate and the ground is 0.36.)

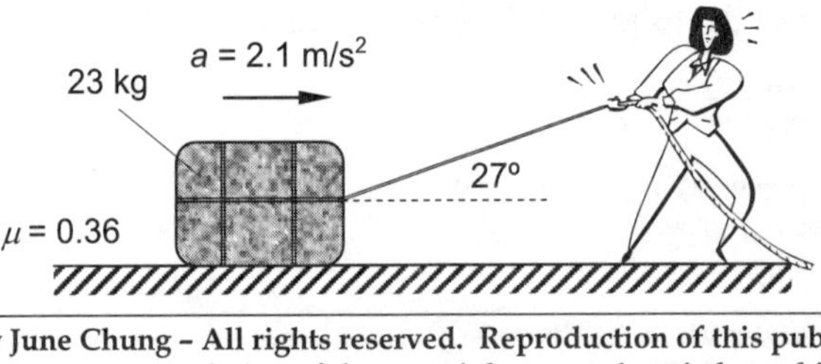

33. An expert skier, starting from rest, slides down an 18° slope and travels 120 m before becoming airborne. Determine the horizontal distance he flies before falling 11 m to the ground below. (The coefficient of friction between the skier and the surface is 0.13.)

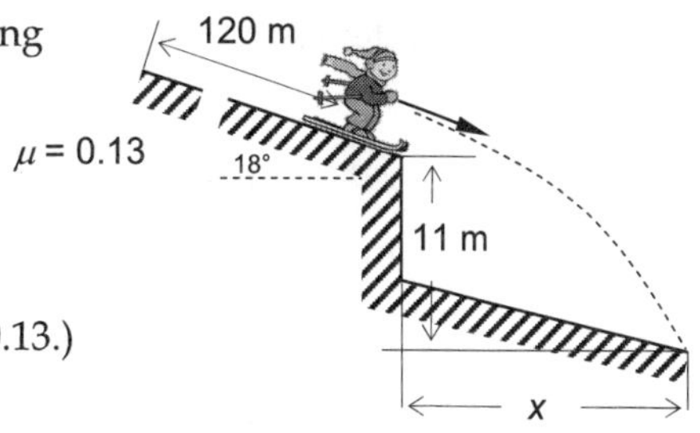

34. A block is moving up an inclined surface with the initial speed of 16 m/s. After 5.0 seconds it reaches the maximum height and then slides down the incline. If the angle of the inclination is 15°, (a) what is the coefficient of friction, and (b) how long would it take for the block to come back to the original position?

35. An 81-kg person pulls himself upward using the pulley system as shown right. If he pulls the rope down with an 870 N force, what is the magnitude and direction of his acceleration?

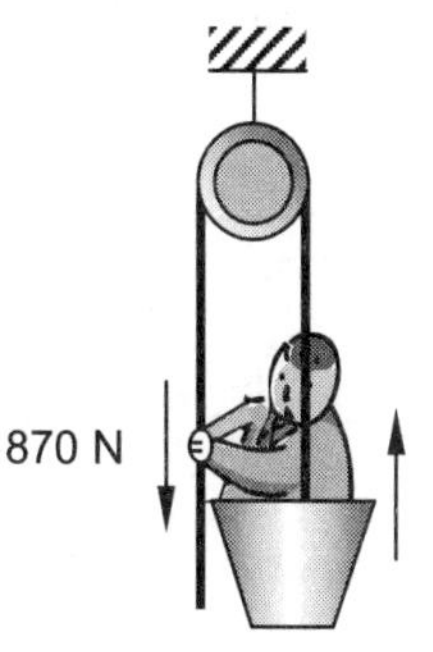

36. A system of masses, when released from the position shown, remains stationary. Express the magnitude of the friction force on M in terms of m, M, θ, and g.

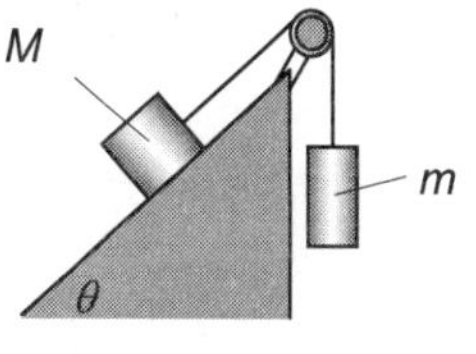

37. A system of masses is connected by a light string as shown below. If the system remains at rest when released, find the angle of the left incline, θ_l. (Assume no friction between the masses and the surface.)

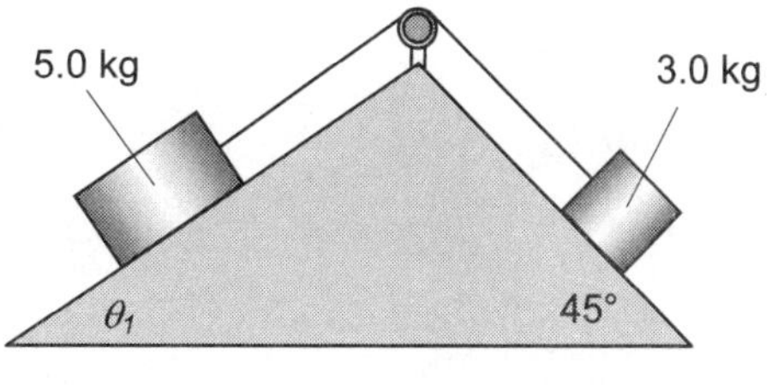

38.* Consider the system of masses in Exercise 3-21. Find the acceleration of the system and the magnitude of the tension force if the coefficient of friction is 0.12.

39. A 2.9-kg can of paint is projected up an inclined plane with an initial speed of 16 m/s, as shown.

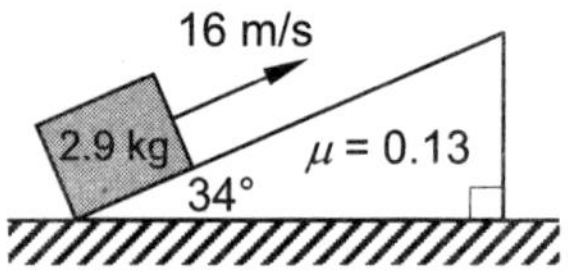

a) Determine how far up the slope the can will slide before stopping.

b) Determine the final velocity of the can when it slides back down to the bottom of the slope.

40. A wooden chest is pushed across a level floor by a 30-N force as shown. The coefficient of friction between the chest and the floor is 0.40, and the chest accelerates at 4.2 m/s².

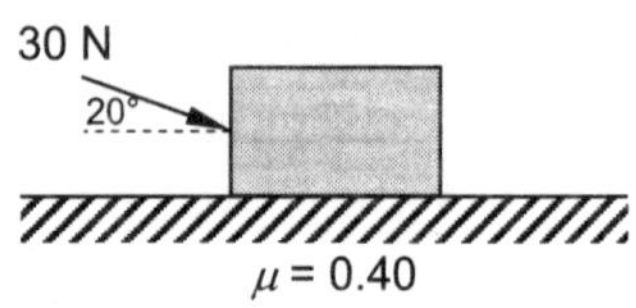

a) What is the mass of the chest?

b) If the chest were to be pulled across the floor with a 30-N force at an angle of 20° above the horizontal, determine the new acceleration of this chest.

c) Compare these two accelerations. If they are different, explain why.

CHAPTER 4

WORK AND ENERGY

So far we have analyzed motion using vector quantities such as velocity, acceleration, and force, but motion can also be analyzed using scalar quantities like work and energy. Energy is closely related to motion; as an object gains more energy, its motion becomes faster and/or the potential for faster motion increases. In a way, this principle applies to us as well. When we are starved, we have no energy to move or work, but after eating, we become livelier and can use the energy in the food to move about, or store the energy in our cells for later actions.

When solving motion problems, you must decide which approach to use - vector or scalar. For example, if you need to determine the direction of an object's motion, you should use vectors. If, on the other hand, the object's speed is required, the scalar approach is shorter and simpler. It usually does not make much difference for simple one-dimensional motions, but for two-dimensional and more complex motions, the correct choice of approach will save you a lot of work.

In this chapter, you will:

- ✓ review work and energy from Physics 11
- ✓ test your knowledge of work and energy from Physics 11
- ✓ learn more about work
- ✓ learn more about work-energy principle
- ✓ learn how to solve work-energy problems
- ✓ solve exercise problems involving work and energy
- ✓ test your knowledge of work and energy from Physics 12

Review of Physics 11: Work and Energy

1. The Law of Conservation of Energy

- Energy can be neither created nor destroyed. It can merely change from one form to another.
- The total amount of energy inside any isolated system is conserved.

An isolated system is one in which no energy is transferred to or from its surroundings.

2. Work (W)

- **Work** indicates a transfer of energy into or out of an object.
- If the direction of the force on an object is the same as that of the object's displacement,

$$W = F\,d \quad [\text{N·m} = \text{J}] \qquad \ldots (4.1)$$

  Then the work done by the force is positive, and an amount of energy equal to the work goes into the object.

- If the direction of the force on an object is perpendicular to that of the object's displacement, the force does no work on the object, and the amount of energy within the object remains constant.
- If the direction of the force on an object is opposite to that of the object's displacement,

$$W = -F\,d \qquad \ldots (4.2)$$

  Then the work done by the force is negative, and an amount of energy equal to the work comes out of the object.

- The **net work** on an object is the sum of all the work done by the individual forces acting on the object.

  Example: Suppose a 20-N object is being pulled to the right by a 15-N force as shown. The friction force is 10 N and acts against the object's motion. The work done by each force and the net work are shown below.

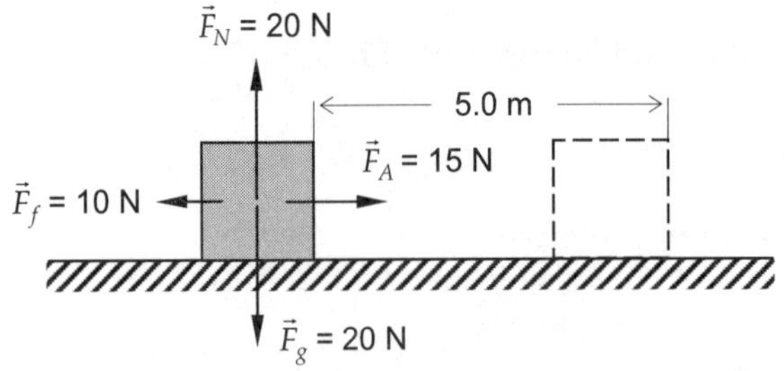

$$W_{\text{by } F_A} = F_A d = 15\,\text{N}(5.0\,\text{m}) = 75\,\text{J}$$

$$W_{\text{by } F_f} = -F_f d = -10\,\text{N}(5.0\,\text{m}) = -50\,\text{J}$$

$$W_{\text{by } F_g} = W_{by\,F_N} = 0\,\text{J}$$

$$W_{\text{net}} = 75\,\text{J} + (-50\,\text{J}) + 0\,\text{J} + 0\,\text{J} = 25\,\text{J}$$

Figure 4-1: Work done by various forces

3. Kinetic Energy (E_K)

- **Kinetic energy** is the amount of energy required to accelerate an object of mass m from rest to speed v and is defined as

$$E_K = \frac{1}{2} m v^2 \quad [\text{kg·m}^2/\text{s}^2 = \text{J}] \qquad \ldots (4.3)$$

Example: If a 10-kg object is moving at 2.0 m/s, the kinetic energy of the object is

$$E_K = \frac{1}{2}(10\,\text{kg})(2.0\,\text{m/s})^2 = 20\,\text{J}$$

4. Work-Energy Principle

- The **work-energy principle** states that the net work done on an object causes a change in its kinetic energy:

$$W_{\text{net}} = \Delta E_K = \frac{1}{2} m \left(v_f^2 - v_i^2\right) \qquad \ldots (4.4)$$

- If the net work done on an object is positive, the object gains energy, which is used to increase its speed. On the other hand, if negative work is done on an object, that much energy is lost by the object, whose speed then decreases.

Example: A 5.0-kg object is initially moving at 3.0 m/s. If 40 J of net work is done on the object, its final speed can be calculated using the work-energy principle as

$$40\,\text{J} = \frac{1}{2}(5.0\,\text{kg})\left[v_f^2 - (3.0\,\text{m/s})^2\right] \implies v_f = 5.0\,\text{m/s}$$

5. Gravitational Potential Energy (E_P)

h is the altitude of the object above some reference level, which may be set arbitrarily. Choose a convenient level, and make all calculations based on that reference level.

- **Potential energy** is stored energy associated with the position of a body within the field of another body and is defined as

$$E_P = m g h \qquad \ldots (4.5)$$

- The value of gravitational potential energy depends on your choice of the reference level (where h = 0) and may be positive, zero, or even negative! However, the actual value of the energy, whether it be kinetic or potential, is not what counts; it is the change in the energy that counts. As you will see in the following example, the change in gravitational potential energy is independent of your choice of reference level.

Example: A 10-kg object, initially 6.0 m above the ground, falls to the ground as shown. Find the change in its gravitational potential energy using (a) the ground as the reference level and (b) 6.0 m above the ground as the reference level.

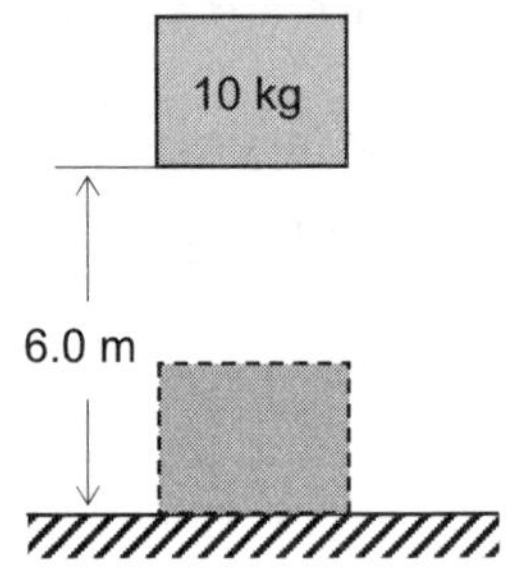

Figure 4-2: Calculation of ΔE_P

(a) The object is initially 6.0 m above the ground, so $h_i = 6.0\,\text{m}$. Thus,

$$E_{P_i} = (10\,\text{kg})(9.8\,\text{m/s}^2)(6.0\,\text{m}) = 588\,\text{J}$$

The object ends up being on the ground, so $h_f = 0\,\text{m}$.

$$E_{P_f} = 0\,\text{J}$$

Hence, the change in gravitational potential energy is

$$\Delta E_P = E_{P_f} - E_{P_i} = 0\,\text{J} - 588\,\text{J} = -588\,\text{J}$$

Regardless of where the reference level is, the change in gravitational potential energy is the same.

(b) This time, the reference level is at 6.0 m above the ground. Thus, $h_i = 0$ m and $h_f = -6.0\,\text{m}$. The final height is -6.0 m because it is 6.0 m below the reference level. The change in gravitational potential energy is

$$\Delta E_P = E_{P_f} - E_{P_i} = (10\,\text{kg})(9.8\,\text{m/s}^2)(-6.0\,\text{m}) - (10\,\text{kg})(9.8\,\text{m/s}^2)(0\,\text{m}) = -588\,\text{J}$$

6. Conservation of Mechanical Energy

- The **mechanical energy** of an object is defined as the sum of its potential and kinetic energies.
- According to the law of conservation of mechanical energy, the mechanical energy of an object is conserved during its motion in the absence of any friction or applied force. This law is expressed mathematically as

$$(E_P + E_K)_{\text{initial}} = (E_P + E_K)_{\text{final}} \quad \ldots (4.6)$$

or alternatively,

$$mgh_i + \frac{1}{2}mv_i^2 = mgh_f + \frac{1}{2}mv_f^2 \quad \ldots (4.7)$$

Example: Consider the 10-kg object shown in Fig. 4-2. It is dropped from 6.0 m above the ground. Calculate the impact speed of the object.

There is no friction or applied force on the object between the initial state (just after the object is dropped) and the final state (just before it hits the ground). Therefore, the mechanical energy of the object is conserved, and we can use Eq. (4.7) to obtain the final speed of the object:

The ground is set as the reference level.

$$(10\,\text{kg})(9.8\,\text{m/s}^2)(6.0\,\text{m}) + \frac{1}{2}(10\,\text{kg})(0\,\text{m/s})^2 = (10\,\text{kg})(9.8\,\text{m/s}^2)(0\,\text{m}) + \frac{1}{2}(10\,\text{kg})v_f^2 \implies v_f = 11\,\text{m/s}$$

7. Power (P) and Efficiency (η)

- **Power** is the rate of doing work, or the rate at which energy is transferred. It is given as

You need to use the force exerted by an object, not the force acting on it, if you want to calculate the power delivered by the object.

$$P = \frac{\text{Work}}{t} = \frac{Fd}{t} = Fv \quad [\text{J/s} = \text{N·m/s} = \text{W}] \qquad \dots (4.8)$$

Example: Suppose you exert 15 N to pull the 20-N object in Fig. 4-1 to the right. If the object moves 5.0 m in 3.0 s, the power delivered by you is

$$P = \frac{(15\,\text{N})(5.0\,\text{m})}{3.0\,\text{s}} = 25\,\text{W}$$

Kilowatt-hour [kW·h] is a unit of energy, not power. 1 kW·h = 3.6×10^6 J

- The **efficiency** of a machine, η, is defined as

$$\eta = \frac{W_{out}}{W_{in}} = \frac{P_{out}}{P_{in}} \qquad \dots (4.9)$$

P_{in} = the power input to the machine.
P_{out} = the power output of the machine.

- Efficiency is always less than or equal to 1.

Example: A 60-W bulb that is 30% efficient generates power equal to

$$P_{out} = P_{in} \times \eta = (60\,\text{W})(0.3) = 18\,\text{W}$$

8. How to Solve Work-Energy Problems

1. Read the problem statement carefully and identify the initial and final states of motion. Draw a diagram to show the starting point, the journey, and the final point, and fill in the data on the diagram. Do not jump to the next step until you are sure what the initial and final states are.
2. Choose an appropriate point to be your reference level. Write down everything you know about the initial and final states such as height and speed. If there is no mention of height in the problem, you might be able to assume the object is moving horizontally. Also, write down the mass and the displacement of the object if they are known.
3. Observe whether a friction force or applied force is involved. If any one of them is, apply the work-energy principle in Eq. (4.4). If not, use conservation of mechanical energy and set up Eq. (4.6) or Eq. (4.7).
4. Substitute the numbers into the equation. Check that there is only one unknown in the equation. Solve for the unknown and compare your answer with your own experience.

Test Your Knowledge of Physics 11: Work and Energy

(Answer Key on p. 368)

1 - 10. True/False questions

1. _____ If the work done on an object is -50 J, it means 50 J of energy has been lost by the object.

2. _____ The work done by the gravitational force on a moving object is always zero.

3. _____ The net work done on an object is zero if the net force on it is zero.

4. _____ An object's kinetic energy is directly proportional to its speed.

5. _____ An object gains speed when the net work done on it is positive.

6. _____ An object gains altitude when the net work on it is positive.

7. _____ An object above the ground has positive gravitational potential energy relative to the ground.

8. _____ For a projectile, a decrease in its gravitational potential energy means that its kinetic energy decreases by the same amount.

9. _____ If one machine can generate more power than another, its efficiency is higher than the efficiency of the other.

10. _____ The efficiency of a machine can never be greater than one.

11 - 20. Multiple choice questions

11. An object is being pulled up by a string as shown. Which of the forces acting on the object does positive work on the object?

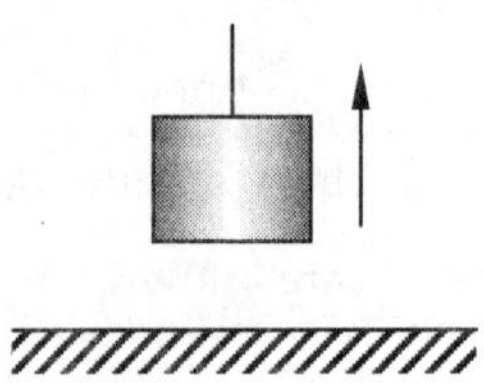

a) only the gravitational force
b) only the tension force
c) both the gravitational and tension forces
d) neither the gravitational nor tension force

12. Which of the following is <u>not</u> a unit of energy?

a) J b) kW·h

c) N·m d) W

13. A moving object is gaining speed. Which of the following statements must be true?

    a) Net work being done on the object is increasing.
    b) The object's kinetic energy is increasing.
    c) The object's potential energy is increasing.
    d) The object's mechanical energy is increasing.

For Problems 14 to 16, refer to the statements below:

I.	The object's gravitational potential energy decreases.
II.	The object's kinetic energy increases.
III.	The object's mechanical energy remains constant.

14. Which of the above statements is true for an object being pulled on level ground at constant speed?

    a) I only　　b) II only
    c) III only　　d) II and III only

15. Which of the above statements is true for an object being pushed up an incline at constant speed?

    a) I and II only　　b) II and III only
    c) I, II, and III　　d) None of the statements is true.

16. Which of the above statements is true for a dropped object falling towards the ground?

    a) I and II only　　b) I and III only
    c) II and III only　　d) All of the statements are true.

17. Three balls are thrown with the same initial speed but at different angles from the top of a tall building as shown. Which ball has the greatest speed at impact with the ground?

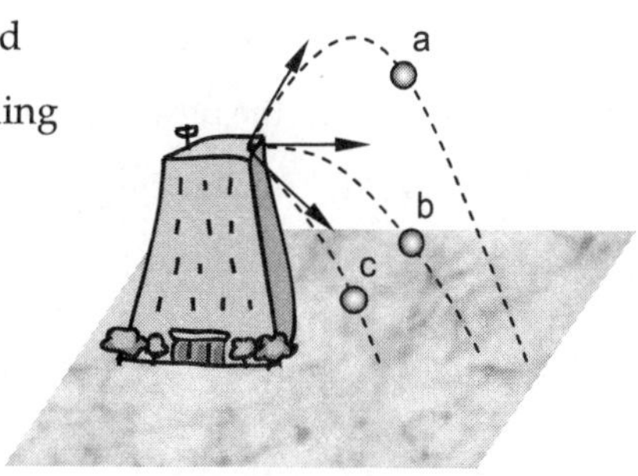

    a) ball A　　b) ball B
    c) ball C
    d) They all have the same speed at impact.

18. An object is thrown vertically upward from the ground. Which of the following graphs best represents the object's potential and kinetic energy during the flight?

a) 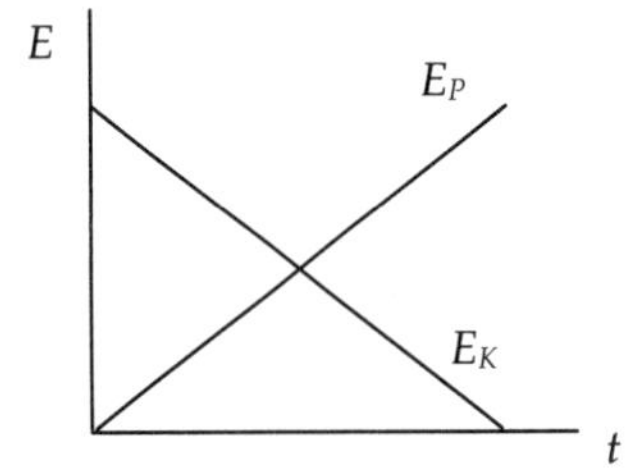

b) 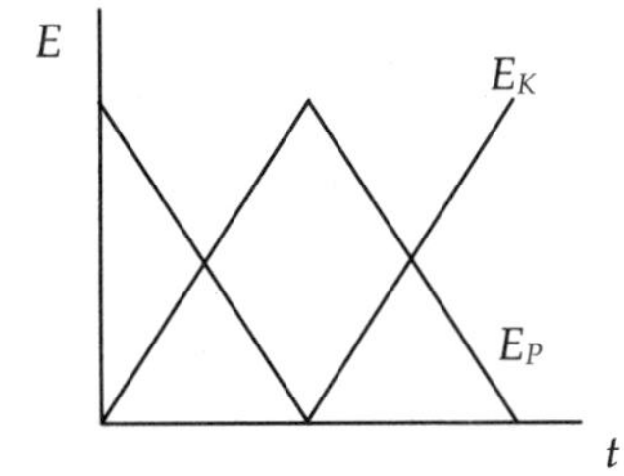

c) 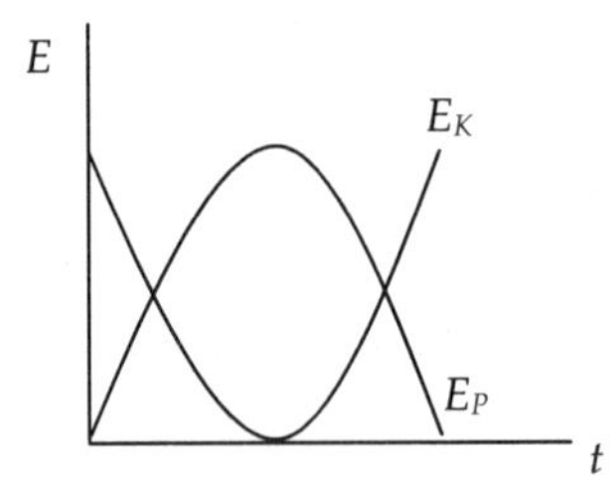

d) 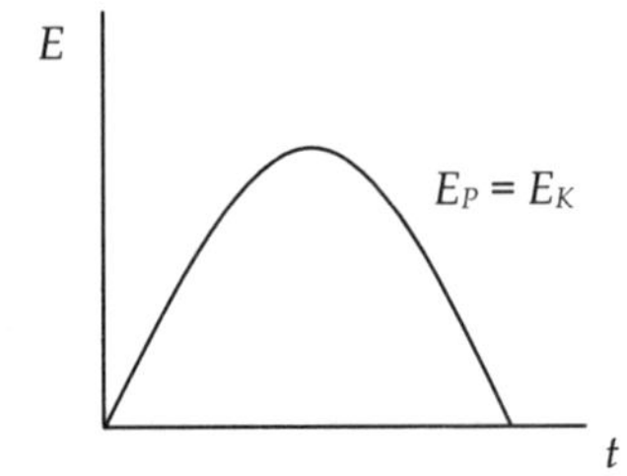

19. An object is sliding down an inclined surface as shown. Which of the following statements could be false?

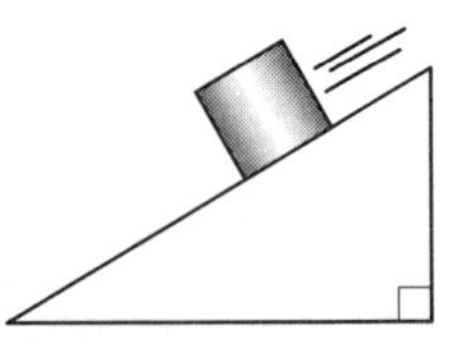

a) The work done by the normal force is zero.
b) The work done by the friction force is negative.
c) The kinetic energy of the object is increasing.
d) The gravitational potential energy of the object is decreasing.

20. A crate is pulled up an inclined surface at constant speed by a motor as shown. If the crate had been pulled up at twice the original speed, then

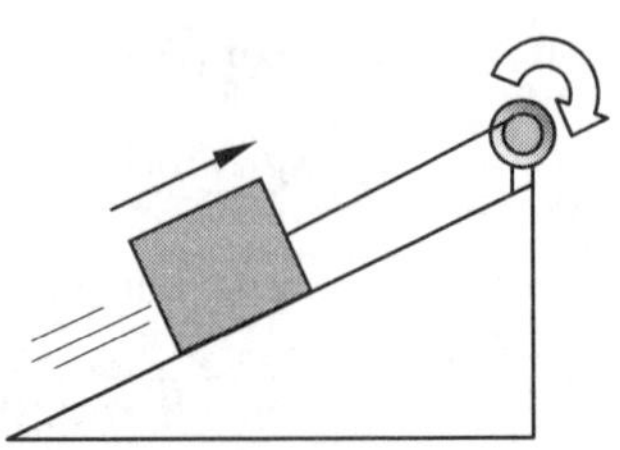

a) the crate's kinetic energy would have doubled.
b) the work done by the motor would have doubled.
c) the power exerted by the motor would have doubled.
d) the efficiency of the motor would have doubled.

21 - 35. Written-response questions

21. A 2.8-kg sled is pulled across level ground by a 62-N force and moves 3.2 m. If the coefficient of friction is 0.27, draw the free-body diagram of the sled and find the work done by each force, and the net work done on the sled.

22. How much work must be done by the friction force if a 100-kg wagon initially sliding along a horizontal road at 8.5 m/s is brought to rest?

23. Joshua needs to accelerate a 10.3-kg wagon from rest to 7.5 m/s while he pushes it 9.0 m on a flat, frictionless floor. How much force should he exert on the wagon?

24. An object at rest accelerates to 10 m/s when 6.8 J of work is done on it. If the same object needs to reach a speed of 30 m/s from rest, how much work must be done on it?

25. A 1.5-kg ball is thrown horizontally from the roof of a 28-m tower with an initial speed of 16 m/s. (a) Find the speed of the ball when it is halfway to the ground; (b) How far above the ground is the ball when its speed reaches 25 m/s?

26. A 6.0-kg object is dropped from a height of 20 m above the ground. At what height above the ground is the object's kinetic energy twice its gravitational potential energy? (Set the ground as the reference level.)

27. A 0.68-kg pendulum bob is released 30° from the vertical, as shown. What is the speed of the bob as the pendulum swings by the vertical position (indicated by the dotted line)? How will the answer change if the mass of the bob is doubled?

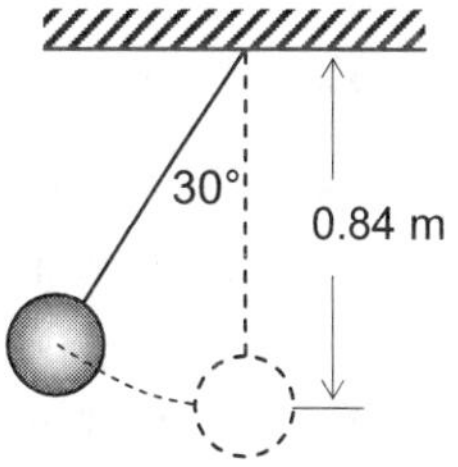

28. A 750-g object is initially moving along level ground at 6.5 m/s, and we must accelerate it to 10.2 m/s in 3.2 s. If the surface is frictionless, calculate (a) the magnitude of the horizontal force we need to apply on the object, (b) the work done on the object while we exert the applied force, and (c) the average power we need to exert on the object.

29. A roller coaster car, initially moving at an unknown speed at point A, passes through point B at the speed of 18 m/s. Find the car's speed at A and at C. (Ignore friction.)

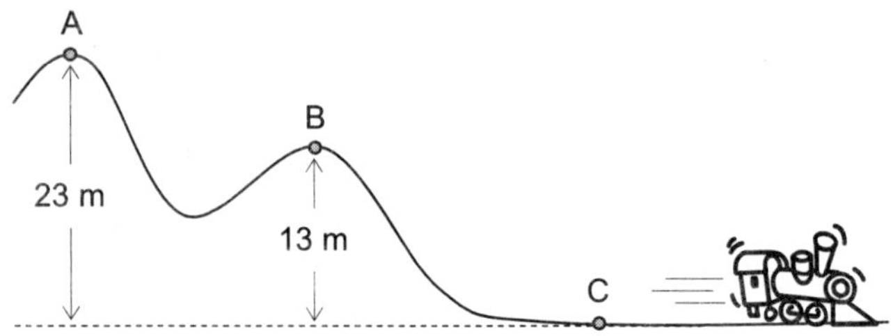

30. A 2.2-kg object is released from rest on an inclined surface as shown.

(a) If there is no friction, calculate the object's speed at the bottom of the slope.

(b) If the object's speed at the bottom of the incline is 2.6 m/s, find the amount of mechanical energy lost to friction.

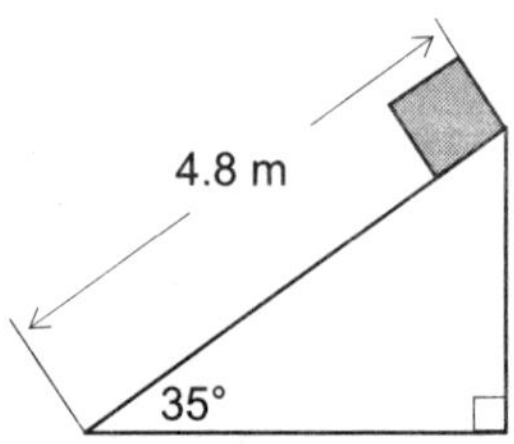

31. A 1200-kg elevator ascends a vertical height of 410 m with an average speed of 2.8 m/s. What is the average power supplied by the motor in lifting the elevator car?

32. A skater increases her kinetic energy from 320 J to 1800 J by skating 32 m in 4.5 s. (a) Find the minimum power of the skater; (b) If she expends 3100 J during that 4.5-s period, find the efficiency of her acceleration.

33. An elevator driven by a 21-kW motor is carrying five occupants. The total mass of the elevator including the occupants is 1450 kg. If the elevator is pulled up at a constant speed of 0.85 m/s by the motor, what is the efficiency of the motor?

34. A machine can lift a 90-kg object from the ground to the top of a 23-m building in 30 seconds. If its efficiency is 45%, determine the power output of the machine and the power input to the machine.

35. On average, the appliances in Andrew's home are 47% efficient, and his monthly electricity consumption from using those appliances is 1200 kW·h. If he upgrades his appliances and improves their average efficiency to 62%, how much will he save per month? (Assume the power company charges seven cents per kilowatt-hour.) (1 kW·h = 3.6×10^6 J)

What's New in Physics 12?

4-1. More on Work

Work done on an object by a force depends on the angle between the force and displacement vectors. Having investigated special cases where the angle is 0°, 90°, or 180° in Physics 11, we now generalize the formula for work that considers situations where the angle can be anything between 0° and 180°.

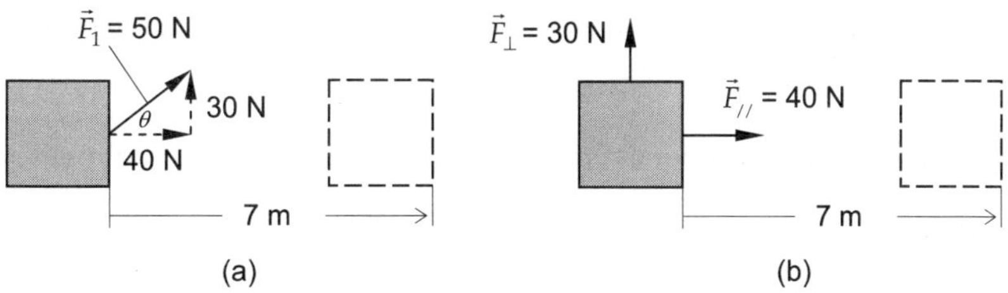

Figure 4-3: Work done by a two-dimensional force

Consider the 50-N force acting on a block shown in Fig. 4-3 (a). This 50-N force, $\vec{F}_1$, has exactly the same effect on the object's motion as the combination of two forces shown in Fig. 4-3 (b), $\vec{F}_\perp$ and $\vec{F}_{//}$, which are $\vec{F}_1$'s components perpendicular and parallel to the object's displacement. That is,

$$\vec{F}_1 = \vec{F}_\perp + \vec{F}_{//} \qquad \ldots (4.10)$$

Thus, the work done by $\vec{F}_1$ equals the sum of the work done by its components:

$$W_{\text{by}\,\vec{F}_1} = W_{\text{by}\,\vec{F}_\perp} + W_{\text{by}\,\vec{F}_{//}} \qquad \ldots (4.11)$$

Since the work done by $\vec{F}_\perp$ is zero, and the work done by $\vec{F}_{//}$ is 280 J, the work done by $\vec{F}_1$ is equal to 280 J.

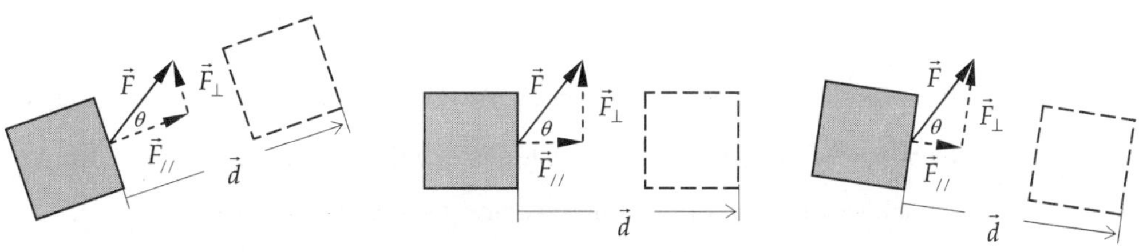

Figure 4-4: Resolving a force into components perpendicular and parallel to displacement

Note that any force can be resolved into components perpendicular and parallel to the object's displacement, no matter what the direction of the displacement, as shown in Fig. 4-4. In any case, the work done by a force is always equal to the work done by its component parallel to the displacement. Since

$$F_{//} = F\cos\theta \qquad \ldots (4.12)$$

we can express work done by a force as

$$W = (F\cos\theta)d = Fd\cos\theta \qquad \ldots (4.13)$$

θ is the angle between the force vector and the displacement vector.

According to Eq. (4.13), work done by a force is positive when the angle between the force and the object's displacement is less than 90° and negative when the angle is greater than 90°.

Now consider an object being pushed up an incline by a horizontal force, $\vec{F}_A$, as shown in Fig. 4-5 (a). To calculate the amount of work done by that force, we first need to find its component parallel to the displacement, $\vec{F}_{//}$. From Fig. 4-5 (b), we find that

$$F_{//} = 35\,\mathrm{N}\cdot\cos 30° = 30.3\ \mathrm{N}$$

Since $\vec{F}_{//}$ is in the same direction as the displacement, the work done by $\vec{F}_A$ is

$$W = F_{//}d = 30.3\,\mathrm{N}\cdot(1.6\,\mathrm{m}) \cong 48\ \mathrm{J}$$

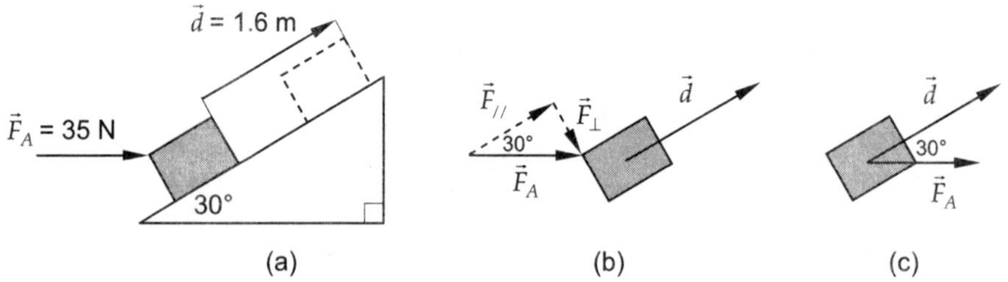

Figure 4-5: Work done on a body pushed up an incline by a horizontal force

Alternatively, we could have found the angle between $\vec{F}_A$ and $\vec{d}$ as shown in Fig. 4-5 (c) and used Eq. (4.13):

$$W = Fd\cos\theta = 35\,\mathrm{N}\cdot(1.6\,\mathrm{m})\cos 30° \cong 48\ \mathrm{J}$$

Be careful not to say the angle between $\vec{F}_A$ and $\vec{d}$ is 150°. When finding an angle between two vectors, you must slide one of the vectors so that its starting point coincides with that of the other vector, as shown in Fig. 4-6.

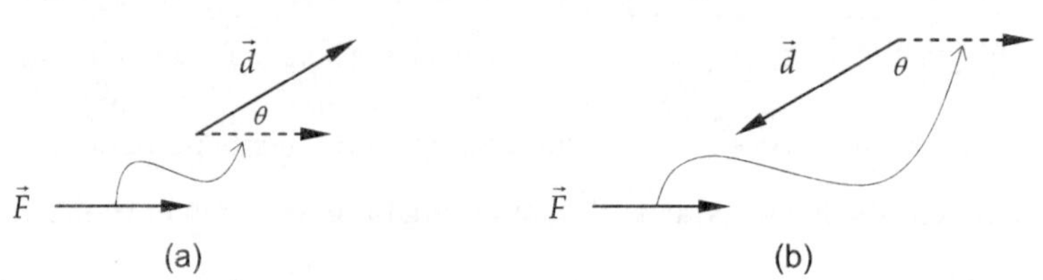

Figure 4-6: Finding angle between two vectors

4-2. Work Done by a Varying Force

Thus far we have only considered work done by a constant force, whose magnitude and direction do not change relative to the displacement. What if a force varies in magnitude and/or direction as an object moves?

Consider an object pulled along level ground by a force applied at a constant 35° angle relative to the ground. Its magnitude is doubled after the object moves 2.5 m, as shown in Fig. 4-7. The total work done on the object is the sum of the work done during the first 2.5 m and the last 3.5 m. That is,

$$W_{\text{total}} = W_1 + W_2 = F_1\,d_1\cos\theta + F_2\,d_2\cos\theta$$
$$= 22\,\text{N}\cdot(2.5\,\text{m})\cos 35^\circ + 44\,\text{N}\cdot(3.5\,\text{m})\cos 35^\circ \cong 170\,\text{J}$$

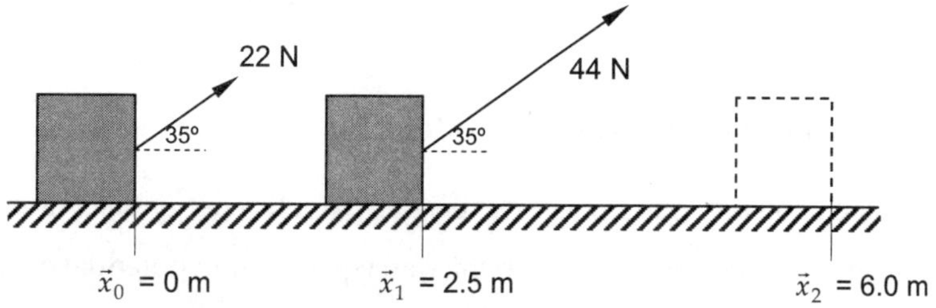

Figure 4-7: Work done by a varying force

We can also obtain this total work of 170 J graphically as follows. Draw the graph of $F\cos\theta$ (which is $\vec{F}_{//}$ according to Eq. (4.12)) versus the position of the object ($\vec{x}$) as shown in Fig. 4-8. In this example, we can find the total area under the graph by adding the areas of the two rectangles. The width of each rectangle is the magnitude of the displacement, and its height is the force component parallel to the displacement. Thus, the area of the small rectangle represents the amount of work done on the object during the first 2.5 m, and the area of the large rectangle is the work done during the last 3.5 m.

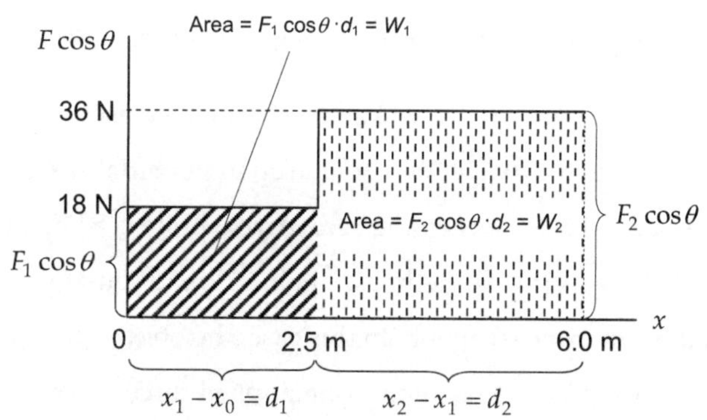

Figure 4-8: Work and the area under the graph of $F\cos\theta$ vs. x

This graphical way of calculating work is applicable to more complex situations. For instance, if the magnitude of the force on an object continually changes with the object's position, as shown in Fig. 4-9, we can no longer use Eq. (4.13) to find the work done by the force. Instead, we should calculate the area under the graph, which represents the total work done during an object's displacement.

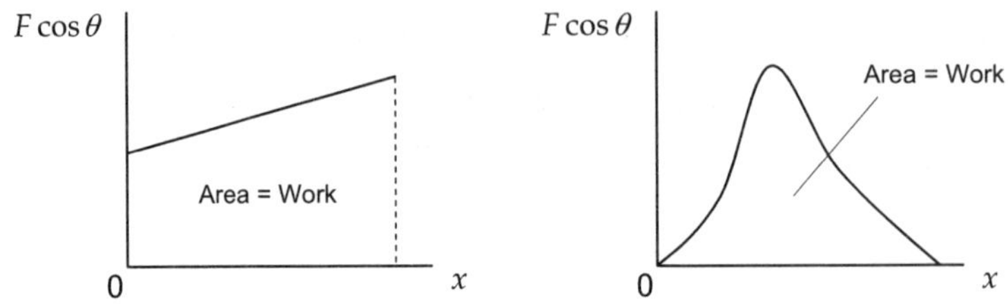

Figure 4-9: Work done by varying forces

4-3. More on Work-Energy Principle

When a moving object loses its speed, its kinetic energy decreases. Sometimes this drop in kinetic energy accompanies an equal gain in potential energy so that the object's total mechanical energy is conserved. An example is a ball thrown up in the air, as shown in Fig. 4-10 (a). On other occasions, however, when an object slows down, its kinetic energy is converted to heat and other forms of energy that, once released into the surroundings, cannot be recovered. In this case, the object's total mechanical energy decreases. An object losing speed due to friction, as shown in Fig. 4-10 (b), is one such example.

Figure 4-10: Conversion of kinetic energy to other forms of energy

When an object gains speed, its kinetic energy increases. Sometimes this gain in kinetic energy accompanies an equal drop in potential energy so that the object's total mechanical energy is conserved. An example is a ball falling in the air, as shown in Fig. 4-11 (a). Sometimes, a gain in kinetic energy is attributed to energy supplied by an external agent. In this case, the object's total mechanical energy increases. An object accelerated by an applied force, as shown in Fig. 4-11 (b), is an example.

(a) (b)

Figure 4-11: Increase in kinetic energy

Comparing the situations in Figs. 4-10 and 4-11, we find that mechanical energy is conserved for the balls in Figs. 4-10 (a) and 4-11 (a) while it is not for the objects in Figs. 4-10 (b) and 4-11 (b). So, how do you determine whether mechanical energy is conserved in a particular situation or not? In order to understand this, we need to learn about conservative and non-conservative forces.

Table 4-1: Conservative and non-conservative forces

Force	Conservative	Non-conservative
Gravitational force ($\vec{F}_g$)	✓	
Normal force ($\vec{F}_N$)		✓
Tension force ($\vec{F}_T$)		✓
Friction force ($\vec{F}_f$)		✓
Applied force ($\vec{F}_A$)		✓
Electric force ($\vec{F}_E$)	✓	
Magnetic force ($\vec{F}_M$)	✓	

Don't be concerned about these forces for now. (Electric force, Magnetic force)

In Chapter 3 we classified forces into contact and non-contact forces for the purpose of drawing free-body diagrams. We can also classify forces into **conservative** and **non-conservative forces**. If work done by a force changes an object's mechanical energy, the force is non-conservative. As you can see in Table 4-1, all the contact forces we covered in Chapter 3 are non-conservative forces. On the other hand, work done by a conservative force has no effect on an object's mechanical energy. That is, positive work done by a conservative force raises an object's kinetic energy but at the expense of an equal decrease in its potential energy. By the same token, negative work done by a conservative force decreases the kinetic energy of an object but increases its potential energy by an equal amount. This relationship between the work done by a conservative force (W_C) and a change in potential energy is expressed mathematically as

$$W_C = -\Delta E_P \qquad \text{... (4.14)}$$

Recall that the net work done on an object is the sum of the work done by individual forces, some of which are conservative and the rest of which non-conservative. Hence, we may write net work as the sum of work done by conservative forces and non-conservative forces (W_{NC}):

$$W_{\text{net}} = W_C + W_{NC} \qquad \text{... (4.15)}$$

Then the work-energy principle in Eq. (4.4) can be re-written as

$$W_C + W_{NC} = \Delta E_K \qquad \text{... (4.16)}$$

Finally, we use Eq. (4.14) to obtain a different form of the work-energy principle:

$$W_{NC} = \Delta E_K + \Delta E_P = \Delta(E_K + E_P) = \Delta\left(\frac{1}{2}mv^2 + mgh\right) \qquad \text{... (4.17)}$$

The right-hand side of Eq. (4.17) represents a change in an object's mechanical energy. Therefore, if work done by non-conservative forces is zero, an object's mechanical energy is conserved, and if not, the mechanical energy changes.

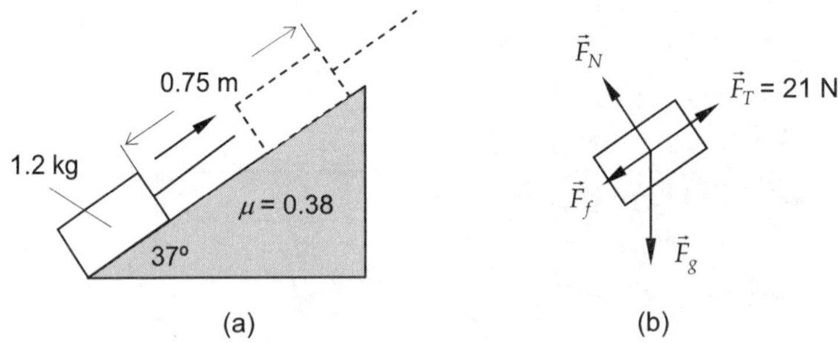

Figure 4-12: An object pulled up an inclined surface

Consider a 1.2-kg object pulled up an inclined surface by a 21-N tension force, as shown in Fig. 4-12 (a). There are four different forces acting on the object: $\vec{F}_g$, $\vec{F}_N$, $\vec{F}_f$, and $\vec{F}_T$, shown Fig. 4-12 (b). The tension force is 21 N, and the magnitudes of the other forces are

$$F_g = 1.2\,\text{kg}\cdot(9.8\,\text{m/s}^2) = 11.8\,\text{N}$$

$$F_N = 1.2\,\text{kg}\cdot(9.8\,\text{m/s}^2)\cos 37° = 9.4\,\text{N}$$

$$F_f = \mu F_N = 0.38(9.4\,\text{N}) = 3.6\,\text{N}$$

Using Eq. (4.13), we can calculate work done by each force:

$$W_{\text{by}\,\vec{F}_g} = 11.8\,\text{N}\cdot(0.75\,\text{m})\cos 127° = -5.3\,\text{J} \quad (\text{or, } -\underbrace{11.8\,\text{N}\cdot\sin 37°}_{\text{component parallel to surface}}(0.75\,\text{m}) = -5.3\,\text{J})$$

(negative because the parallel component is in opposite direction of displacement)

$$W_{\text{by}\,\vec{F}_N} = 9.4\,\text{N}\cdot(0.75\,\text{m})\cos 90° = 0\,\text{J}$$

$$W_{\text{by } \vec{F}_f} = 3.6\,\text{N} \cdot (0.75\,\text{m}) \cos 180° = -2.7\,\text{J}$$

$$W_{\text{by } \vec{F}_T} = 21\,\text{N} \cdot (0.75\,\text{m}) \cos 0° = 15.8\,\text{J}$$

Apply the work-energy principle:

$$W_{\text{net}} = \underbrace{W_{\text{by } \vec{F}_g}}_{} + \underbrace{W_{\text{by } \vec{F}_N} + W_{\text{by } \vec{F}_f} + W_{\text{by } \vec{F}_T}}_{}$$

$$= W_C \quad + \quad W_{NC} = \Delta E_K$$

$$= -5.3\,\text{J} \quad + \quad 13.1\,\text{J} = 7.8\,\text{J}$$

Rewrite the above equation as

$$W_{NC} = \Delta E_K - W_C = \Delta E_K + \Delta E_P$$

Kinetic energy increases by 7.8 J

Potential energy increases by 5.3 J

Mechanical energy increases by 13.1J → $13.1\,\text{J} = 7.8\,\text{J} - (-5.3\,\text{J}) = 7.8\,\text{J} + 5.3\,\text{J}$

The above equation shows that the object's mechanical energy increases by 13.1 J because of the work done by the non-conservative forces, and 5.3 J of it is taken away by the gravitational force and stored as gravitational potential energy. The rest, 7.8 J, is used to raise the object's kinetic energy.

Note that Eqs. (4.4) and (4.17) are the same equation, and either of them may be used to solve work-energy problems in this chapter. Although they are completely interchangeable, it might be conceptually easier to apply Eq. (4.17) in certain work-energy problems, so it is recommended that you practice using Eq. (4.17) for problems here on.

4-4. How to Solve Work-Energy Problems

The work-energy principle is the most important concept in this chapter. It connects all the quantities we learned so far in work and energy into one equation, and allows us to quantify changes in an object's kinetic, potential, and overall mechanical energies brought about by work done by different forces acting on the object. You will need to apply the work-energy principle for most of the problems in this chapter, so follow the solution procedure summarized in these four steps:

1. From the problem statement, identify the initial and final states of motion. It is critical that you identify them correctly because otherwise you will probably obtain a wrong answer.

2. Choose an appropriate point as your reference level for *h*, and write down the values of all known quantities in Eq. (4.17). If there is no mention of height in the problem, you might be able to assume $h_i = h_f$.

3. Identify the variable you need to solve for in Eq. (4.17). If the unknown variable is W_{NC}, you should be given the values of $m, v_i, v_f, h_i,$ and h_f, and be able to calculate the kinetic and potential energies at both the initial and final states. On the other hand, if the unknown is in the right-hand side, you need to examine the existence of non-conservative forces acting on the object and, if they exist, calculate work done by them using Eq. (4.13) or graphically.

4. Make sure you have the values of all quantities except for one unknown variable. Substitute the numbers into Eq. (4.17) and solve for the unknown variable. Compare your answer with your own experience.

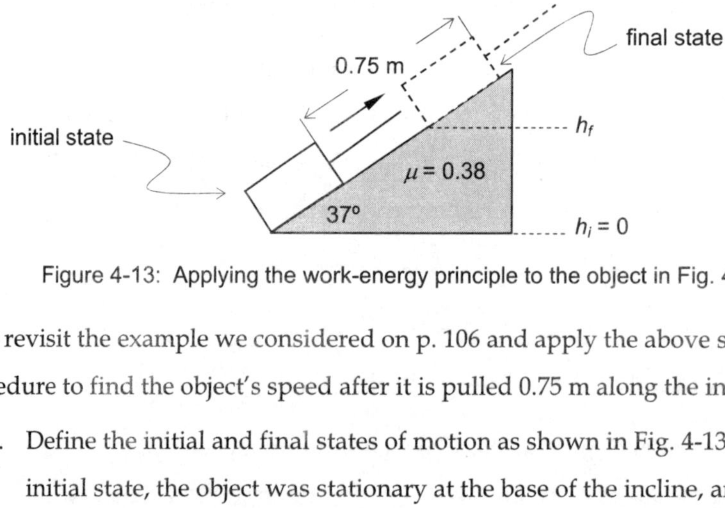

Figure 4-13: Applying the work-energy principle to the object in Fig. 4-12

Let's revisit the example we considered on p. 106 and apply the above solution procedure to find the object's speed after it is pulled 0.75 m along the incline.

1. Define the initial and final states of motion as shown in Fig. 4-13. At the initial state, the object was stationary at the base of the incline, and, at the final state, it has moved 0.75 m up the incline.

2. Set the base of the incline to be the reference level. Thus,

$$h_i = 0 \text{ m}$$

$$h_f = 0.75 \text{ m} \cdot \sin 37° = 0.45 \text{ m}$$

Write down other known quantities as well:

$$m = 1.2 \text{ kg}$$

$$v_i = 0 \text{ m/s}$$

$$g = 9.8 \text{ m/s}^2$$

3. We need to find v_f, which is in the right-hand side of Eq. (4.17). We have found all other quantities in Step 2 except for W_{NC}. Since there are friction and tension forces acting on the object, we do need to find W_{NC}. Previously, we have calculated it to be 13.1 J (see p. 107).
4. Substitute the numbers into Eq. (4.17):

$$W_{NC} = \left(\frac{1}{2}mv_f^2 + mgh_f\right) - \left(\frac{1}{2}mv_i^2 + mgh_i\right)$$

$$13.1\,\text{J} = \left(\frac{1}{2}(1.2\,\text{kg})v_f^2 + (1.2\,\text{kg})(9.8\,\text{m/s}^2)(0.45\,\text{m})\right) - \left(\frac{1}{2}(1.2\,\text{kg})(0\,\text{m/s})^2 + (1.2\,\text{kg})(9.8\,\text{m/s}^2)(0\,\text{m})\right)$$

Solve the above equation for v_f, and we find that it is 3.6 m/s.

In the next example, Mary is driving a 1000-kg car at a speed of 20 m/s when she sees the traffic light ahead turning red. She steps on the brake, and the car skids to a stop. Knowing that the car experienced a constant 5000-N force of friction during the deceleration, we want to determine the stopping distance of the car.

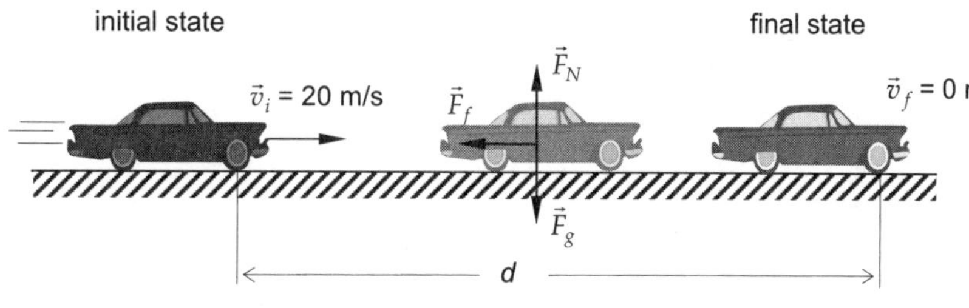

Figure 4-14: Applying the work-energy principle to a car skidding to a stop

When only the word description of a situation is given as above, it helps to visualize the problem by drawing a diagram as shown in Fig. 4-14. Then we begin the solution process as follows:

1. The initial state is just before Mary stepped on the brake, and the final state is when the car came to a complete stop.
2. Since there is no mention of height in the problem statement, we assume the car traveled horizontally during the deceleration. So, $h_i = h_f = 0$ m. Other known quantities are

$$m = 1000\,\text{kg} \qquad v_i = 20\,\text{m/s}$$

$$F_f = 5000\,\text{N} \qquad v_f = 0\,\text{m/s}$$

3. The unknown is the stopping distance, *d*, which is in the left-hand side of Eq. (4.17) because $W = F\,d\cos\theta$. While the car is braking, the non-conservative forces acting on the car are the normal and friction forces. The work done by the normal force is zero because the force is perpendicular to the direction of motion. Therefore, W_{NC} is the work done by the friction force and can be expressed as $W_{NC} = -F_f d$ (negative because cos 180° = -1).

4. Substitute the numbers into Eq. (4.17):

$$-F_f d = \left(\frac{1}{2}mv_f^2 + mgh_f\right) - \left(\frac{1}{2}mv_i^2 + mgh_i\right)$$

$$-5000\,\text{N}\cdot d = \frac{1}{2}(1000\,\text{kg})(0\,\text{m/s})^2 - \frac{1}{2}(1000\,\text{kg})(20\,\text{m/s})^2$$

Solve the above equation for *d*; the stopping distance is 40 m.

PHYSICS 12 ANALYZED: WORK AND ENERGY

4-1. More on Work

- Work done by a constant force is given by $W = F\,d\cos\theta$
- If the angle between the force vector and the displacement vector is…
    - less than 90°, the work done is positive ⟹ The object gains energy!
    - equal to 90°, no work is done by the force
    - greater than 90°, the work done is negative ⟹ The object loses energy!

4-2. Work Done by a Varying Force

- To find the work done by a varying force, draw the graph of $F\cdot\cos\theta$ versus the position (x), and find the area under the graph.

4-3. More on Work-Energy Principle

- Work done by a non-conservative force changes an object's mechanical energy.
- Work done by a conservative force does not change an object's mechanical energy; it merely changes the form of energy between kinetic and potential energy.
- Another form of the work-energy principle: $W_{NC} = \Delta(E_K + E_P)$

4-4. How to Solve Work-Energy Problems

1. Identify the initial and final states of motion.
2. Choose an appropriate point as your reference level for h, and write down the values of all known quantities.
3. Identify the variable you need to solve for in Eq. (4.17) (on p. 106), and make sure you have the values of all quantities expect for one unknown variable.
4. Solve for the unknown and ask yourself whether your answer makes sense.

EXERCISES ON CHAPTER 4: WORK AND ENERGY

Exercise 4-1

A 250-N force is applied to a 48-kg object as shown. The coefficient of friction between the object and the ground is 0.22.

a) Draw the free-body diagram of the object.

b) Find the amount of work done by each force acting on the object while it moves 13 m along the surface.

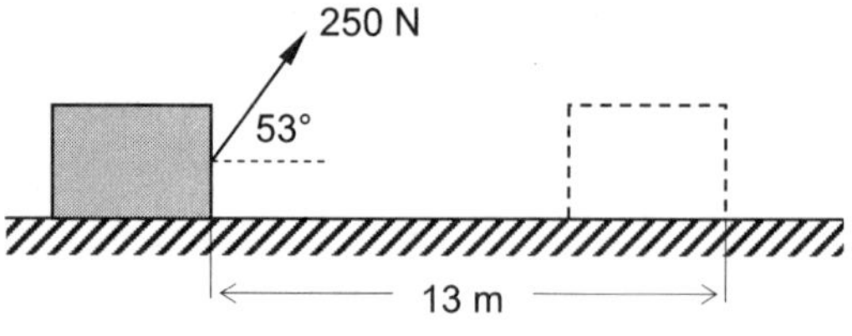

Exercise 4-2

Suppose the object in Exercise 4-1 was initially at rest. Find the object's speed after it has moved 13 m along the surface by

a) calculating the object's acceleration first and then applying one of the kinematic equations.

b) calculating the net work done on the object and then applying the work-energy principle.

Exercise 4-3

A rock is thrown horizontally at 22 m/s from the top of a 57-m high cliff. Find the speed of the rock when it is 24 m above the ground by

a) applying the kinematic equations.

b) applying the work-energy principle.

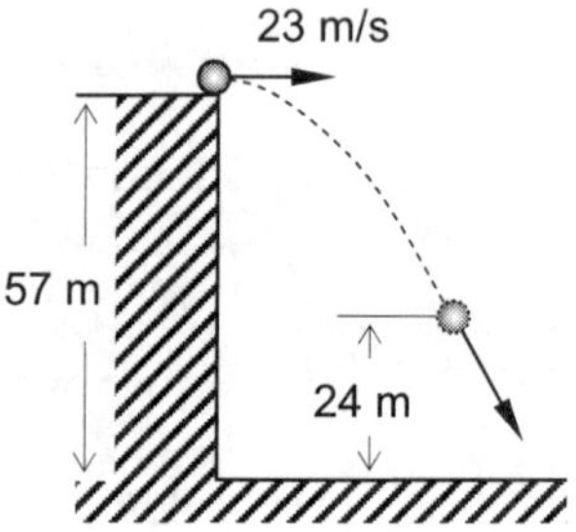

Exercise 4-4

A 75,000-kg rocket, initially at rest, is launched vertically and accelerates to 200 km/h over a distance of 130 m. Find the average force applied to the rocket using

a) the kinematic equations and Newton's second law.

b) the work-energy principle.

Exercise 4-5

A hockey puck is moving at 11 m/s when it starts climbing an incline. It travels 5.3 m from the base of the incline before sliding back down the slope, as shown. Find the coefficient of friction between the puck and the incline by

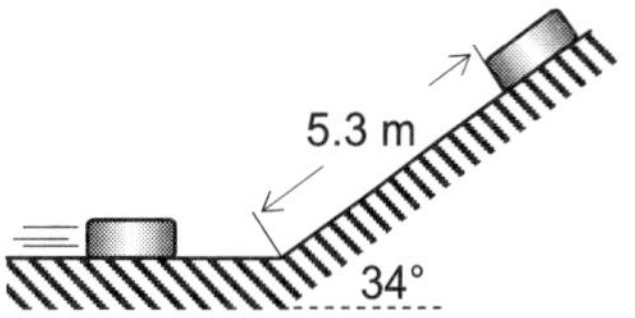

a) finding the puck's acceleration and applying Newton's second law.

b) applying the work-energy principle.

Exercise 4-6

A 35-kg crate of salt, initially at the base of an icy ramp, is pulled up to the top at constant speed by a 130-W motor. The ramp is 15 m long and inclined 22° to the horizontal. Ignore friction.

a) Find the speed of the crate.

b) Find the work done by the motor.

Exercise 4-7

a) A 6.1-kg object is launched vertically at 12 m/s, and 32 J of energy is lost to heat while the object is rising. Find the maximum height reached by the object.

b) A 6.1-kg object is launched at 12 m/s, 41° above the horizontal. If the maximum height reached by the object is 2.9 m, find the amount of energy lost to air resistance.

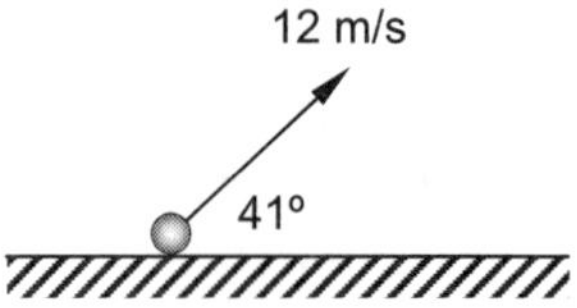

Exercise 4-8

A 2.2-kg object, initially at rest, is pulled horizontally by an applied force on a frictionless surface. The graph below shows the magnitude of the applied force versus the position of the object.

a) Calculate the kinetic energy of the object at $x = 20$ m.

b) Find the object's speed at $x = 40$ m.

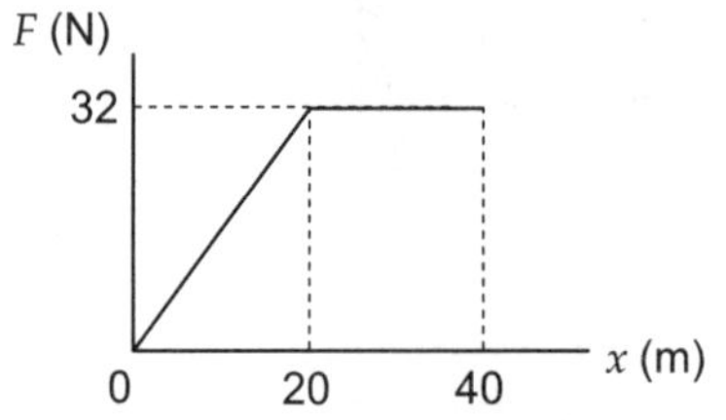

Exercise 4-9

An object, initially moving at 15 m/s on level ground, is pushed by an applied force as shown. The graph shows the magnitude of the applied force versus the position of the object. At $x = 14$ m, the object's speed is 23 m/s. Find the mass of the object.

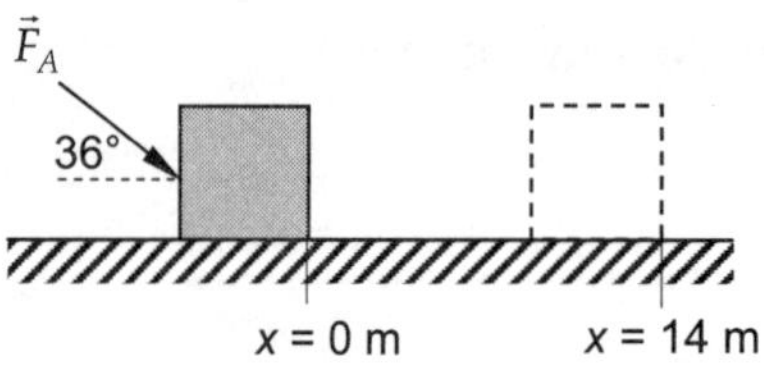

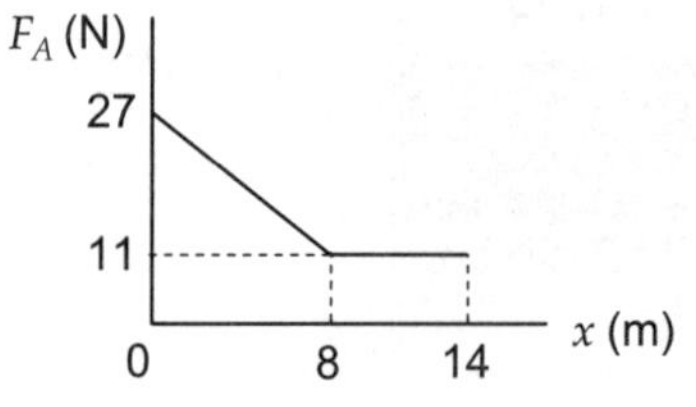

Exercise 4-10

A 990-kg car is cruising on level ground at 17 m/s when the driver steps on the gas pedal. The graph shows the applied force provided by the engine versus the position of the car. If the coefficient of friction between the car and the road is 0.36, find the speed of the car after it has traveled 440 m.

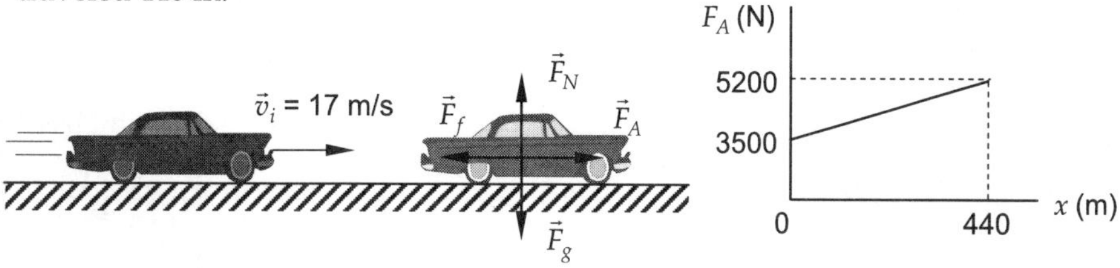

Exercise 4-11

A 0.85-kg object, initially traveling at 1.3 m/s, comes in contact with a spring at x = 0 m as shown and starts to slow down. The graph shows the magnitude of the force by the spring as it is compressed. Find the distance the object travels before coming to a stop. (Ignore friction.)

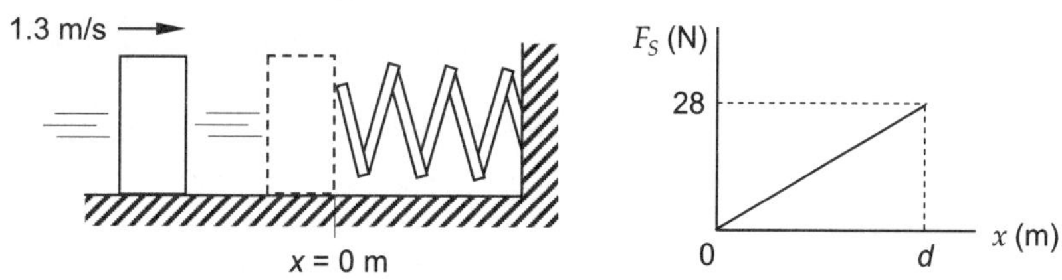

Exercise 4-12

A 7.2-kg object, initially traveling at 3.5 m/s on a smooth, horizontal surface, is acted upon by a horizontal force that varies with the object's position as shown below. Find the final speed of the object.

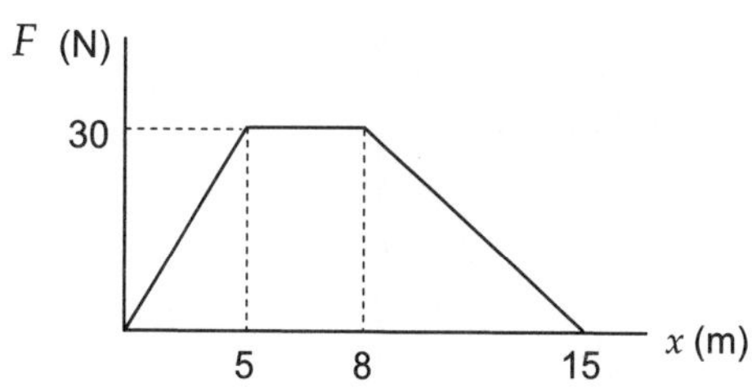

Exercise 4-13

A 45-g bullet moving horizontally at 335 m/s strikes a wooden wall. Upon impact, the bullet slows down and stops 11 cm from the front of the wall. Find the average force the wall exerts on the bullet. (The vertical motion of the bullet in this problem is negligible, so we can ignore the weight of the bullet and assume it travels horizontally until it comes to a full stop.)

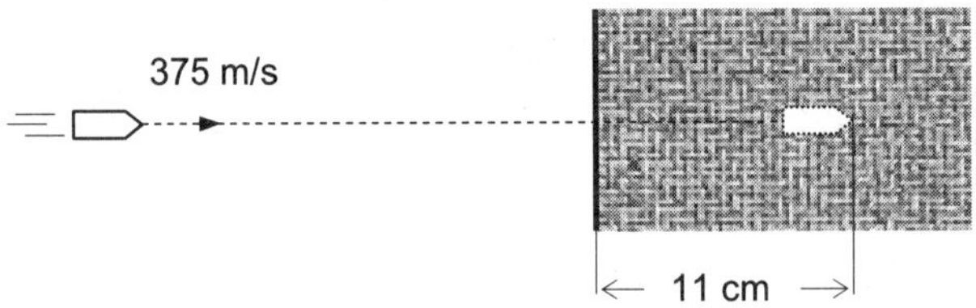

Exercise 4-14

A 64-kg skier, initially moving at 7.6 m/s, descends to the bottom of the slope and travels up a hill to a maximum height of h, as shown.

a) If the skier's speed at the bottom is 13.5 m/s, how much energy has been lost to friction between A and B?

b) If 2500 J of energy is lost to heat between B and C, find the maximum height the skier can reach.

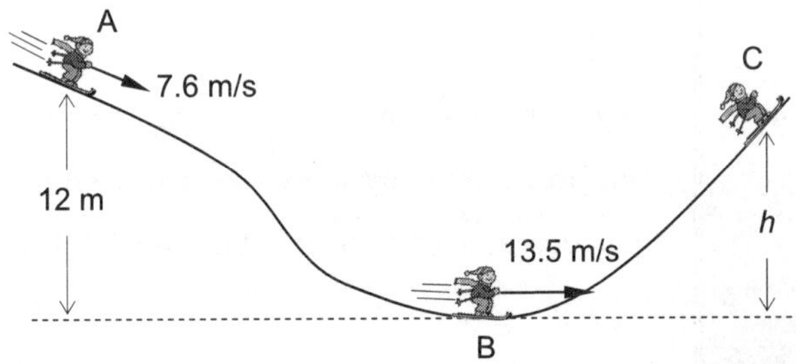

Exercise 4-15

A 17-kg object, initially at rest, slides down a 28° slope to the bottom of the slope, as shown, and reaches a speed of 8.6 m/s.

a) Find the amount of energy lost to friction.

b) Find the average force of friction that acted on the object.

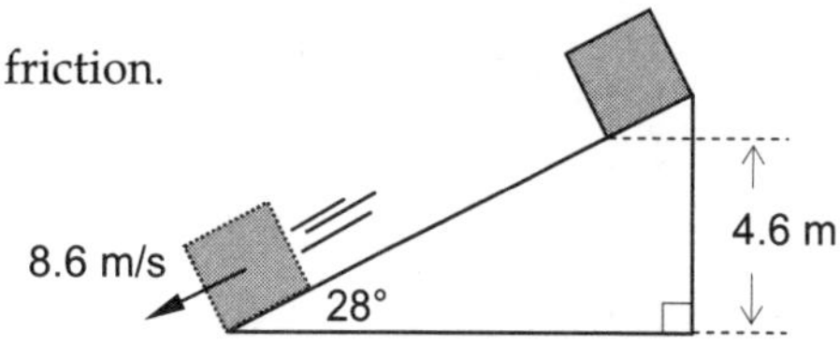

Exercise 4-16

A ball, initially at rest, is let go from h m above the base of a frictionless track. The ball slides down and leaves the track horizontally, hitting the ground at a distance d as shown. If you want to double the range of the ball to $2d$, how high from the base of the track should the ball be let go?

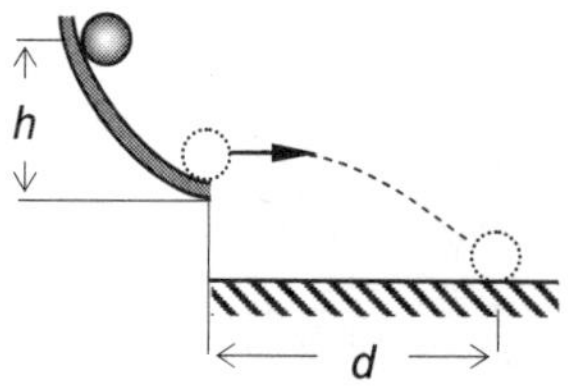

Exercise 4-17

A 0.93-kg ball is dropped 75 cm from the base of a bowl as shown. It travels across the base and rises straight up. If 1.6 J of energy is lost to friction, find the maximum height reached by the ball.

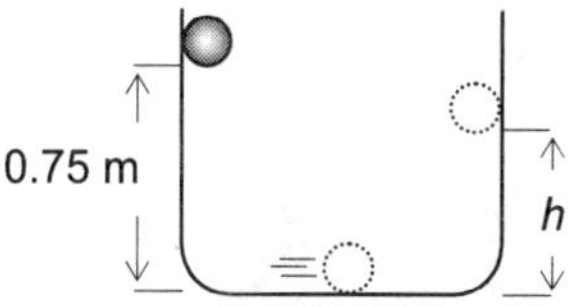

Exercise 4-18

A 3.7-kg object slides down a track from rest, 1.2 m above the base of the track. The object leaves the track horizontally and strikes the ground 0.23 s later as shown. Find the amount of energy lost to heat.

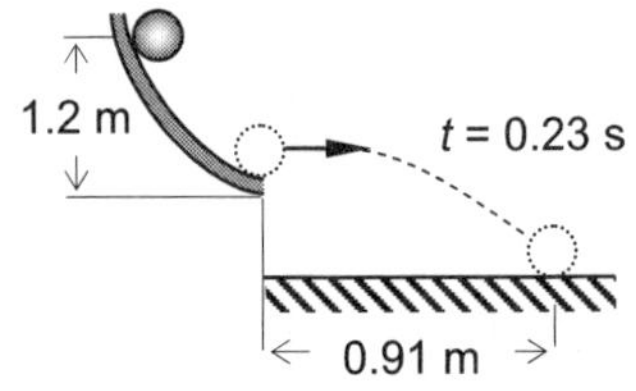

Exercise 4-19

A 7400-W motor is used to pull a 450-kg block along a horizontal surface. The coefficient of friction between the block and the surface is 0.55. The block travels at 1.6 m/s. Find the efficiency of the motor.

Exercise 4-20

A 12-kg object at rest is pulled up an incline by a motor. The graph shows the magnitude of the force exerted by the motor as the object moves 7.5 m up the incline. If 220 J of heat energy is generated through friction, find the object's speed at $x = 7.5$ m.

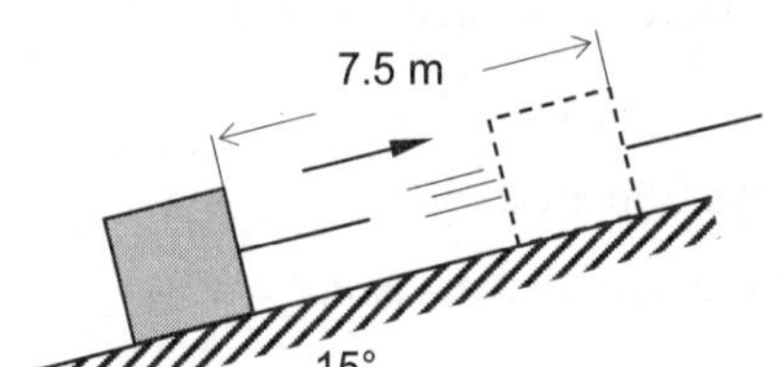

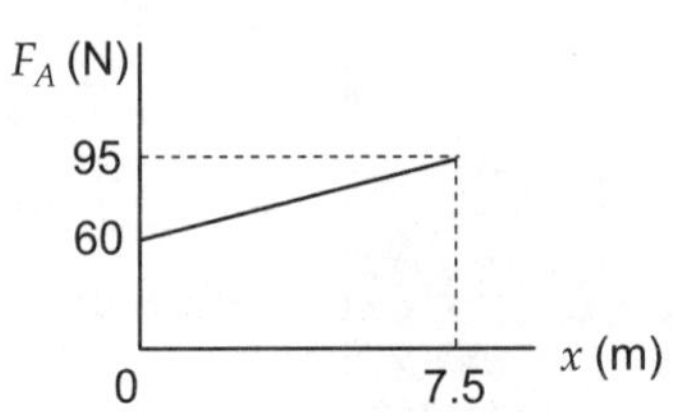

PROBLEMS ON CHAPTER 4: WORK AND ENERGY

(Answer Key on p. 368)

1 – 7. Fill in the blanks

1. __________ done on an object indicates the transfer of energy into or out of it.
2. The __________-__________ __________ states that the net work done on an object is equal to the change in its kinetic energy.
3. __________ __________ energy is stored energy associated with the position of a body within the gravitational field of another body.
4. The mechanical energy of an object is the sum of its __________ __________and __________ __________.
5. Work done by __________ forces does not change an object's mechanical energy; it leads to a change in its potential energy.
6. __________ is defined as the rate of doing work, or as the rate at which energy is transferred.
7. The __________ of a machine is defined as the power output over the power input.

8 – 17. True/False questions

8. _____ When a force acts on an object, only its component parallel to movement contributes to work done on the object.
9. _____ An object gains kinetic energy if the angle between the net force and the object's displacement is greater than 90°.
10. _____ Work done by a force can be found by calculating the area under the graph of force versus time.
11. _____ Tension force acting on a stationary object is a conservative force.
12. _____ In the absence of other forces, positive work done by a conservative force causes an object's kinetic energy to increase.
13. _____ If there are no non-conservative forces acting on a body, its mechanical energy is conserved.
14. _____ When an object's kinetic energy increases, its potential energy must decrease.

15. _____ We are allowed to use $W_{net} = \Delta E_K$ only when there is no change in an object's potential energy.

16. _____ The work-energy principle may be used even when an object's acceleration changes with time.

17. _____ The work-energy principle may <u>not</u> be used if the net force acting on an object changes with position.

18 – 28. Multiple choice questions

18. Two objects of the same shape but different masses are dropped from the same height. Which of the following statements is true?

    a) The heavier object gains more kinetic energy than the lighter object.
    b) The objects have the same kinetic energy at impact.
    c) The heavier object hits the ground first.
    d) The lighter object gains kinetic energy more quickly than the heavier object.

19. An object moves up an incline as shown but, because of friction and gravity, slows down before momentarily coming to a stop. Which graph best represents the object's potential (E_P) and mechanical energy (E_T) as a function of time?

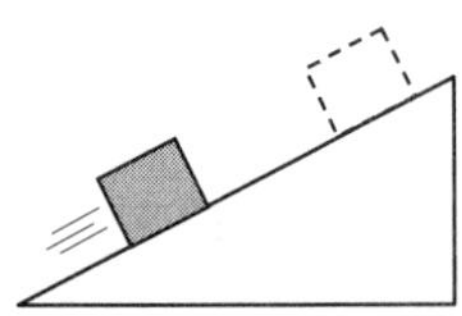

a)

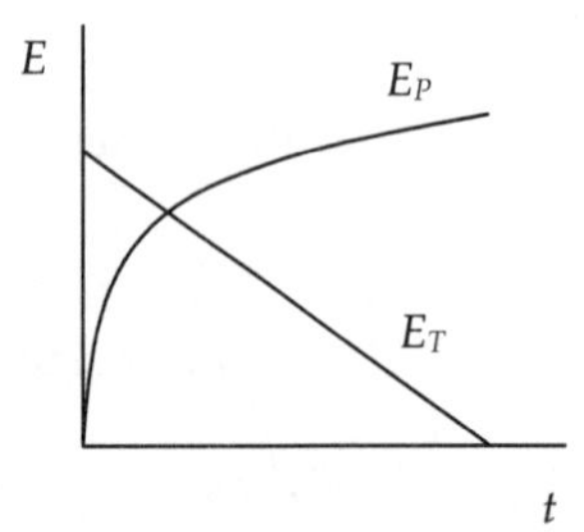

b)

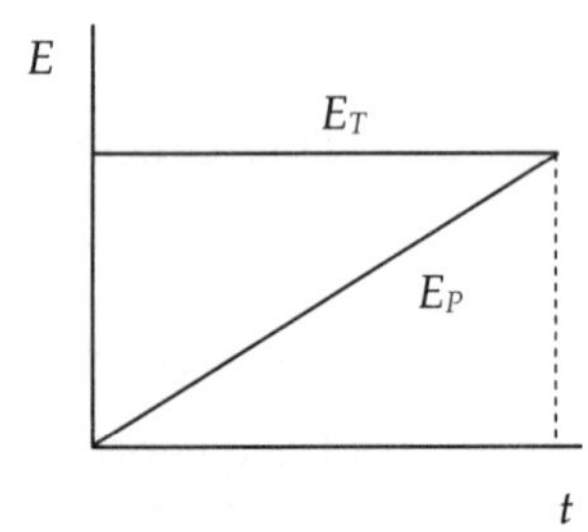

c)

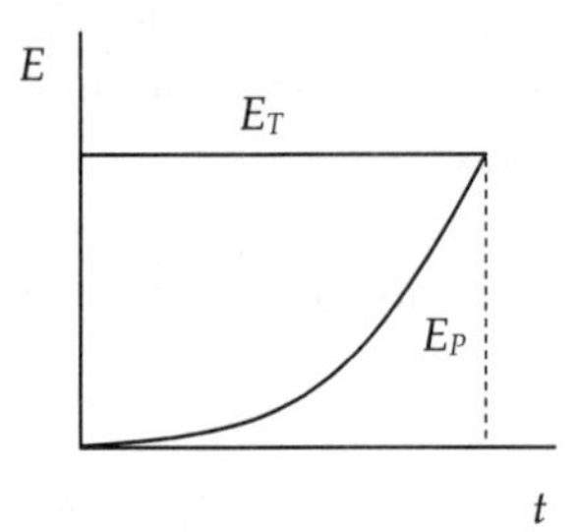

d)

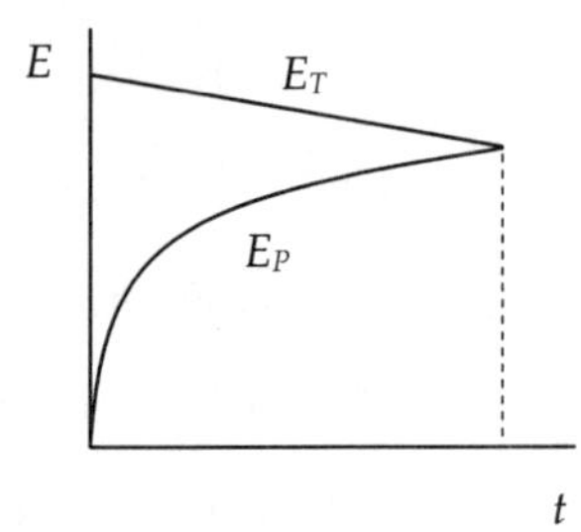

20. An object is acted upon by two forces, one a conservative and the other a non-conservative force. If the work done by the conservative force is 50 J and that by the non-conservative force is -60 J, by how much does the object's kinetic energy change?

a) E_K decreases by 10 J.
b) E_K increases by 50 J.
c) E_K decreases by 60 J.
d) E_K decreases by 110 J.

21. An object is pulled along level ground by a horizontal applied force whose magnitude changes with the object's position. Which of the following graphs shows the greatest amount of work done by the force?

a)
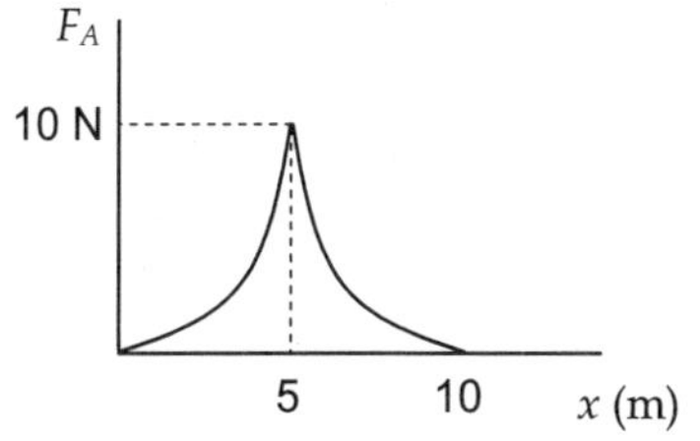

b)
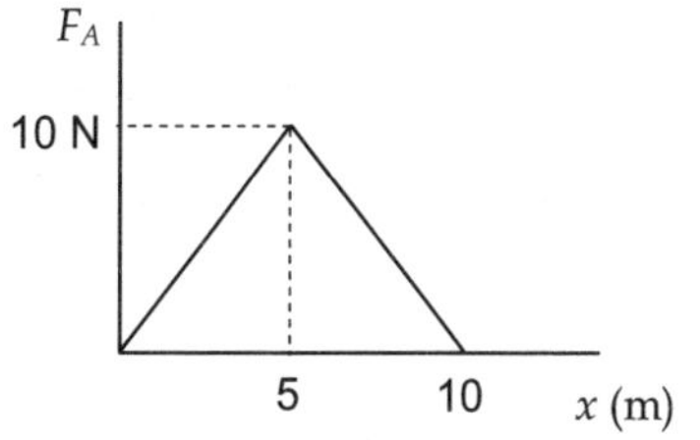

c)
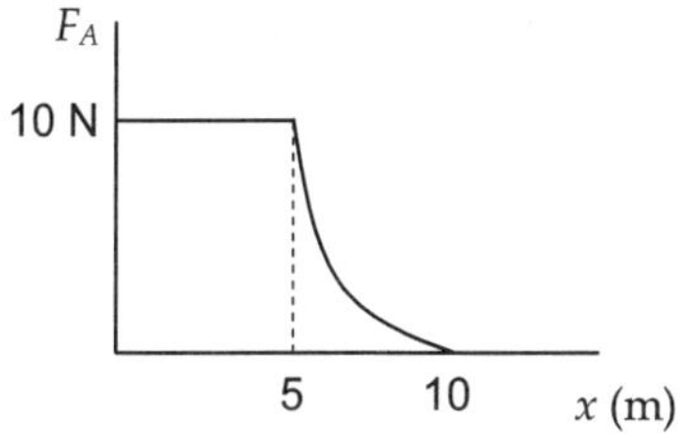

d)
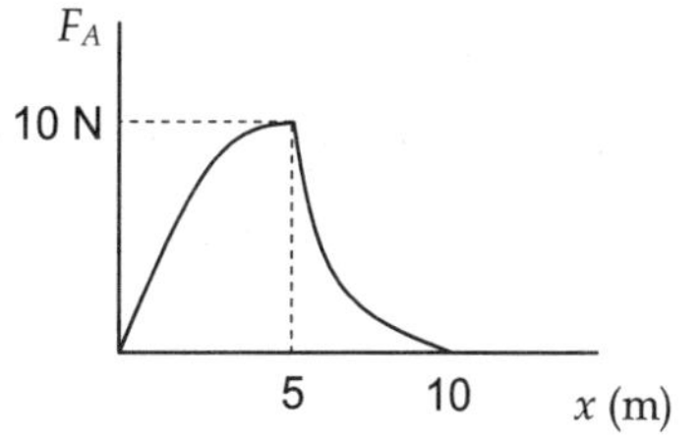

22. An object initially at rest is let go from a height of h on a frictionless slope, as shown. If you repeat the same experiment with a heavier mass, you expect its speed at the bottom of the slope to be

a) greater than the speed of the lighter object.
b) less than the speed of the lighter object.
c) the same as the speed of the lighter object.
d) We cannot compare their speeds unless we know the angle of the slope.

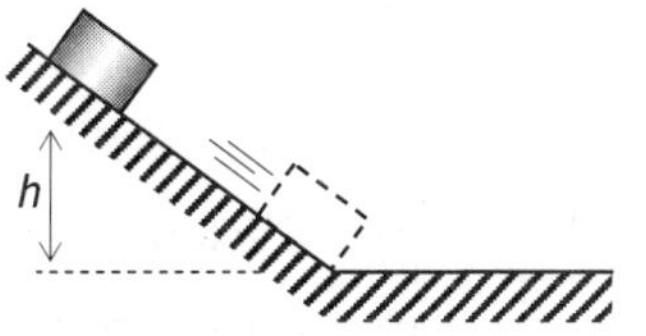

23. In Problem 22, if there is friction between the object and the slope, which object will have the greater speed at the bottom of the slope?

a) the heavier mass
b) the lighter mass
c) They both have the same speed.
d) We cannot compare their speeds because we do not know the coefficient of friction.

24. An object initially at rest is pulled along level ground by a horizontal applied force. The graph of the force's magnitude versus the object's position is shown right. At which position does the object attain the maximum speed?

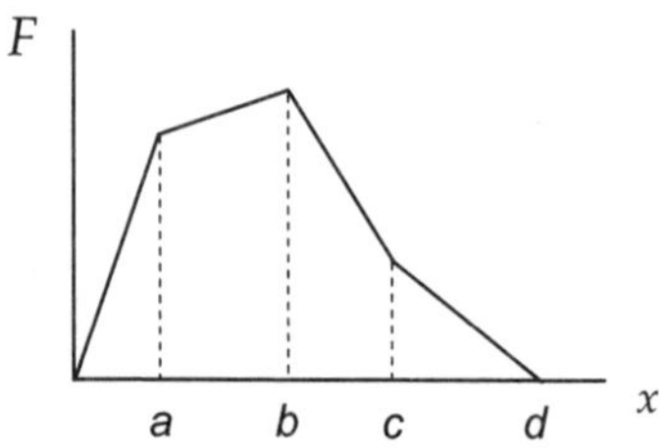

a) *a* b) *b*
c) *c* d) *d*

25. A car initially traveling at constant speed of v_0 comes to a stop in 40 m when its brakes are applied. In order to reduce the stopping distance in half to 20 m with the same braking force, the car must reduce its speed to

a) $\frac{v_0}{4}$ b) $\frac{v_0}{2}$
c) $\frac{v_0}{\sqrt{2}}$ d) $v_0 - 20$

26. An object is moving in a circle with constant speed as shown. Which of the following statements must be correct?

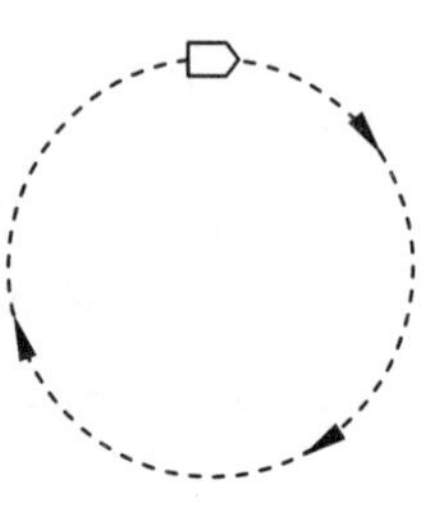

a) The object's acceleration is zero.
b) The forces acting on the object are balanced.
c) There is no net work done on the object.
d) There are only conservative forces acting on the object.

27. An object initially at rest is pulled up an incline through a string by a machine. The graph of the tension force versus the object's position is shown right. The work output by the machine increases

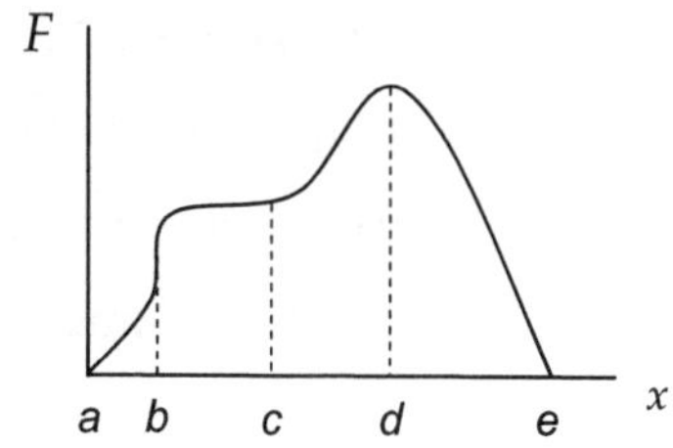

a) only between *a* and *b*
b) only between *c* and *d*
c) only between *a* and *d*
d) between *a* and *e*

28. Three identical objects are released from the same vertical height but along different slopes, as shown. The coefficient of friction is non-zero and is the same for all three slopes. Which object will attain the greatest speed at A, and which object will arrive at A first?

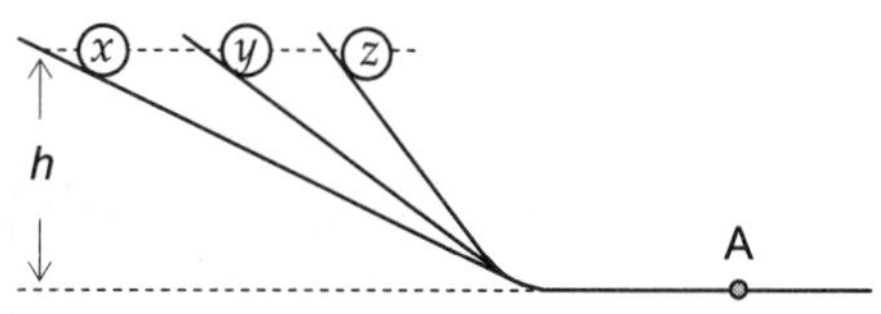

a) *z* has the greatest speed at A and arrive there first.
b) *z* has the greatest speed at A, but they all arrive at A at the same time.
c) They all have the same speed at A, but *z* takes arrives at A first.
d) They all have the same speed at A and arrive at A at the same time.

29 – 40. Written-response questions

29. A string is used to let a 3.5-kg object slide down a smooth incline at constant speed. Find the work done on the object by the tension force when the object has traveled 2.2 m along the surface.

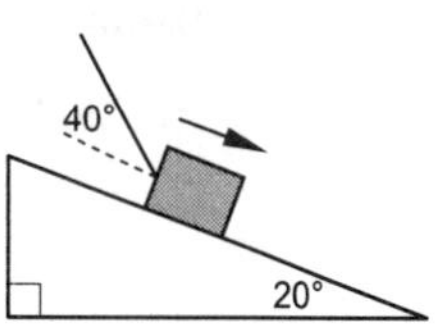

30. Suppose the object in Problem 29 experiences friction but still slides down the incline at constant speed. If the coefficient of friction is 0.25, find the work done by the tension force when the object has traveled 2.2 m along the surface.

31. A 1.2-kg pendulum is released from the position shown right. Indicate the point at which the pendulum bob attains the maximum speed, and find the maximum speed. (Ignore air resistance.)

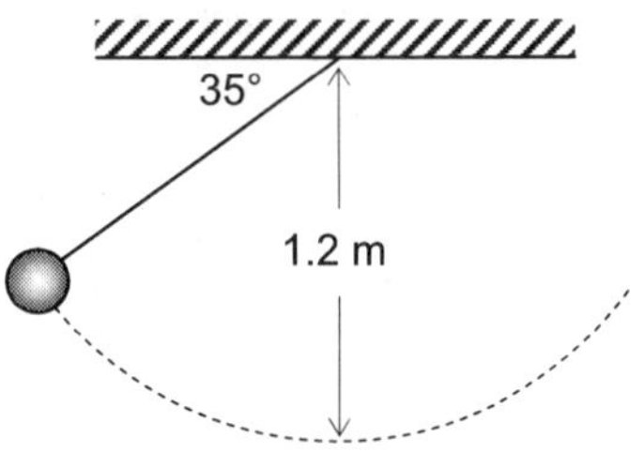

32. In Problem 31, assume the pendulum bob loses 0.45 J of energy per cycle due to air resistance. Find the angle between the pendulum and the ceiling after one complete cycle.

33. A string is used to pull a stationary 14-kg object up a smooth incline as shown. If the magnitude of the tension force varies with the object's position as shown in the graph below, find the speed of the object after it has moved 1.5 m along the incline.

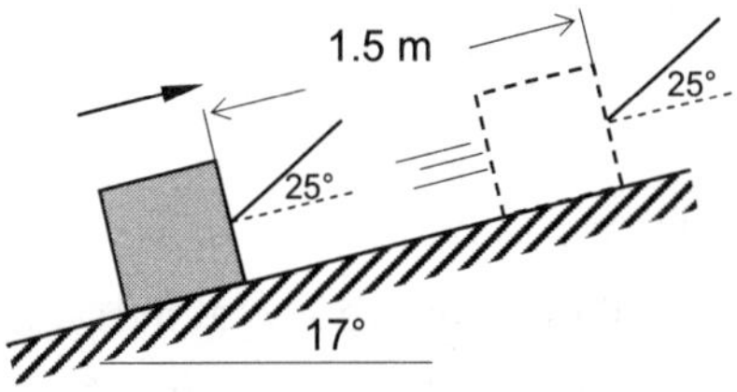

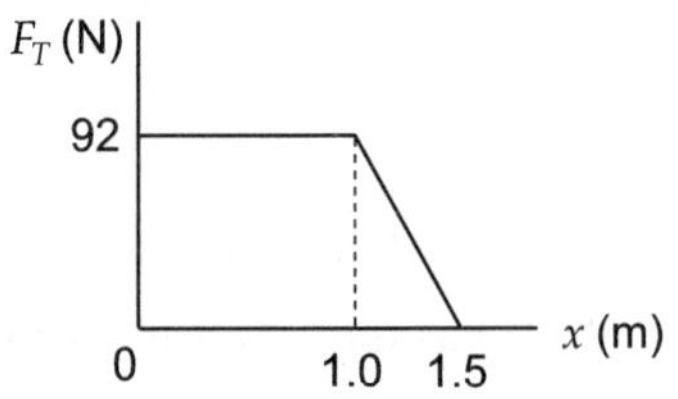

34. A 1400-kg roller-coaster car, initially traveling at 2.5 m/s at the top of the first hill, glides down the hill but barely reaches the top of the second hill as shown. If the distance of travel between a and b is 280 m, find the magnitude of the average friction force on the car.

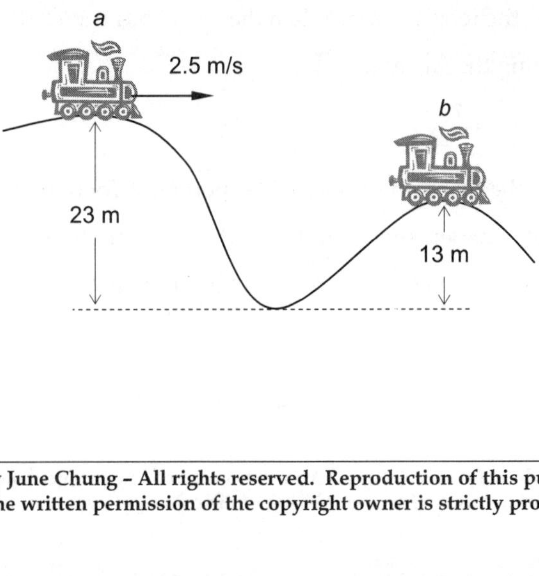

35. A 6.2-kg object is given an initial speed of 12 m/s down a 35° slope. While it travels 25 m down the slope, 450 J of energy is lost to friction. If the object's speed is to be kept constant throughout its movement, how much work has to be supplied to or taken away from the object?

36. Two objects connected by a string are released from the position shown. Find the speed of the 11-kg mass when the 15-kg mass hits the ground by (a) finding their acceleration and then using the kinematic equations and (b) applying the work-energy principle to each mass and then solving the resulting system of equations.

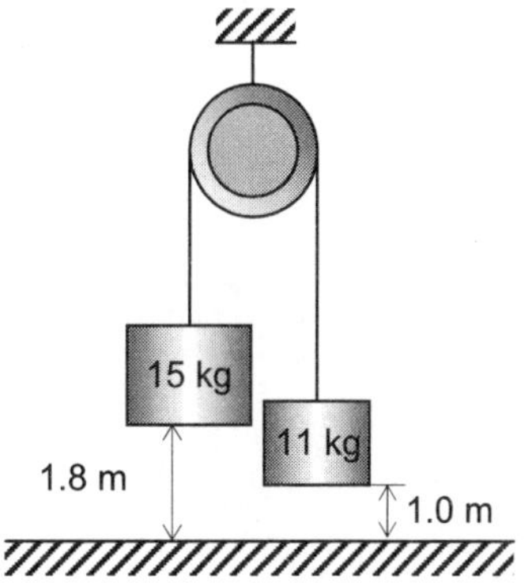

37. A system of masses is connected by a light string as shown. The system is initially at rest and starts to accelerate when released from the given position. We want to find the speed of the system when the 7.2-kg mass falls 0.35 m. (Assume no friction between the 9.0-kg mass and the surface.)

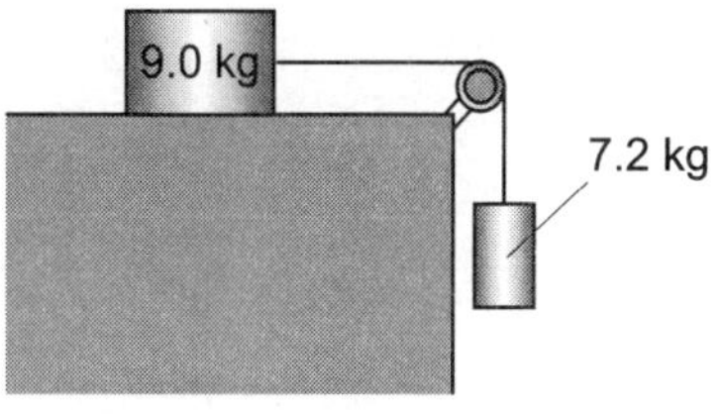

a) Apply the work-energy principle to each mass and set up a system of two equations. (You should have two unknowns in the equations.)

b) Solve the system of equations for the speed of the system. (Use the fact that the masses have the same speed.)

38.* Repeat Problem 37 assuming the coefficient of friction between the 9.0-kg mass and the surface is 0.67.

39. A system of masses is connected by a light string as shown. The system is initially at rest and starts to accelerate when released from the given position. Find the speed of the system when the system has moved 1.2 m. Assume no friction. (Note that the 2.5-kg mass could either rise or fall.)

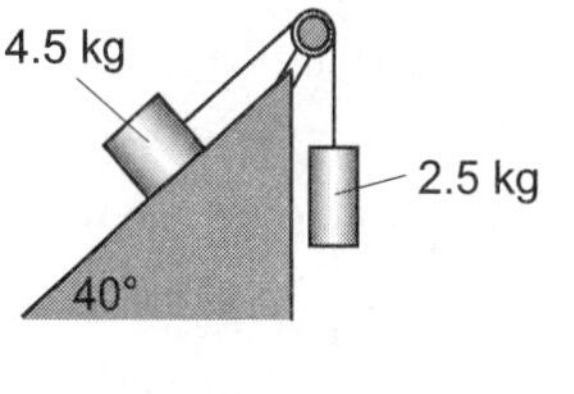

40. A machine known to be 65% efficient is used to lift a 42-kg crate along an inclined surface shown at right. If the machine is to lift the crate 1.2 m vertically in 2.1 seconds, how much power needs to be given to it? (Assume no friction.)

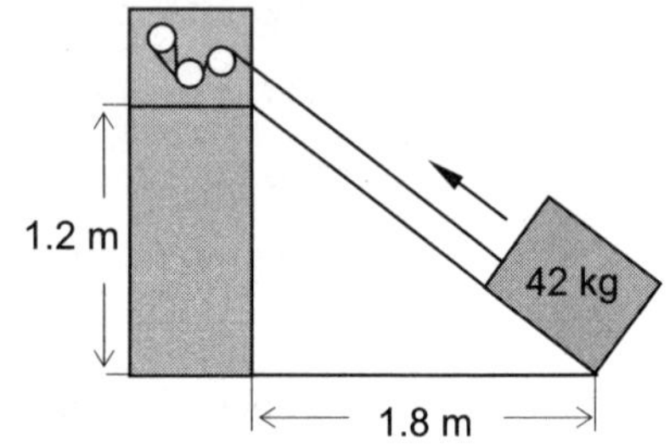

CHAPTER 5

MOMENTUM

You have already learned the core concepts of this chapter in Physics 11: momentum, impulse, and the law of conservation of momentum. No new concepts are introduced in Physics 12. Instead, we will use the vector analysis technique we learned in Chapter 1 to investigate two-dimensional collision and explosion problems. (Investigating two-dimensional motion problems is what we have mainly been doing in Chapters 2 and 3.) Make sure you know how to resolve vectors into components and find a vector's magnitude and direction from its components. If not, it is strongly suggested that you review Chapter 1. Remember that vector calculation is the key to solving the problems in this chapter, and pay attention to the <u>direction</u> of vectors!

In this chapter, you will:

- ✓ review momentum from Physics 11
- ✓ test your knowledge of momentum from Physics 11
- ✓ learn how to obtain impulse given by a force changing with time
- ✓ learn how to solve two-dimensional collision and explosion problems
- ✓ test your knowledge of momentum from Physics 12

Review of Physics 11: Momentum

1. Momentum ($\vec{p}$)

- **Momentum** is the quantity of motion an object possesses and is defined as

The direction of an object's momentum is always the same as that of its velocity.

$$\vec{p} = m\vec{v} \quad [\text{kg·m/s}] \qquad \dots (5.1)$$

Example: The momentum of a 2.5-kg object moving east at 3.0 m/s is

$$\vec{p} = 2.5\,\text{kg}\cdot(3.0\,\text{m/s}) = 7.5\,\text{kg·m/s, east}$$

- Then a change in momentum is

$$\Delta\vec{p} = \vec{p}_f - \vec{p}_i = m(\vec{v}_f - \vec{v}_i) \qquad \dots (5.2)$$

- Net force acting on a body can be defined as the rate of change in its momentum:

$$\vec{F}_{\text{net}} = \frac{\Delta\vec{p}}{\Delta t} \quad [\text{kg·m/s}^2] \qquad \dots (5.3)$$

Example: A 2.5-kg object is initially moving east at 3.0 m/s. After 2.0 s, the object's velocity has become 5.0 m/s, west. Find the average net force during the 2.0-s interval.

Find the initial and final momenta:

$$\vec{p}_i = 2.5\,\text{kg}\cdot(3.0\,\text{m/s}) = 7.5\,\text{kg·m/s, east}$$

$$\vec{p}_f = 2.5\,\text{kg}\cdot(5.0\,\text{m/s}) = 12.5\,\text{kg·m/s, west}$$

Set east as positive, calculate the change in momentum, and divide it by 2.0 s:

$$\Delta\vec{p} = \vec{p}_f - \vec{p}_i = -12.5\,\text{kg·m/s} - 7.5\,\text{kg·m/s} = -20\,\text{kg·m/s}$$

$$\vec{F}_{\text{net}} = \frac{\Delta\vec{p}}{\Delta t} = \frac{-20\,\text{kg·m/s}}{2.0\,\text{s}} = -10\,\text{kg·m/s}^2\text{, or 10 N, west}$$

Both momentum and impulse are vectors!

2. Impulse ($\vec{J}$)

- **Impulse** is defined as the product of the <u>average</u> net force applied to a body and the time period over which the force was applied:

The direction of impulse on a body is always the same as that of net force on the body.

$$\vec{J} = (\vec{F}_{\text{net}})_{\text{avg}}\cdot\Delta t \quad [\text{N·s} = \text{kg·m/s}] \qquad \dots (5.4)$$

Example: The impulse on the 2.5-kg object in the previous example is

$$\vec{J} = (\vec{F}_{\text{net}})_{\text{avg}}\cdot\Delta t = 10\,\text{N}\cdot(2.0\,\text{s}) = 20\,\text{N·s, west}$$

- Using Eq. (5.3), we may express impulse as the change in momentum:

$$\vec{J} = \left(\vec{F}_{\text{net}}\right)_{\text{avg}} \cdot \Delta t = \Delta\vec{p} = m\left(\vec{v}_f - \vec{v}_i\right) \quad \text{... (5.5)}$$

Example: In the previous example, we could have obtained the impulse on the 2.5-kg object by calculating the change in its momentum:

$\Delta\vec{p} = \vec{p}_f - \vec{p}_i = -12.5\,\text{kg}\cdot\text{m/s} - 7.5\,\text{kg}\cdot\text{m/s} = -20\ \text{kg}\cdot\text{m/s}$, or 20 kg·m/s, west

kg·m/s is equivalent to N·s.

3. The Law of Conservation of Momentum

- The **law of conservation of momentum** states that, when two or more objects interact in any way, the sum of their momenta is conserved during the interaction, as long as the system is isolated. (You may assume all systems are isolated unless told otherwise.)

- For all types of collisions and explosions, you may apply the law of conservation of momentum, which is expressed mathematically as

$$\left(m_A\vec{v}_A + m_B\vec{v}_B\right)_{\text{before}} = \left(m_A\vec{v}_A + m_B\vec{v}_B\right)_{\text{after}} \quad \text{... (5.6)}$$

This equation is derived from multiplying the time of interaction to Newton's third law of motion: $\underbrace{\vec{F}_{\text{on }A\text{ by }B}\cdot\Delta t}_{\Delta\vec{p}_A} = \underbrace{-\vec{F}_{\text{on }B\text{ by }A}\cdot\Delta t}_{\Delta\vec{p}_B}$

4. Collisions and Explosions

- There are three types of collisions: perfectly elastic, inelastic, and perfectly inelastic.

- In perfectly elastic collisions, the total kinetic energy of colliding objects is conserved as well as the sum of their momenta. Thus, in addition to Eq. (5.6), you can use the following equation:

If a collision is perfectly elastic, you will be told that it is.

$$\left(\frac{1}{2}m_Av_A^2 + \frac{1}{2}m_Bv_B^2\right)_{\text{before}} = \left(\frac{1}{2}m_Av_A^2 + \frac{1}{2}m_Bv_B^2\right)_{\text{after}} \quad \text{... (5.7)}$$

- In inelastic and perfectly inelastic collisions, the total kinetic energy of colliding objects is not conserved. Hence, you may use only the law of conservation of momentum in Eq. (5.6). In perfectly inelastic collisions, the colliding objects become locked and move away together.

Example: A 6.0-kg object, initially traveling east at 2.0 m/s, collides with a 9.0-kg object at rest. After the collision, the 6.0-kg object continues moving east at 1.2 m/s. Find the velocity of the 9.0-kg object.

Set east as positive, and apply the law of conservation of momentum:

$$\left(6.0\,\text{kg}\cdot(2.0\,\text{m/s}) + 9.0\,\text{kg}\cdot(0\,\text{m/s})\right) = \left(6.0\,\text{kg}\cdot(1.2\,\text{m/s}) + 9.0\,\text{kg}\cdot v_{\text{after}}\right)$$

Solve the above equation for the velocity, and we obtain $v_{\text{after}} = 0.53\,\text{m/s}$, east.

- Think of an explosion as the reverse of a perfectly inelastic collision. The pieces of an exploding object are initially locked together before the explosion and separate afterwards.

  Example: A 10-kg object, initially at rest, explodes into two pieces. A 6.0-kg piece flies north at 6.0 m/s. Find the velocity of the smaller piece.

  Set north as positive and apply the law of conservation of momentum:

  $$\left(6.0\,\text{kg}\cdot(0\,\text{m/s})+4.0\,\text{kg}\cdot(0\,\text{m/s})\right)=\left(6.0\,\text{kg}\cdot(6.0\,\text{m/s})+4.0\,\text{kg}\cdot v_{\text{after}}\right)$$

  Solve the above equation for the velocity: $v_{\text{after}} = -9.0$ m/s, or 9.0 m/s, south.

5. How to Solve Collision and Explosion Problems

1. Read the problem statement carefully. Make sure two objects are colliding or two objects initially joined together are separating.
2. Draw a diagram showing both objects before and after the collision (or before and after the separation). An example is shown below.

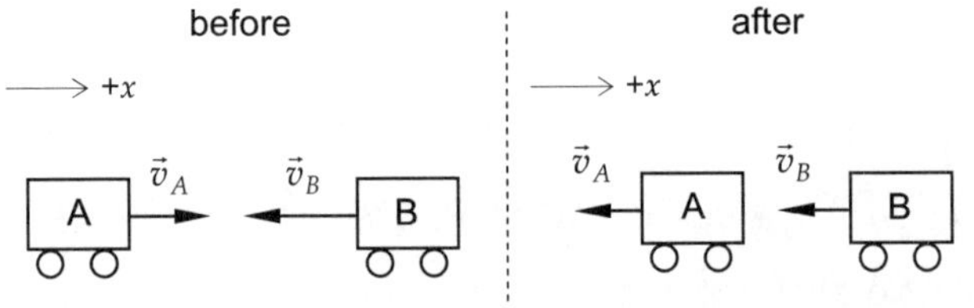

Figure 5-1: Diagram showing the objects before and after collision

3. Choose the positive direction, and calculate the momentum of each object before and after the collision (or before and after the separation) based on that direction.
4. Set up Eq. (5.6). Check that there is only one unknown in the equation. Solve for the unknown, and think about whether your answer makes sense.

Test Your Knowledge of Physics 11: Momentum

(Answer Key on p. 369)

1 - 10. True/False questions

1. _____ Momentum is a vector quantity.

2. _____ Impulse is a scalar quantity.

3. _____ While an object is moving south, its momentum is also directed south.

4. _____ While an object is moving with constant velocity, the impulse on it is zero.

5. _____ To double the magnitude of an object's momentum, you need to double the magnitude of the net force on the object.

6. _____ When an object receives a non-zero impulse, its momentum changes.

7. _____ 1 kg·m/s is equivalent to 1 N·s.

8. _____ For two colliding objects, the sum of their momenta is conserved only if the collision is perfectly elastic.

9. _____ The law of conservation of momentum does not apply to an exploding object unless it was at rest before the explosion.

10. _____ When an object is thrown vertically upward and explodes into two pieces at the maximum height, the pieces must travel in opposite directions.

11 - 20. Multiple choice questions

11. An object traveling on a smooth surface is acted on by a constant force. Which of the following expressions is the change in the object's momentum?

a) $\vec{F}_{net}d$ b) $\vec{F}_{net}v$

c) $\vec{F}_{net}\Delta t$ d) ma

12. Objects A and B are moving on a frictionless surface. If the magnitude of the impulse required to stop A is twice that required to stop B, which of the following statements must be true?

a) A has twice the mass of B.
b) A was initially moving twice as fast as B.
c) The magnitude of A's initial momentum was twice that of B's.
d) A has the greater mass than B but was initially moving slower than B.

13. A ball is thrown at a window. In which case does the ball experience the greatest magnitude of impulse?

    a) The ball bounces back at the same speed.
    b) The ball bounces back at a reduced speed.
    c) The ball sticks to the window and does not bounce back.
    d) The ball breaks the window and continues moving at a reduced speed.

14. An object initially traveling north receives an impulse due south. Which of the following statements is true?

    a) The object continues moving north even after receiving the impulse.
    b) The object stops moving after receiving the impulse.
    c) The impulse reverses the direction of the objects motion to south.
    d) Any of the above could happen.

For Problems 15 to 17, refer to the statements below:

I.	The total mass is conserved.
II.	The sum of the colliding bodies' momenta is conserved.
III.	The sum of the colliding bodies' kinetic energies is conserved.
IV.	The colliding objects have the same velocity after the collision.

15. Which of the above statements is true for all collisions between two objects?

    a) I and II only b) II only
    c) II and III only d) I, II, and III only

16. Which of the above statements can be used to determine whether a collision is perfectly elastic?

    a) I b) II
    c) III d) IV

17. Which of the above statements can be used to determine whether a collision is perfectly inelastic?

    a) I b) II
    c) III d) IV

18. A 1.0-kg object moves to the left at 9.0 m/s and collides with a stationary 2.0-kg object. After the collision, the 1.0-kg object heads to the right at 3.0 m/s, and the 2.0-kg object moves to the left at 6.0 m/s. Which type of collision is it?

    a) perfectly elastic collision
    b) inelastic collision
    c) perfectly inelastic collision
    d) The above collision cannot occur as described.

19. A car and a bicycle collide head-on. Which of the following is true?

    a) The bicycle receives the greater magnitude of impulse.
    b) They receive the same magnitude of impulse.
    c) The momentum of the car is conserved.
    d) The bicycle sticks to the car.

20. When you fire a gun at a target in front of you, the gun recoils back toward you. Why is the gun's speed less than that of the bullet?

    a) Because the gun receives less magnitude of impulse.
    b) Because the gun has less magnitude of momentum.
    c) Because the gun has the greater mass.
    d) Because the gun receives less magnitude of force.

21 - 35. Written-response questions

21. A 3.5-kg object is thrown vertically upward at 14 m/s. What is the object's momentum after 1.8 seconds?

22. A 12-kg ball traveling north at 40 m/s collides with a 15-kg ball heading south. If the 12-kg ball comes to a stop immediately after the collision, find the impulse on the 15-kg ball.

23. A 330-kg object moving south at 7.1 m/s is subjected to an impulse equal to 1200 N·s, north. Find the final velocity of the object.

24. A 0.51-kg object is sliding across a smooth horizontal surface when a 1.7-N force directed west acts on it for 3.5 s. The object is seen moving directly west at 1.3 m/s. Find the initial velocity of the object.

25. A 57-g tennis ball is traveling horizontally in air at 17 m/s when it is struck by a racket. The ball bounces back in the opposite direction at 23 m/s. If the contact between the ball and the racket lasted 0.51 ms, find the magnitude of the average force applied by the racket on the ball.

26. An applied force causes a 9.5-kg object to accelerate uniformly from rest on a horizontal, smooth surface. If this object moves 5.5 meters in 2.0 seconds, how much impulse has the applied force exerted on the object?

27. A 2.5-kg ball, dropped 7.2 m above the ground, hits the ground and rebounds up to 5.7 m above the ground. Find the impulse on the ball by the ground.

28. A 4.0-kg object traveling north at 7.8 m/s collides with a stationary object whose mass is unknown. After the collision, the 4.0-kg object heads south at 0.50 m/s, and the other object moves north at 6.5 m/s. Find the change in momentum of each object.

29. A 35000-kg rail car traveling at 1.8 m/s collides with a 40000-kg car moving in the same direction at 1.1 m/s. If the two cars become locked during the collision, find their speed after the collision.

30. A 1900-kg truck is coasting through a forest at 8.3 m/s when three men suddenly jump out of trees and land on the truck bed safely. If the average mass of the men is 85 kg, find the final speed of the truck. (Assume the men lands on the truck bed vertically.)

31. A 1.5-kg object moving east at 2.9 m/s collides with a 0.90-kg object traveling west at 4.2 m/s. After the collision, the 0.90-kg object heads east at 2.3 m/s. Find the velocity of the 1.5-kg object.

32. A 2.5-kg cart moving at 4.8 m/s across a horizontal smooth surface collides with a stationary 6.5-kg cart. If the 2.5-kg cart rebounds at 1.3 m/s, how much kinetic energy is lost during this collision process?

33. A 47-kg object is traveling horizontally in air when it suddenly explodes into two pieces. A 33-kg piece comes to a complete stop immediately after the explosion and falls vertically toward the ground. Express the speed of the other piece in terms of the initial speed before the explosion.

34. An 80-kg astronaut was working on a lab experiment in outer space. At one point during the experiment, he found himself all alone, just floating in space and going nowhere. Fortunately, he was holding a 10-kg spherical object, and when he threw it as hard as he could, he felt his body moving towards the space station at 0.63 m/s. What is the speed of separation between the astronaut and the 10-kg object?

35. Bobby is teaching his little sister, Katie, how to skate. They initially stand far apart on ice, holding each end of a rope. Bobby starts pulling on the rope, and they both start moving towards each other. When Katie's speed reaches 1.1 m/s, they let go of the rope. If Bobby's mass is 55-kg and Katie's mass is 45-kg, find the total work done. (Ignore friction.)

What's New in Physics 12?

5-1. Impulse Given by a Varying Force

Consider a 5.0-kg ball thrown vertically up in the air. Since the only force acting on the ball is its own weight, the impulse the ball receives in any 0.30-s period is the product of its weight and the time interval of 0.30 s:

$$\vec{J} = \vec{F}_{net}\,\Delta t = m\,\vec{g}\,\Delta t = (5.0\,\text{kg})(9.8\,\text{m/s}^2)(0.30\,\text{s}) \cong 15\,\text{N·s, downward}$$

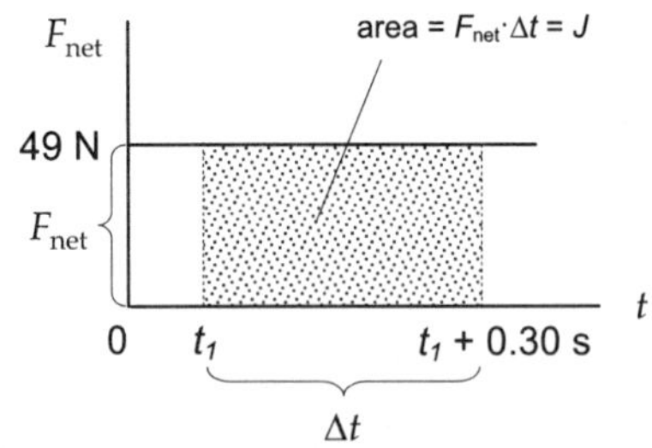

Figure 5-2: Plot of net force vs. time for a ball thrown vertically

Figure 5-2 shows the graph of the magnitude of the net force acting on the ball versus time. The net force on the ball, which is its own weight, does not change with time, so the graph is a horizontal line. Then the area of the shaded rectangle is the product of F_{net} and Δt, which is the magnitude of the impulse on the ball.

For the ball in the above example, it is not necessary to plot a graph of the net force versus time when you are finding the impulse on it; the use of Eq. (5.4) (on p. 128) will do. However, objects often experience net force varying with time during collisions and produce graphs of net force versus time similar to the one drawn in Fig. 5-3. Examples are a tennis ball smashed by a racket, a golf ball struck by a golf club, and a baseball hit by a bat, and you may not apply Eq. (5.4) for such objects. Instead, you can either use Eq. (5.5) (on p. 129) or find the magnitude of impulse graphically by calculating the area under the graph of the net force versus time.

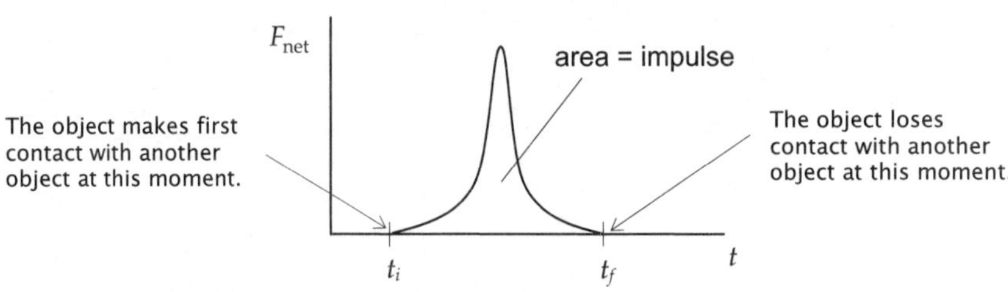

Figure 5-3: Plot of net force varying with time

So far, we have learned three physical quantities that can be represented by the area under the graph. The first was displacement, obtained from the graph of velocity versus time, and the second was work, obtained from the graph of force versus position. And the last is impulse that we discussed in this section. Pay attention to which variables are on the x- and y-axes, and make sure you can name the quantity represented by the area under a particular graph.

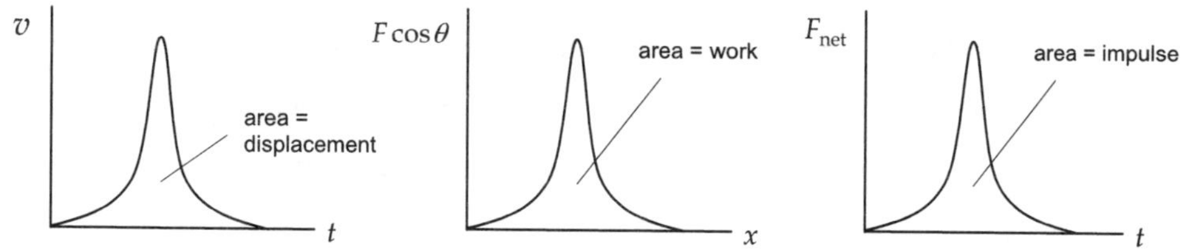

Figure 5-4: Physical quantities represented by the areas under various graphs

5-2. Two-Dimensional Collisions and Explosions

The law of conservation of momentum holds true for all types of collisions and explosions. However, applying the law to solve two-dimensional collision and explosion problems is not as easy as doing the same for one-dimensional problems; for two-dimensional collisions and explosions, you need to use the vector analysis technique covered in Chapter 1.

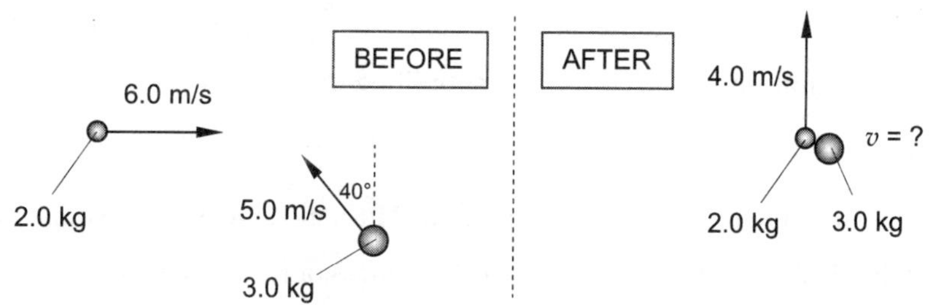

Figure 5-5: Diagram showing two-dimensional collision

Consider the following two-dimensional collision. A 2.0-kg object moving directly east at 6.0 m/s collides with a 3.0-kg object traveling at 5.0 m/s, 40° west of north, and heads north at 4.0 m/s after the collision, as shown in Fig. 5-5. Let's find the velocity of the 3.0-kg object after this collision. First, denote the 2.0-kg object as A and the 3.0-kg object as B. Then we can find three of the four vectors in the equation for the conservation of momentum:

$$(m_A\vec{v}_A + m_B\vec{v}_B)_{\text{before}} = (m_A\vec{v}_A + m_B\vec{v}_B)_{\text{after}}$$

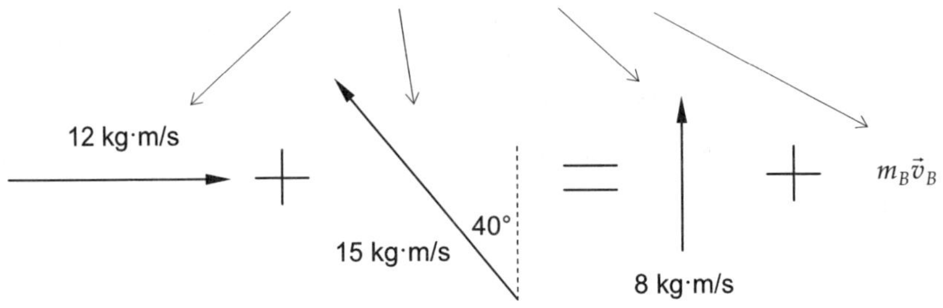

Move the final momentum of *A* to the other side. Then finding the final momentum of *B* is just a matter of performing two-dimensional vector addition and subtraction, after which we obtain $(m_B\vec{v}_B)_{\text{after}} = 4.2\,\text{kg·m/s}$, 56° N of E.

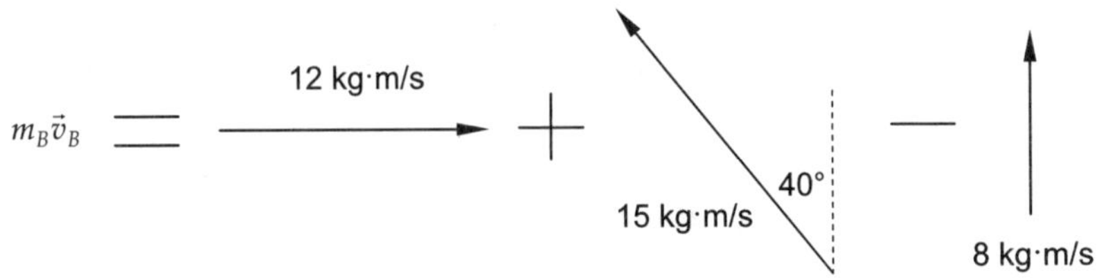

Now divide the final momentum of *B* by its mass to find its velocity:

$$\vec{v}_{B\text{after}} = \frac{(m_B\vec{v}_B)_{\text{after}}}{m_B} = \frac{4.2\,\text{kg·m/s}}{3.0\,\text{kg}} = 1.4\,\text{m/s, 56° N of E}$$

If the collision in the previous example were perfectly inelastic, the two objects would move locked together after the collision. The equation for the conservation of momentum for such a collision is simplified to

$$(m_A\vec{v}_A + m_B\vec{v}_B)_{\text{before}} = (m_A + m_B)\vec{v}_{B\text{after}}$$

Knowing all the momentum vectors in the left-hand side, we can simply add them and divide the resultant vector by the total mass of the objects to find their velocity after the collision.

As we have seen in this section, solving two-dimensional collision problems is not hard at all if you can do the vector operations you practiced in Chapter 1.

5-3. How to Solve Two-Dimensional Momentum Problems

The problems in this chapter are roughly divided into two groups. One group involves a collision between two objects or an explosion of an object into several pieces. Such problems require application of the law of conservation of momentum, as we have seen in the previous section. Those in the other group focus on the motion of one object, whether it is involved in a collision or not. This kind of problems usually asks for impulse, average net force, duration of impact, change in momentum, and so on. For such problems, you may not be able to use the law of conservation of momentum because either the second object's motion is unknown or insufficient information is given. Problems about a ball bouncing against a wall or an object receiving an impulse imparted by an applied force fall into this second category. So, be sure you know which type of problem you are solving.

If you see two colliding objects or an object exploding into pieces, use the following solution procedure:

1. Read the problem statement and draw a diagram showing both objects before and after the collision (or separation) as in Fig. 5-5.
2. Check that you know five of the following six variables appearing in Eq. (5.6) (on p. 129): m_A, m_B, $(\vec{v}_A)_{\text{before}}$, $(\vec{v}_B)_{\text{before}}$, $(\vec{v}_A)_{\text{after}}$, and $(\vec{v}_B)_{\text{after}}$. (In case of a perfectly inelastic collision, you must know four of the following five variables: m_A, m_B, $(\vec{v}_A)_{\text{before}}$, $(\vec{v}_B)_{\text{before}}$, $\vec{v}_{\text{after}} = (\vec{v}_A)_{\text{after}} = (\vec{v}_B)_{\text{after}}$.) If you have more than one unknown, you cannot use the conservation of momentum. Calculate the momentum vectors using known quantities and draw the vectors below the diagram as shown in Fig. 5-6.

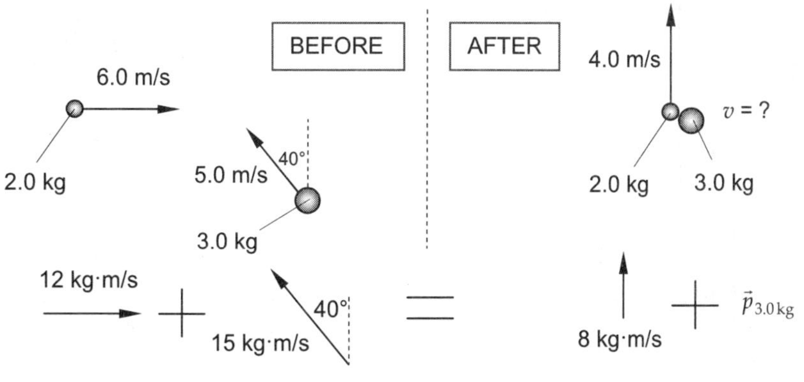

Figure 5-6: Diagram showing colliding objects and their momentum vectors

3. Choose the +x- and +y-direction, and resolve each momentum vector into its x- and y-components.

4. Apply the law of conservation of momentum to solve for the unknown. Usually the unknown is the velocity of one of the colliding objects. Then you can solve for the components of the unknown velocity from the following set of equations:

   Conservation of momentum in x-direction:

   $$(m_A\vec{v}_{Ax} + m_B\vec{v}_{Bx})_{\text{before}} = (m_A\vec{v}_{Ax} + m_B\vec{v}_{Bx})_{\text{after}} \quad \text{... (5.8)}$$

   Conservation of momentum in y-direction:

   $$(m_A\vec{v}_{Ay} + m_B\vec{v}_{By})_{\text{before}} = (m_A\vec{v}_{Ay} + m_B\vec{v}_{By})_{\text{after}} \quad \text{... (5.9)}$$

   Once the components are obtained, you can find the magnitude and direction of the unknown velocity using Eqs. (1.3) and (1.4) (on p. 14).

5. Check the answer against your own intuition as you re-enact the collision in your head. Especially ask yourself these questions: 'Does the speed seem too fast or too slow?', 'Is the object heading in the right direction?'

Let's apply this solution procedure to the following collision problem. Suppose a 3.7-kg object traveling at 4.5 m/s, 34° S of E, collides with a 2.2-kg object moving at 5.0 m/s, 62° N of E. Knowing that the 3.7-kg object's velocity changes to 2.8 m/s, 49° N of E, after the collision, we can find the velocity of the 2.2-kg object as follows:

1. Draw a diagram as shown below to help visualize the collision process.

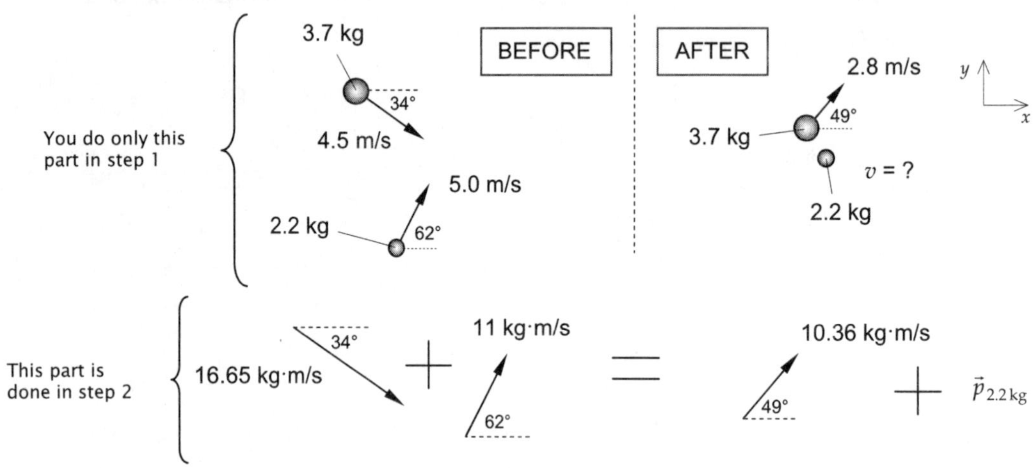

Figure 5-7: Diagram showing colliding objects and their momentum vectors

2. We are given five of the six variables, and the only unknown is the final velocity of the 2.2-kg object. Thus, we can calculate the momentum vectors of both objects before the collision and that of the 3.7-kg object after the collision. Letting the 3.7-kg object be *A* and the 2.2-kg object be *B*, they are

$$(m_A\vec{v}_A)_{\text{before}} = (3.7\,\text{kg})(4.5\,\text{m/s}) = 16.65\,\text{kg·m/s, } 34°\text{ S of E}$$

$$(m_B\vec{v}_B)_{\text{before}} = (2.2\,\text{kg})(5.0\,\text{m/s}) = 11.00\,\text{kg·m/s, } 62°\text{ N of E}$$

$$(m_A\vec{v}_A)_{\text{after}} = (3.7\,\text{kg})(2.8\,\text{m/s}) = 10.36\,\text{kg·m/s, } 49°\text{ N of E}$$

Draw the momentum vectors below the diagram you drew in the previous step (see Fig. 5-7).

3. Set up the coordinate system as shown at the top right corner of Fig. 5-7, and resolve each momentum vector:

$$(m_A\vec{v}_{Ax})_{\text{before}} = (16.65\,\text{kg·m/s})\cos 34° = 13.80\,\text{kg·m/s}$$

$$(m_A\vec{v}_{Ay})_{\text{before}} = (-16.65\,\text{kg·m/s})\sin 34° = -9.31\,\text{kg·m/s}$$

$$(m_B\vec{v}_{Bx})_{\text{before}} = (11.00\,\text{kg·m/s})\cos 62° = 5.16\,\text{kg·m/s}$$

$$(m_B\vec{v}_{By})_{\text{before}} = (11.00\,\text{kg·m/s})\sin 62° = 9.71\,\text{kg·m/s}$$

$$(m_A\vec{v}_{Ax})_{\text{after}} = (10.36\,\text{kg·m/s})\cos 49° = 6.80\,\text{kg·m/s}$$

$$(m_A\vec{v}_{Ay})_{\text{after}} = (10.36\,\text{kg·m/s})\sin 49° = 7.82\,\text{kg·m/s}$$

Note that the *y*-component of the 3.7-kg object's momentum before the collision is negative.

4. Use Eqs. (5.8) and (5.9) to solve for the components of the unknown velocity:

x-direction: $13.80\,\text{kg·m/s} + 5.16\,\text{kg·m/s} = 6.80\,\text{kg·m/s} + 2.2\,\text{kg}\cdot(\vec{v}_{Bx})_{\text{after}}$

$$\Longrightarrow (\vec{v}_{Bx})_{\text{after}} = 5.53\,\text{m/s}$$

y-direction: $-9.31\,\text{kg·m/s} + 9.71\,\text{kg·m/s} = 7.82\,\text{kg·m/s} + 2.2\,\text{kg}\cdot(\vec{v}_{By})_{\text{after}}$

$$\Longrightarrow (\vec{v}_{By})_{\text{after}} = -3.37\,\text{m/s}$$

With the components found, we can calculate the magnitude and direction of the velocity:

$$(v_B)_{\text{after}} = \sqrt{(5.53\,\text{m/s})^2 + (3.37\,\text{m/s})^2} \cong 6.5\,\text{m/s};\quad \theta = \tan^{-1}\left(\frac{3.37\,\text{m/s}}{5.53\,\text{m/s}}\right) \cong 31°\ \text{S of E}$$

Thus, the 2.2-kg object's velocity after the collision is 6.5 m/s, 31° S of E.

For problems other than those involving collisions and explosions, you still go through the same solution procedure except that the main equation to use is Eq. (5.5) (on p. 129). You might not be using all the variables in the equation because you do not need to. For instance, suppose a 13-N force directed east acts for 2.3 s on an 8.0-kg object initially moving at 3.5 m/s, 50° S of W. You can find the final velocity of the object as follows:

1. Draw a diagram as shown below to help visualize the situation.

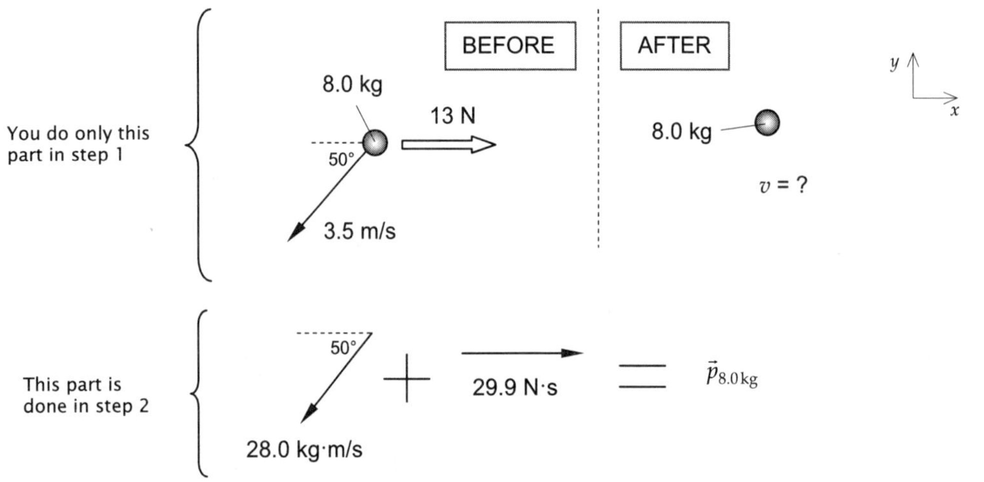

Figure 5-8: Diagram showing an object receiving impulse

2. In Eq. (5.5), we know four of the five variables needed to solve the problem.

$$\vec{J} = \left(\vec{F}_{\text{net}}\right)_{\text{avg}} \cdot \Delta t = m\left(\vec{v}_f - \vec{v}_i\right) = \vec{p}_f - \vec{p}_i \qquad \text{... (5.10)}$$

13 N, east   2.3 s   8.0 kg   3.5 m/s, 50° S of W

Calculate the impulse and the initial momentum vector:

$$\vec{J} = \left(\vec{F}_{\text{net}}\right)_{\text{avg}} \cdot \Delta t = (13\,\text{N})(2.3\,\text{s}) = 29.9\ \text{N·s, east}$$

$$\vec{p}_i = m\vec{v}_i = (8.0\,\text{kg})(3.5\,\text{m/s}) = 28.0\ \text{kg·m/s, 50° S of W.}$$

Draw these vectors below the diagram you drew in the previous step.

3. Resolve each vector according to the coordinate system shown at the top right corner of Fig. 5-8:

$$\vec{J}_x = 29.9\,\text{N·s} \qquad \vec{p}_{ix} = (-28.0\,\text{kg·m/s})\cos 50° = -18.0\,\text{kg·m/s}$$

$$\vec{J}_y = 0\,\text{N·s} \qquad \vec{p}_{iy} = (-28.0\,\text{kg·m/s})\sin 50° = -21.4\,\text{kg·m/s}$$

4. According to Eq. (5.10), the final momentum of the object is the sum of its initial momentum and the impulse vector:

$$\vec{p}_f = \vec{p}_i + \vec{J} \qquad \text{... (5.11)}$$

Add the two vectors by components, and we obtain

$$\vec{p}_{f_x} = 11.9 \text{ kg·m/s and } \vec{p}_{f_y} = -21.4 \text{ kg·m/s}$$

Thus, the final momentum of the object is

$$p_f = \sqrt{(11.9\,\text{kg}\cdot\text{m/s})^2 + (21.4\,\text{kg}\cdot\text{m/s})^2} \cong 25 \text{ kg·m/s}$$

$$\theta = \tan^{-1}\left(\frac{21.4\,\text{kg}\cdot\text{m/s}}{11.9\,\text{kg}\cdot\text{m/s}}\right) \cong 61^\circ \text{ S of E}$$

Finally, the final velocity of the object is

$$\vec{v}_f = \frac{\vec{p}_f}{m} = \frac{25\,\text{kg}\cdot\text{m/s}}{8.0\,\text{kg}} = 3.1\text{ m/s, } 61^\circ \text{ S of E}$$

As you have seen from the examples above, an accurate before-and-after diagram of the interaction is very helpful in solving these kinds of problems. Once a diagram is drawn and all necessary vectors are found, your ability to work with vectors will be the key to obtaining the right answer in any momentum problem.

PHYSICS 12 ANALYZED: MOMENTUM

5-1. Impulse Given by a Varying Force

- To find the magnitude of impulse on an object, find the area under the graph of net force versus time.

5-2. Two-Dimensional Collisions and Explosions

- Solving two-dimensional collision and explosion problems requires two-dimensional vector addition.

5-3. How to Solve Two-Dimensional Momentum Problems

- If you see two objects colliding or separating, do the following:
  1. Draw a diagram showing both objects before and after the collision or explosion.
  2. Make sure you have the values for the five of the following six variables: m_A, m_B, $(\vec{v}_A)_{\text{before}}$, $(\vec{v}_B)_{\text{before}}$, $(\vec{v}_A)_{\text{after}}$, and $(\vec{v}_B)_{\text{after}}$. The unknown variable has to be what the problem is asking you to find.
  3. Set the $+x$- and $+y$-direction, and resolve each momentum vector.
  4. Apply the following equations to solve for the unknown:

$$(m_A\vec{v}_{Ax} + m_B\vec{v}_{Bx})_{\text{before}} = (m_A\vec{v}_{Ax} + m_B\vec{v}_{Bx})_{\text{after}}$$

$$(m_A\vec{v}_{Ay} + m_B\vec{v}_{By})_{\text{before}} = (m_A\vec{v}_{Ay} + m_B\vec{v}_{By})_{\text{after}}$$

- When the motion of one particular object is given special attention (e.g., a basketball bouncing off a wall, a baseball struck by a bat, etc.), use the following equation:

$$\vec{J} = (\vec{F}_{\text{net}})_{\text{avg}}\Delta t = \Delta\vec{p} = m(\vec{v}_f - \vec{v}_i)$$

Do the necessary vector addition or subtraction to find the answer.

EXERCISES ON CHAPTER 5: MOMENTUM

Exercise 5-1

A 350-g rock is thrown horizontally at 4.7 m/s from the top of a 23-m high cliff.

a) Find the momentum of the rock when it is 8.0 m above the ground.

b) Find the impulse the rock receives while it is traveling in the air.

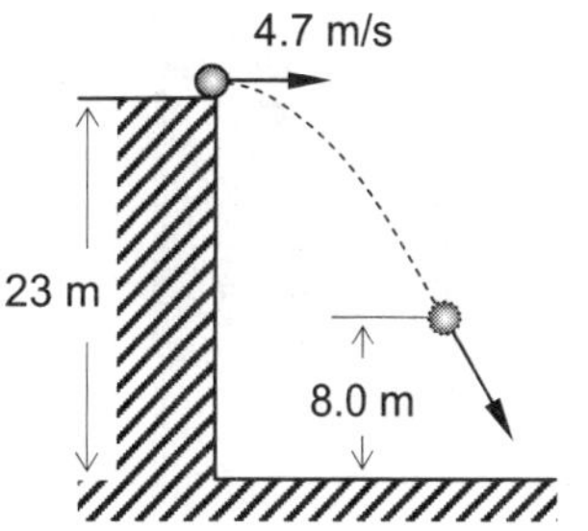

Exercise 5-2

A 5.3-kg cart traveling at 0.35 m/s, 25° S of E, bumps against a wall and moves 0.21 m/s, 13° N of E. Find the impulse received by the cart.

Exercise 5-3

A 52-kg object initially traveling east at 4.1 m/s experiences an applied force whose magnitude changes with time as shown. If the applied force is directed north, find (a) the impulse on the object and (b) the object's final velocity.

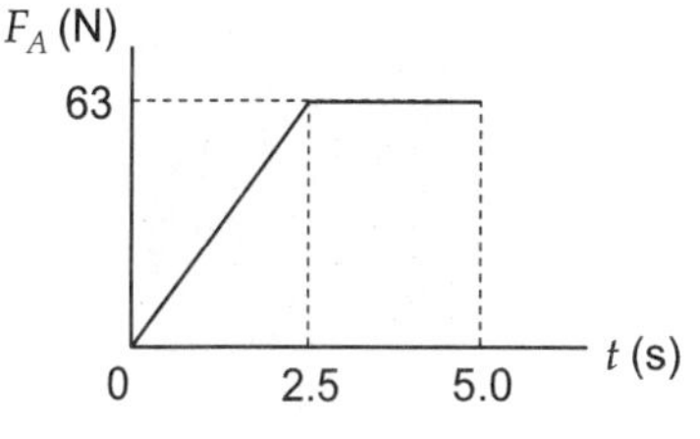

Exercise 5-4

A 10-kg object moving at 3.3 m/s, 27° N of E, is acted on by a 55-N force. If the object's velocity changes to 7.8 m/s, 3.5° N of E, how long and in what direction did the force act on the object?

Exercise 5-5

An 8.3-kg object initially traveling west receives 95-N force directed at 49° N of E for 1.5 s and changes the direction to north. Find the final speed of the object.

Exercise 5-6

A 350-g tennis ball initially moving at 18 m/s, 10° above the horizontal, is struck by a racket swung horizontally. The graph below shows the force applied by the racket on the ball as a function of time. Find (a) the average force on the ball and (b) the final velocity of the ball.

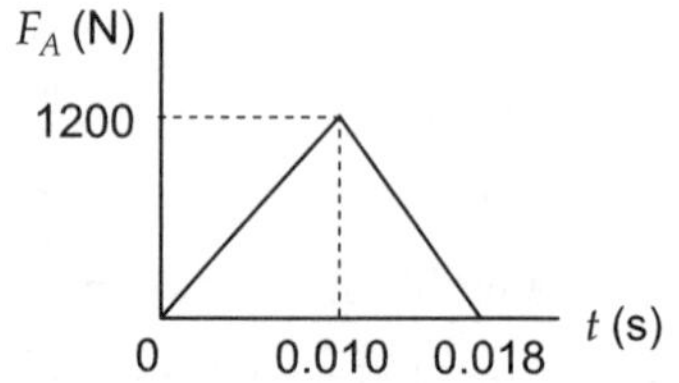

Exercise 5-7

A 1600-kg car traveling east at 60 km/h collides with a 2300-kg pickup truck heading south at 45 km/h at an intersection. After the collision, the wrecked vehicles move together. Find their speed immediately after the collision.

Exercise 5-8

A 2.5-kg ball traveling at 8.3 m/s, 42° N of E, collides with a 3.0-kg ball moving at 5.8 m/s, 36° S of E. After the collision, the 2.5-kg ball changes its direction to east but continues moving at the same speed of 8.3 m/s. Find (a) the velocity of the 3.0-kg ball after the collision and (b) the loss of kinetic energy during the collision.

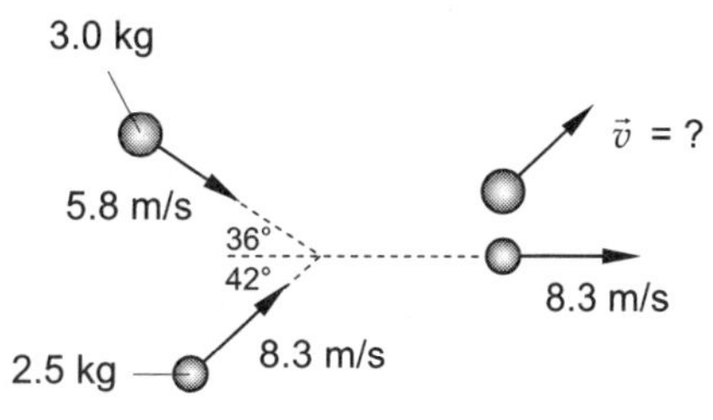

Exercise 5-9

A 42-kg cart moving at 2.5 m/s, 66° S of W, collides with a 23-kg cart whose initial velocity is unknown. After the collision, they move together at 2.3 m/s, 12° W of S. Find the initial velocity of the 23-kg cart.

Exercise 5-10

A 6.5-kg object traveling at 2.6 m/s collides with an object of unknown mass moving at 1.8 m/s. Upon collision, they stick together and move to the right as shown. Find the mass of the second object.

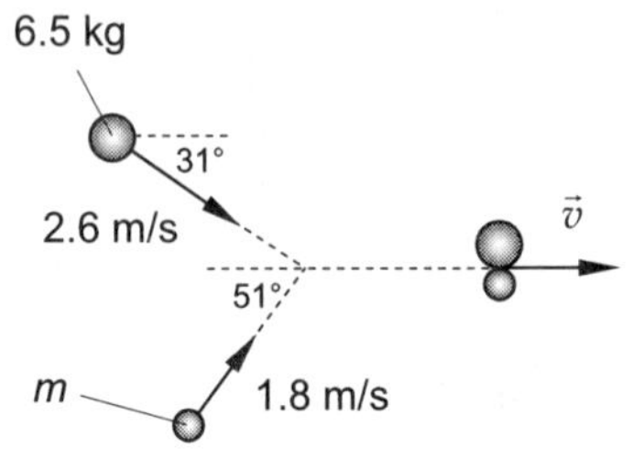

Exercise 5-11

A 73-kg object traveling at 25 m/s, 23° E of N, explodes into two fragments. If the larger 48-kg fragment moves at 17 m/s, 48° E of N, find the velocity of the other fragment.

Exercise 5-12

Three objects travel as shown. If the objects collide simultaneously and come to a stop immediately after the collision, find the initial velocity of the 3.6-kg object.

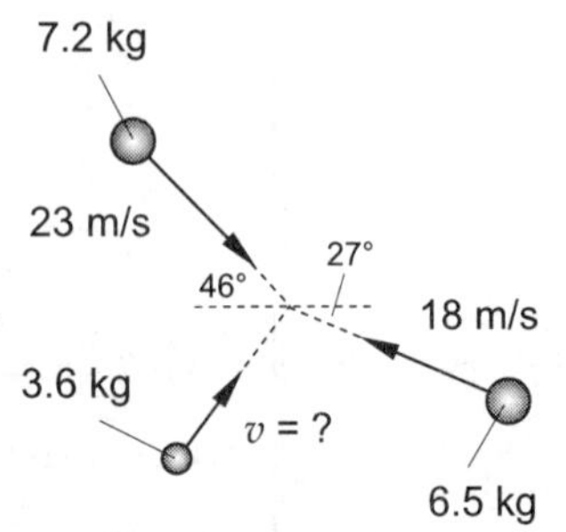

Exercise 5-13

A 19-kg object traveling at 5.9 m/s, east, explodes into three pieces as shown. Find (a) the velocity of the 4.0-kg piece and (b) the amount of energy added during the explosion.

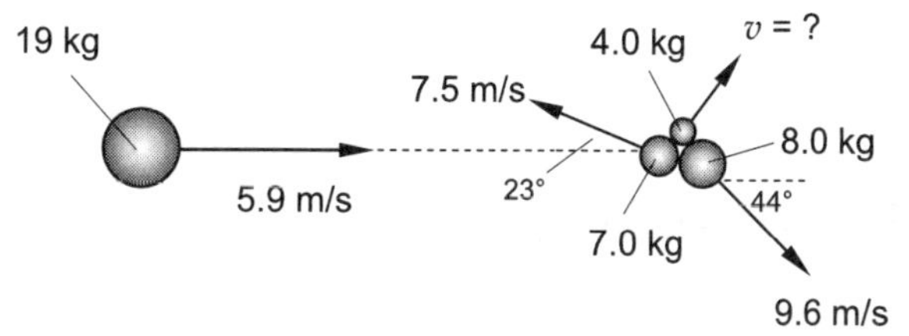

Exercise 5-14

A 37-g bullet is fired horizontally at a 1.3-kg block as shown. The bullet collides with the block at the speed of 330 m/s and gets embedded within it. If the surface is smooth, find the maximum height the block reaches. (Assume the block remains motionless until the bullet gets fully embedded within it.)

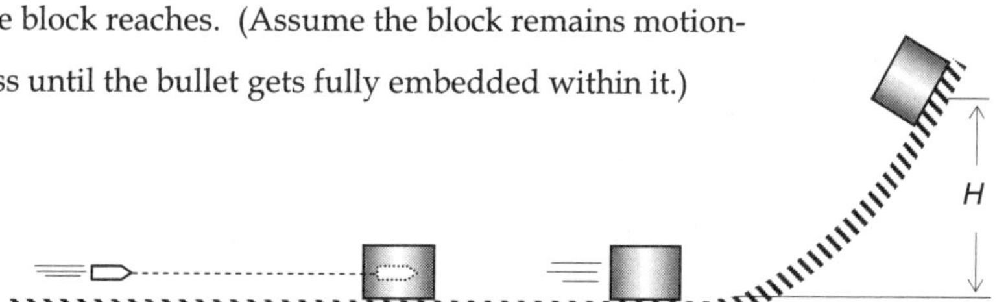

Exercise 5-15

An 85-kg object traveling east at speed v collides with a 57-kg object heading north at speed $2v$. Upon collision, they stick together and move at 12 m/s. Find (a) the initial speed of each object before the collision and (b) the direction of their velocity after the collision.

PROBLEMS ON CHAPTER 5: MOMENTUM

(Answer Key on p. 369)

1 – 6. Fill in the blanks

1. __________ is the amount of motion an object possesses and is defined as the product of the object's mass and __________.

2. The ________ ________ on an object indicates how fast its momentum changes with respect to time.

3. __________ is defined as the product of net force applied to an object and time period over which the force was applied. It is equal to the change in __________.

4. The area under the graph of net force versus time represents the magnitude of __________.

5. Colliding bodies stick together in a __________ __________ collision.

6. The total kinetic energy of colliding bodies is conserved in a __________ __________ collision.

7 – 15. True/False questions

7. _____ A moon orbiting a planet at constant speed has constant momentum.

8. _____ An object receives impulse only through contact with another object.

9. _____ If an object loses speed while traveling in a straight path, it implies that the object received an impulse in the opposite direction of its motion.

10. _____ The direction of an impulse given to an object coincides with its direction of motion.

11. _____ The impulse generated by a force whose magnitude changes with time can be calculated by measuring the change in an object's momentum.

12. _____ Among Newton's laws of motion, the law of conservation of momentum is most closely related to Newton's third law.

13. _____ In a perfectly elastic collision, each object's kinetic energy remains constant.

14. _____ If a ball initially moving with speed v strikes a stationary ball of the same mass and comes to a complete stop, the collision is perfectly elastic.

15. _____ The total kinetic energy before an explosion is always less than that after it.

16 – 25. Multiple choice questions

16. A ball initially heading south collides with an object moving north and turns east, as shown right. Which of the following could be the direction of the impulse given by the ball during the collision?

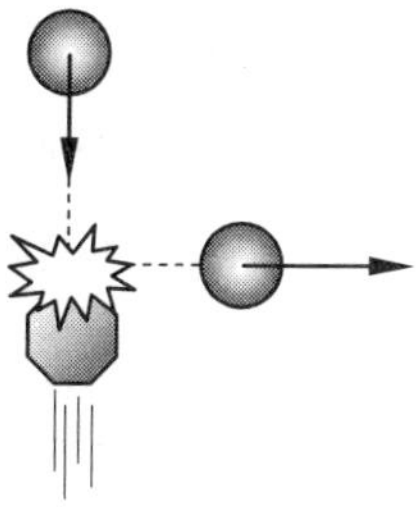

a) 

b) 

c) 

d) 

17. An object initially heading east experiences a force directed northwest as shown. Which of the following could not be the direction of the object's final velocity?

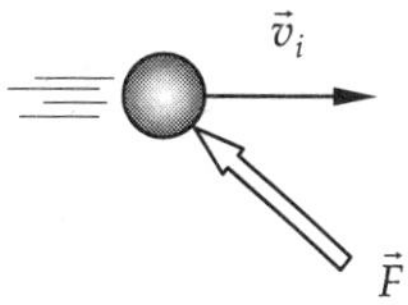

a) 

b) 

c) 

d) 

18. An object initially flying east is split into two pieces, as shown. If the first piece heads directly north after the separation, which of the following could be the direction of the other piece's velocity?

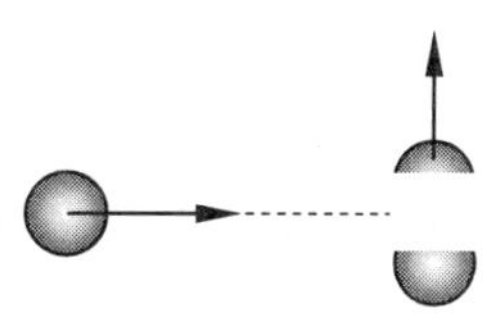

a) 

b)

c) 

d) 

19. A red object heading west collides with a blue object and bounces back to move east. Which of the following statements must be true?

a) The objects have unequal masses.

b) It was a head-on collision.

c) It was a perfectly elastic collision.

d) The blue object received impulse directed west.

20. A ball is thrown at 3.5 m/s, 25° above the horizontal, as shown. Which of the following statements is <u>false</u>?

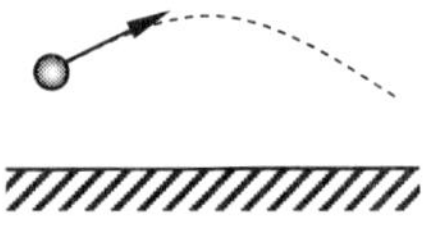

a) The ball receives impulse directed vertically downward.

b) The magnitude of impulse on the ball is the greatest at the maximum height.

c) The magnitude of the ball's momentum is the minimum at the maximum height.

d) The ball constantly receives impulse while it is in the air.

21. An egg does not break when dropped onto a mattress while it does when dropped from the same height onto a concrete floor. This occurs because

a) the egg experiences the greater change in momentum when it hits the concrete floor.

b) the egg receives the greater average net force when it hits the concrete floor.

c) the egg has the greater impact speed when dropped onto the concrete floor.

d) All of the above

22. Hitting a baseball far is easier with a ball coming towards you at high speed than with a stationary ball. Which of the following statements explains the reason?

a) The hitter can exert greater force on a moving ball than on a stationary ball.

b) The hitter can exert force longer on a moving ball than on a stationary ball.

c) The hitter can hit a moving ball higher than a stationary ball.

d) It is actually easier to hit a stationary ball farther than a moving ball.

For Problems 23 to 25, refer to the following problem:

A bullet is fired horizontally at a wooden block as shown. The bullet collides with the block, which then slides along a smooth surface and climbs a frictionless track, as shown. (Assume the block remains motionless until the bullet gets fully embedded within it.)

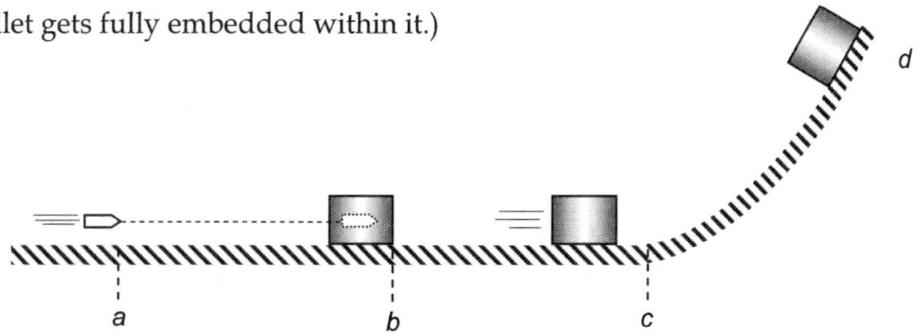

23. Which of the following is conserved between *a* and *b*?
    a) The sum of momenta of the bullet and the block only
    b) The total mechanical energy of the bullet and the block only
    c) Both the sum of their momenta and their total kinetic energy
    d) Neither the sum of their momenta nor their total kinetic energy

24. Which of the following is conserved between *b* and *c*?
    a) The sum of momenta of the bullet and the block only
    b) The total mechanical energy of the bullet and the block only
    c) Both the sum of their momenta and their total kinetic energy
    d) Neither the sum of their momenta nor their total kinetic energy

25. Which of the following is conserved between *c* and *d*?
    a) The sum of momenta of the bullet and the block only
    b) The total mechanical energy of the bullet and the block only
    c) Both the sum of their momenta and their total kinetic energy
    d) Neither the sum of their momenta nor their total kinetic energy

26 – 35. Written-response questions

26. An 8.1-kg object moving at 1.2 m/s, 63° N of E, collides with a 2.7-kg object traveling at 3.2 m/s, 11° N of E. They stick together during the collision. Find (a) their final velocity and (b) the loss of total kinetic energy during the collision.

27. A 12-kg block is accelerated from rest across a horizontal surface by a 52-N force for 4.2 s, as shown. The coefficient of friction between the block and the surface is 0.18. Find the block's final momentum.

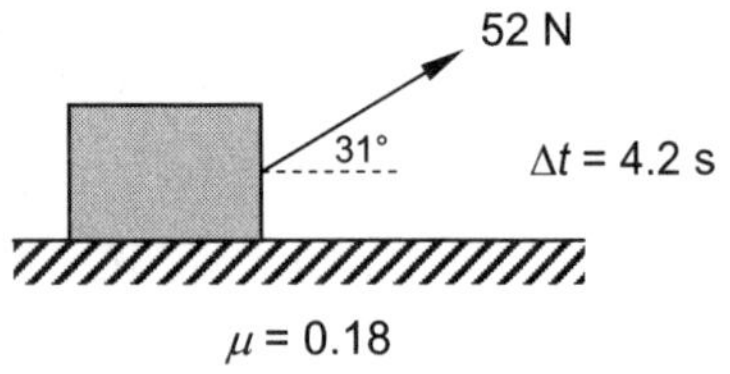

28. Water is ejected from a hose at a rate of 4.4 kg/s. It hits a wall at 11 m/s and is splashed back at the same speed. What is the magnitude of the force exerted by the water on the wall?

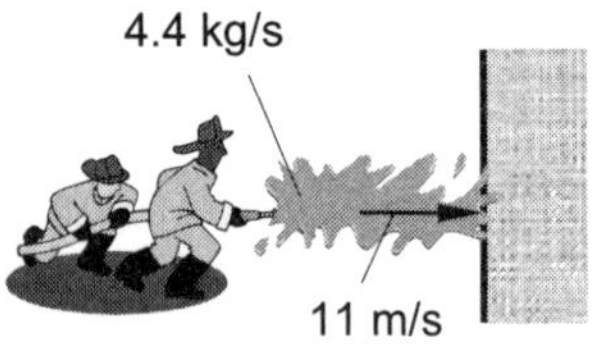

29.* A 130-kg cannon fires a 15-kg cannon ball at 37 m/s, 26° above the horizontal, and recoils back on a smooth surface, as shown. The cannon eventually falls off the 26-m high platform. Find the distance between where the cannon lands and the base of the wall.

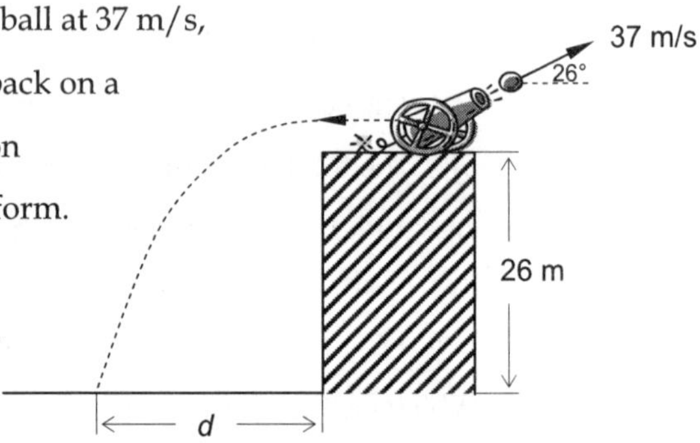

30.* A 3.0-kg object heading east at 2.5 m/s collides with a 9.0-kg object moving west at 1.2 m/s. If the collision is perfectly elastic, find the velocity of each object immediately after the collision.

31. A 150-kg explosive device is launched at 45 m/s, 42° above the horizontal, from the ground, as shown below. The device explodes into four fragments when it reaches the maximum altitude. Immediately after the explosion, a 30-kg fragment heads vertically upward at 23 m/s, a 40-kg fragment moves at 18 m/s, 68° below the horizontal, and a 60-kg fragment comes to a complete stop. Find the velocity of the fourth fragment.

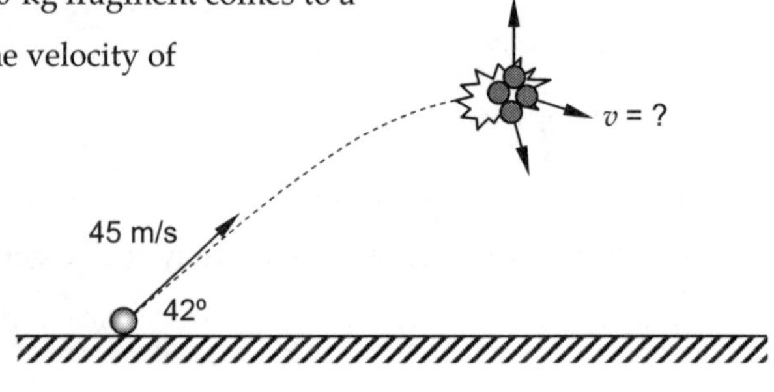

32. A 40-g plastic bullet fired horizontally at 56 m/s strikes a pendulum bob at the moment the pendulum is vertical, as shown. As a result of the collision, they both come to a complete stop. If the pendulum was initially released 49° from the vertical, find the mass of the pendulum bob.

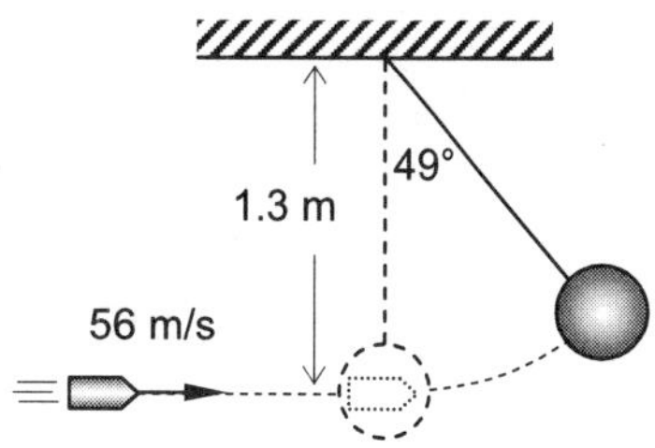

33. A 250-g bullet fired parallel to an inclined surface at 350 m/s strikes a 4.8-kg block at rest. Upon impact, the bullet is embedded within the block. If the coefficient of friction between the block and the ground is 0.45, how far does the block travel before coming to a stop? (Assume the block does not move until the bullet is fully embedded and comes to rest within it.)

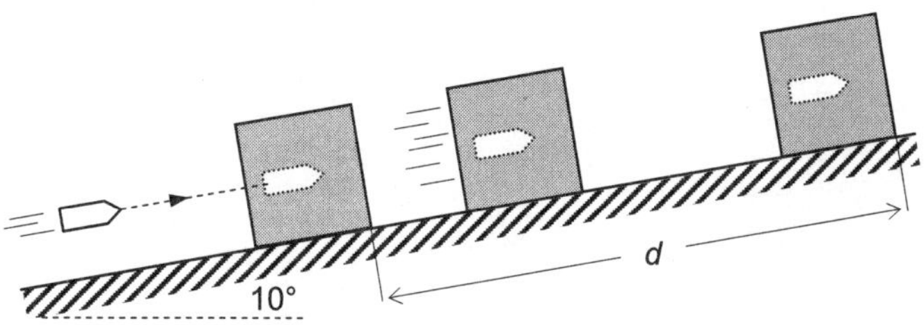

34. A 10-kg object, initially traveling at 60 m/s, 35° S of W, explodes into ten fragments of equal mass. If all the fragments have the same speed, find the sum of their momenta.

35. An object moving at 7.0 m/s collides with another object with the same mass and speed. The collision is perfectly inelastic, and their speed after the collision is reduced in half. Find the angle between the objects' initial directions.

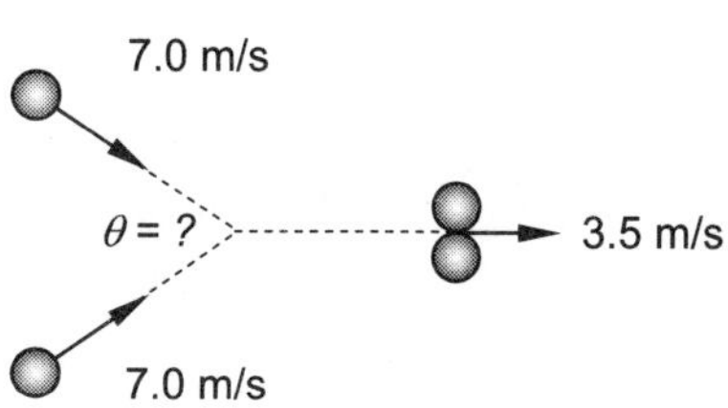

6

CHAPTER 6

STATIC EQUILIBRIUM

Have you ever wondered how the Leaning Tower of Pisa† in Italy can remain stable without toppling? In order to ensure that a structure remains stationary and stable, engineers must be certain that the forces acting on it are balanced. However, that is not their only concern; they must also keep it from rotating! So they need to understand the causes of rotation.

In this chapter, you will be introduced to a new physical quantity called torque. This is what causes a stationary object to rotate. So we can deduce that, when an object is motionless, the torques applied to make it spin clockwise are balanced by those trying to make it spin counterclockwise. Or, simply put, the net torque acting on a stationary object is zero. This condition, together with the one that states that the net force is also zero, allows us to analyze the forces acting on a stationary object and predict stability.

In this chapter, you will:

- ✓ learn about torque
- ✓ learn the two conditions of static equilibrium
- ✓ learn how to solve static equilibrium problems
- ✓ solve exercise problems on static equilibrium
- ✓ test your knowledge of static equilibrium

---

† Ever since it was constructed in the 12th century, the Leaning Tower of Pisa began leaning to one side due to a poorly laid foundation. During the 1990s, structural engineers stabilized the tower with steel cables, buried underground. Now the tower is declared stable for a few hundred years.

6-1. The First Condition of Static Equilibrium

When the motion of an object does not change with time, the object is said to be in equilibrium. If at rest, such an object is said to be in **static equilibrium**, and if it is moving, we say the object is in **dynamic equilibrium**. (We have already discussed dynamic equilibrium in Chapter 3. Objects in dynamic equilibrium move with constant velocity.) Since an object in equilibrium—whether it be static or dynamic—experiences no acceleration, the forces acting on it must be balanced. So, for an object at rest to remain motionless, the net force acting on it has to be zero, and that is the first condition of static equilibrium:

$$\vec{F}_{\text{net}} = 0 \qquad \text{... (6.1)}$$

For two-dimensional problems, both components of the net force are zero:

$$\vec{F}_{\text{net}_x} = 0 \text{ and } \vec{F}_{\text{net}_y} = 0 \qquad \text{... (6.2)}$$

Figure 6-1 shows some objects in static equilibrium and the forces acting on them. By now, it should be clear to you that the net force acting on each object is zero.

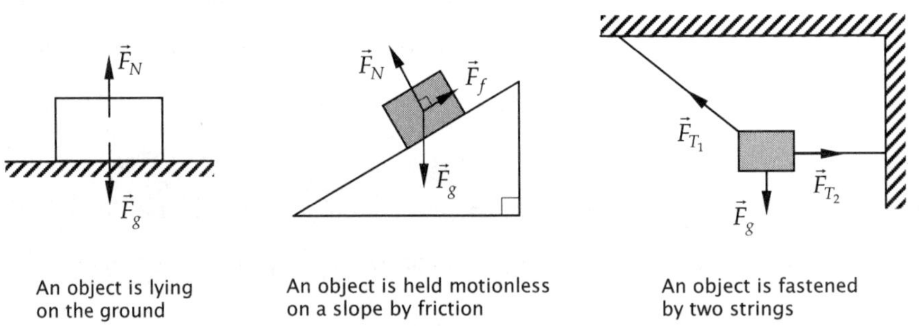

Figure 6-1: Objects in static equilibrium and the forces acting on them

6-2. Torque

If you examine the three objects shown in Fig. 6-1, you will realize that all the forces acting on them could be extended to meet at a single point. When this is the case, the first condition of static equilibrium is the only one that need be applied (since no rotation is possible). However, the situation is very different if the applied forces act at different points on the object and do not meet. Then, even if the net force on the object is zero, rotation may still occur. For example, suppose a pair of 15-N forces is applied simultaneously on each pencil lying on the table as shown in

Fig. 6-2. While the net force acting on each pencil is zero, the one on the right remains motionless, and the other does not. Why do you think that is?

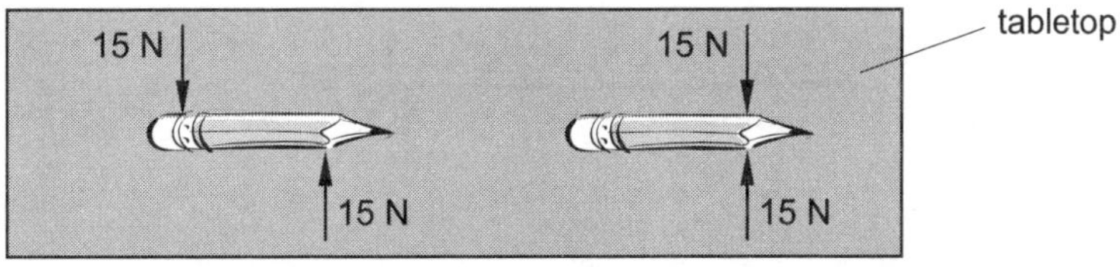

Figure 6-2: A pair of 15-N forces applied on two pencils

In this chapter, we will model long, slender objects as lines. The motion of such objects is complicated as they can not only translate but also <u>rotate</u>. Hence their motion may be purely translational (Fig. 6-3 (a)), purely rotational (Fig. 6-3 (b)), or a combination of both (Fig. 6-3 (c)). You probably knew that the pencil on the left in Fig. 6-2 rotates about a pivot located somewhere between the two 15-N forces, so its motion is purely rotational.

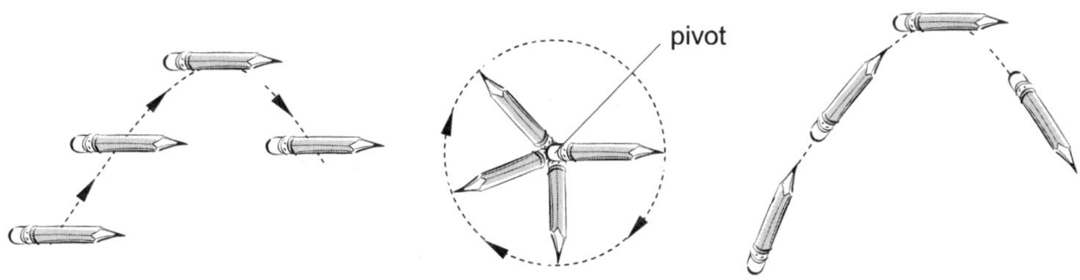

(a) purely translational (b) purely rotational (c) translational and rotational

Figure 6-3: Various types of motion

In order to understand what makes an object turn, you need to know what torque is. **Torque** is defined as the product of the magnitude of a force and the length of its lever arm (or moment arm). The **lever arm** of a force is the perpendicular distance from the **pivot** (or the **axis of rotation**) to the line along which the force acts, as shown in Fig. 6-4. Mathematically, torque is expressed as

> The standard unit of torque is N·m. Although it is equivalent to J [Joule], we use Joules only for work and energy.

$$\tau = F\underbrace{r\sin\theta}_{\text{lever arm}} \quad [\text{N·m}] \qquad \dots (6.3)$$

where r is the distance from the pivot to the point at which the force is applied, and θ is the angle between the force and the object ($0° \le \theta \le 90°$).

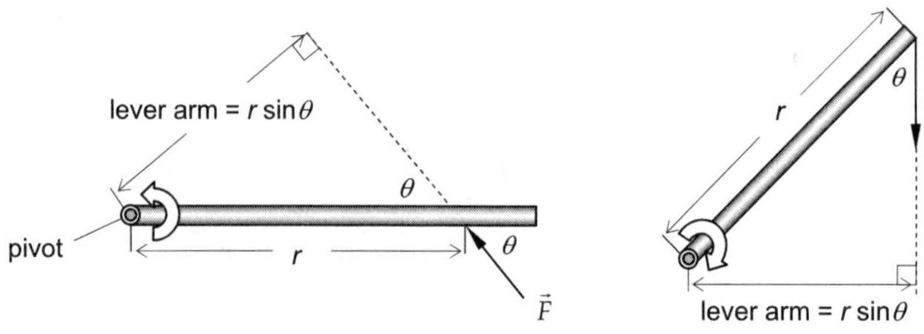

(a) (b)

Figure 6-4: Finding the lever arm

Torque is a vector quantity as it can make an object spin clockwise or counter-clockwise. Determining the direction of a torque should not be difficult. For instance, the torque created by the force in Fig. 6-4 (a) is counterclockwise (CCW), and that created by the force in Fig. 6-4 (b) is clockwise (CW).

(a) $\tau = 0$ because $r = 0$ (b) $\tau = 0$ because $\sin\theta = 0$

Figure 6-5: Torque is zero whenever the lever arm is zero

The torque generated by a force is zero whenever its lever arm is zero, and the lever arm is zero when the force is applied at the pivot, as shown in Fig. 6-5 (a). The lever arm can also be zero if the force is parallel to the object (see Fig. 6-5 (b)), and the line of action of the force goes through the pivot. This leads to the following idea: Only the force's component perpendicular to the object ($F_\perp$) generates torque, and torque may be interpreted alternatively as the product of r and $F_\perp$:

$$\tau = rF_\perp = rF\sin\theta \qquad \text{... (6.4)}$$

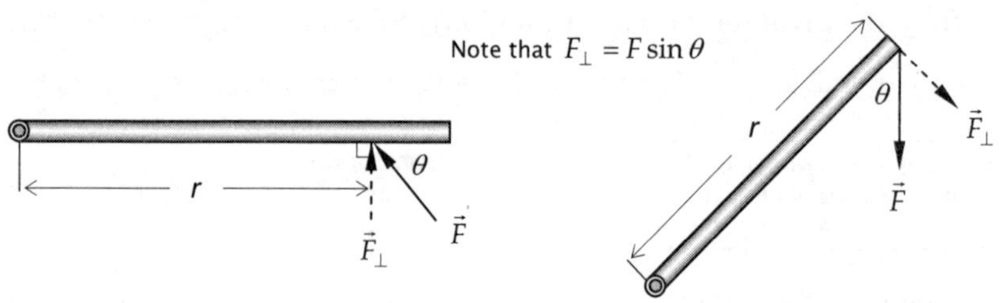

(a) (b)

Figure 6-6: Finding torque by resolving force into components

Thus, instead of finding the lever arm, you can resolve a force vector into components parallel and perpendicular to the object, as shown in Fig. 6-6, and multiply the perpendicular component with the distance from the pivot to the force. The result will be identical to multiplying the lever arm with the magnitude of the force.

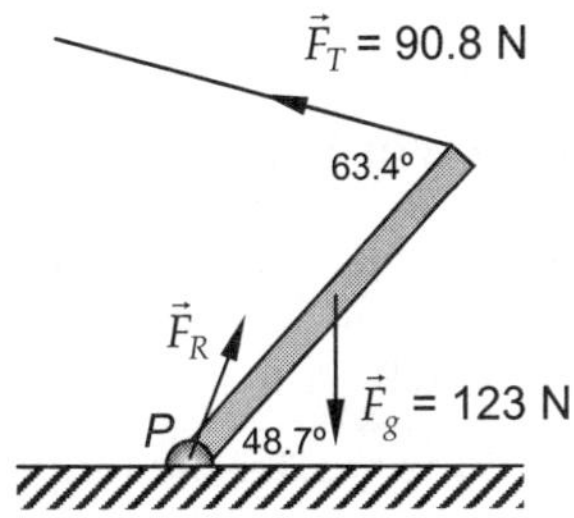

Figure 6-7: Calculating torque generated by each force on a beam

Consider the 2.50-m beam shown in Fig. 6-7. Let's find torque created by each force acting on it. First we need to find the pivot, which, for this beam, is at the base (point *P*). The force on the beam at the hinge, $\vec{F}_R$, generates no torque because it is applied at the pivot. The torque created by the tension force, $\vec{F}_T$, is counter-clockwise, and its magnitude is

$$\tau_{\text{by }\vec{F}_T} = (2.50\text{ m})(90.8\text{ N})\cdot\sin(63.4°) = 203\text{ N}\cdot\text{m (CCW)}$$

The beam's weight, acting at the center of the beam, creates a clockwise torque. To find the magnitude of this torque, we need to find the angle the weight makes with the beam, which is 90.0° - 48.7° = 41.3°. Thus,

$$\tau_{\text{by }\vec{F}_g} = \left(\frac{2.50\text{ m}}{2}\right)(123\text{ N})\cdot\sin(41.3°) = 101\text{ N}\cdot\text{m (CW)}$$

Keep in mind that θ in Eqs. (6.3) and (6.4) must be less than or equal to 90°.

Now that we know all the torques created by the forces acting on the beam, we can combine them to find the net torque. Set the counterclockwise direction as positive. Then

$$\tau_{\text{net}} = \tau_{\text{by }\vec{F}_R} + \tau_{\text{by }\vec{F}_T} + \tau_{\text{by }\vec{F}_g} = 0\text{ N}\cdot\text{m} + 203\text{ N}\cdot\text{m} + (-101\text{ N}\cdot\text{m}) = +102\text{ N}\cdot\text{m}$$

or 102 N·m, counterclockwise.

Note that it is not allowed to use the net force in Eq. (6.3) or (6.4) to find the net torque. That is, never use the following equation:

$$\tau_{net} = r\,F_{net}\sin\theta \quad \Longleftarrow \quad \text{Never do this!}$$

The only way to find the net torque is to compute each torque one by one and add them together.

6-3. The Second Condition of Static Equilibrium

Now that we have learned how to calculate torque, let's investigate how torque is related to rotational motion. You probably have guessed that torque is what makes an object turn and that the greater the torque, the faster it turns. In fact, torque is to rotational motion as force is to translational motion. As you know, a force can make an object accelerate and change its velocity, and when forces acting on an object are balanced, it is either at rest or moving with constant velocity. Likewise, torque can make an object turn faster or slower, and when torques applied to an object are balanced, it is either at rest or rotating at a constant rate. Therefore, for an object in static equilibrium, the net torque as well as the net force must be zero.

Consider the pencils in Fig. 6-2. The left one rotates because the torques are not balanced. On the other hand, the pencil on the right remains stationary because the torques are balanced regardless of where you assume the pivot is on the pencil. Actually, for an object at rest, you may choose any point on it to be the pivot, and the net torque must be zero. (However, the point that simplifies the analysis is the best candidate for the pivot, as you will see in the next section.)

Thus, an object in static equilibrium must satisfy not only the first condition in Eq. (6.1) (on p. 158) but also the following second condition:

$$\tau_{net} = 0 \qquad \text{... (6.5)}$$

The above equation can also be expressed as

$$\Sigma\tau_{CW} = \Sigma\tau_{CCW} \qquad \text{... (6.6)}$$

When asked to analyze the forces acting on a long, slender body, you will probably need to apply this second condition.

6-4. How to Solve Static Equilibrium Problems

In this chapter, you will only see objects in static equilibrium and usually be asked to find the forces acting on them. If all the forces act at a single point, the second condition of static equilibrium is not applicable. Follow the solution procedure in Section 3-2 (p. 72), and you should have no trouble finding the answer.

The following problem is a typical example where only the first condition of static equilibrium is needed to find the answer. Suppose a 25-kg object is hanging from a ceiling as shown in Fig. 6-8, and we are asked to find the tension in each cable.

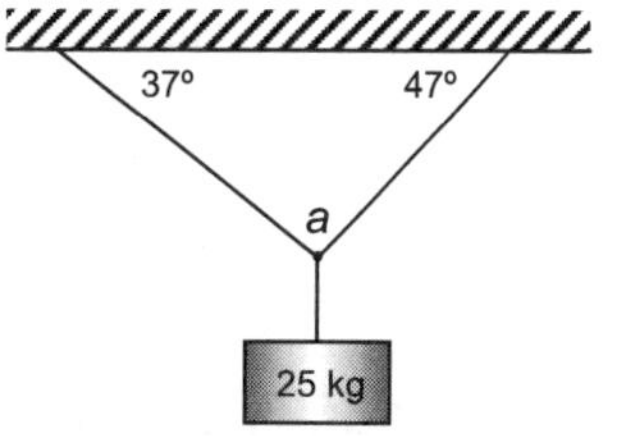

Figure 6-8: An object hanging from a ceiling by cables

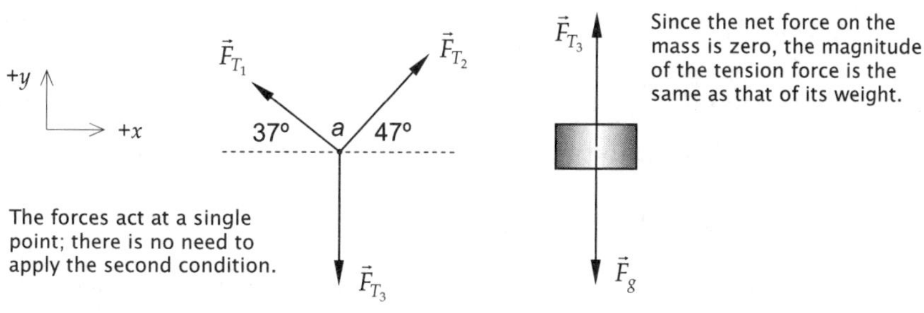

Figure 6-9: The free-body diagrams at point *a* and of the 25-kg mass

First set up the coordinate system and draw the free-body diagram, both of which are shown in Fig. 6-9. Note that the free-body diagram is drawn at the point where the cables meet, instead of the 25-kg object. There are three tension forces, and we resolve each into its x- and y-components:

$$F_{T_{1x}} = -F_{T_1}\cos 37° = -0.799F_{T_1} \qquad F_{T_{1y}} = F_{T_1}\sin 37° = 0.602F_{T_1}$$

$$F_{T_{2x}} = F_{T_2}\cos 47° = 0.682F_{T_2} \qquad F_{T_{2y}} = F_{T_2}\sin 47° = 0.731F_{T_2}$$

$$F_{T_{3x}} = 0 \qquad F_{T_{3y}} = -F_{T_3}$$

Now apply Eq. (6.2) (on p. 158), the first condition of static equilibrium:

$$\vec{F}_{\text{net}_x} = -0.799F_{T_1} + 0.682F_{T_2} + 0 = 0$$

$$\vec{F}_{\text{net}_y} = 0.602F_{T_1} + 0.731F_{T_2} - F_{T_3} = 0$$

We have more unknowns (3) than equations (2), so we need to set up one more equation in order to be able to solve the system of equations. From the free-body diagram of the 25-kg object, we can obtain F_{T_3}, which is equal to the object's weight:

$$F_{T_3} = mg = 25\,\text{kg}\cdot(9.8\,\text{m/s}) = 245\,\text{N}$$

We now have a system of two equations with two unknowns:

$$\left.\begin{aligned} -0.799F_{T_1} + 0.682F_{T_2} &= 0 \\ 0.602F_{T_1} + 0.731F_{T_2} &= 245\,\text{N} \end{aligned}\right\} F_{T_1} = 168\,\text{N},\ F_{T_2} = 197\,\text{N}$$

Therefore, the tension forces are as follows: $F_{T_1} = 170\,\text{N},\ F_{T_2} = 200\,\text{N},\ F_{T_3} = 250\,\text{N}$.

When a problem involves a long, slender object at rest, you almost always need to find the torque created by each force acting on the object and apply both conditions of static equilibrium in order to solve the problem. Your ability to determine torque correctly is the key to obtaining the right solution, so make sure you know how to do it. The solution procedure is outlined as follows:

1. Set up the coordinate system. The *x*-axis may be chosen to be parallel to the body or to the ground. (The *y*-axis is always perpendicular to the *x*-axis.)
2. Draw the free-body diagram of the object. Here are a few tips:
   a. Unless told otherwise, you may assume the force of gravity acts at the center of the object.
   b. A wire, cable, string, etc. is flexible and can only have a force acting along its length, as any force acting at an angle to it would bend it.
   c. A rigid device like a hinge can exert a force in any direction, not just along a beam or up a wall. So, draw a force in the direction you think it acts. If you guess wrong, your calculations will give you a negative value.
   d. Objects leaning against walls or standing on the ground at an angle will be subject to a normal force acting perpendicular to the surface and a static friction force acting to prevent motion. If the surface is 'smooth,' however, no friction force is in effect.
3. Choose a point on the object to be the pivot. A good candidate is the point where one of the unknown forces is applied.

4. Calculate the torque generated by each force in the diagram. (This is also called 'taking moment about a pivot.') Don't forget the direction!

5. Apply the second condition of static equilibrium, Eq. (6.6). Check that there is only one unknown in the equation, and solve for the unknown. If there is more than one unknown, try selecting a different point for the pivot.

6. Apply the first condition of static equilibrium if necessary.

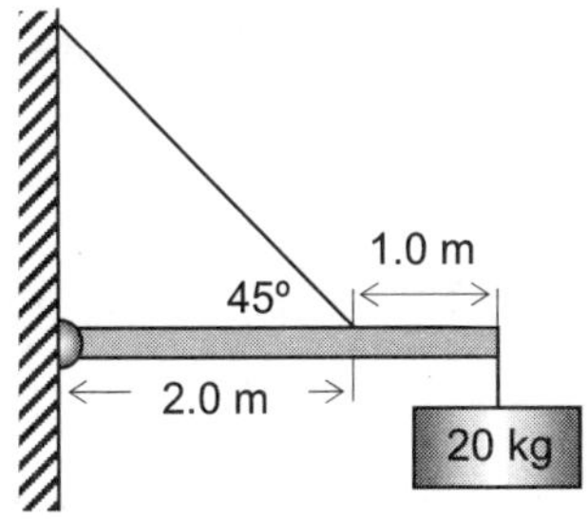

Figure 6-10: A beam supported by a cable

Let's apply the above solution procedure to a couple of example problems. In the first problem, we have a 10-kg uniform beam supported by a cable as shown in Fig. 6-10. One end of the beam is fixed to the wall, and a 20-kg mass hangs by a string at the other end. We need to find all unknown forces acting on the beam.

1. The x- and y-axes are chosen to be parallel and perpendicular to the beam, respectively.

2. The free-body diagram of the beam, together with the coordinate system set up in the previous step, is shown in Fig. 6-11. Note that the weight of the beam acts at its center and that we have two unknown forces in the diagram, the tension in the cable ($\vec{F}_T$) and the force exerted by the wall ($\vec{F}_R$). (Since we are not sure about the direction of $\vec{F}_R$, we may draw the vector in any direction for now.)

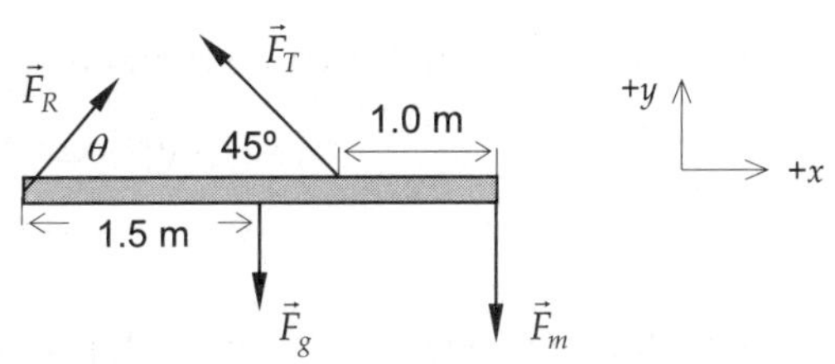

Figure 6-11: Free-body diagram of the beam in Fig. 6-10

3. Choose the center of the beam to be the pivot. We will soon see whether this is a good choice.

4. Calculate the torque generated by each force:

$$\tau_{\text{by}\,\vec{F}_R} = (1.5\,\text{m})F_R \sin\theta\ \text{N·m, CW}$$

$$\tau_{\text{by}\,\vec{F}_g} = (0\,\text{m})(10\,\text{kg})(9.8\,\text{m/s}^2)\sin 90° = 0\ \text{N·m}$$

$$\tau_{\text{by}\,\vec{F}_T} = (0.5\,\text{m})F_T \sin 45°\ \text{N·m, CCW}$$

$$\tau_{\text{by}\,\vec{F}_M} = (1.5\,\text{m})(20\,\text{kg})(9.8\,\text{m/s}^2)\sin 90° = 294\ \text{N·m, CW}$$

5. Set up Eq. (6.6):

$$\Sigma\tau_{\text{CW}} = \tau_{\text{by}\,\vec{F}_R} + \tau_{\text{by}\,\vec{F}_M} \text{ and } \Sigma\tau_{\text{CCW}} = \tau_{\text{by}\,\vec{F}_T}$$

Thus,

$$(1.5\,\text{m})F_R \sin\theta + 294\,\text{N·m} = (0.5\,\text{m})F_T \sin 45°$$

We have too many unknowns in the above equation (F_R, θ, and F_T). Perhaps moving the pivot to a different point may yield a simpler equation. If we choose the left end of the beam to be the pivot, the torques become

$$\tau_{\text{by}\,\vec{F}_R} = (0\,\text{m})F_R \sin\theta = 0\ \text{N·m}$$

$$\tau_{\text{by}\,\vec{F}_g} = (1.5\,\text{m})(10\,\text{kg})(9.8\,\text{m/s}^2)\sin 90° = 147\ \text{N·m, CW}$$

$$\tau_{\text{by}\,\vec{F}_T} = (2.0\,\text{m})F_T \sin 45° = (1.41\,\text{m})\cdot F_T\ \text{N·m, CCW}$$

$$\tau_{\text{by}\,\vec{F}_M} = (3.0\,\text{m})(20\,\text{kg})(9.8\,\text{m/s}^2)\sin 90° = 588\ \text{N·m, CW}$$

Now we have only one unknown, F_T, when we set up Eq. (6.6):

$$147\,\text{N·m} + 588\,\text{N·m} = (1.41\,\text{m})\cdot F_T$$

Solve for the tension force, and we get

$$F_T = 520\ \text{N}$$

6. We can now apply the first condition of static equilibrium to find $\vec{F}_R$. First, resolve the tension force into components:

$$F_{T_x} = -(520\,\text{N})\cos 45° = -367.5\ \text{N}$$

$$F_{T_y} = (520\,\text{N})\sin 45° = 367.5\ \text{N}$$

Find the components of $\vec{F}_R$ using Eq. (6.2):

$$\vec{F}_{\text{net}_x} = F_{R_x} + (-367.5\,\text{N}) = 0$$

$$\vec{F}_{\text{net}_y} = F_{R_y} + F_{T_y} - F_g - F_M = F_{R_y} + 367.5\,\text{N} - 98\,\text{N} - 196\,\text{N} = 0$$

$$\Longrightarrow \quad F_{R_x} = 367.5\,\text{N and } F_{R_y} = -73.5\,\text{N}$$

Thus, $\vec{F}_R$ is 370 N, 11° below the $+x$-axis. Note that its direction is below the $+x$-axis because its y-component is negative.

In the next problem, a 2.0-m ladder is leaning against a frictionless wall as shown in Fig. 6-12. Given that the ladder's mass is 10 kg, we are to analyze the forces acting on the ladder.

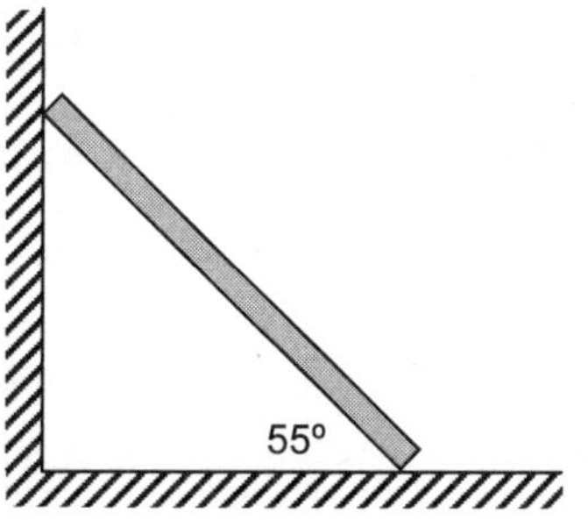

Figure 6-12: A ladder leaning against a wall

1. Set up the coordinate system so that the x-axis is horizontal, as shown in Fig. 6-13.
2. The free-body diagram of the ladder is drawn in Fig. 6-13. There are three unknown forces in the diagram: the normal force exerted by the wall ($\vec{F}_{N_1}$), the one exerted by the ground ($\vec{F}_{N_2}$), and the friction force against the ground ($\vec{F}_f$). (Note that there is no friction on the wall.)
3. Choose the pivot to be where the ladder is touching the ground. This choice eliminates the torques created by $\vec{F}_{N_2}$ and $\vec{F}_f$, two of the unknowns in this problem.

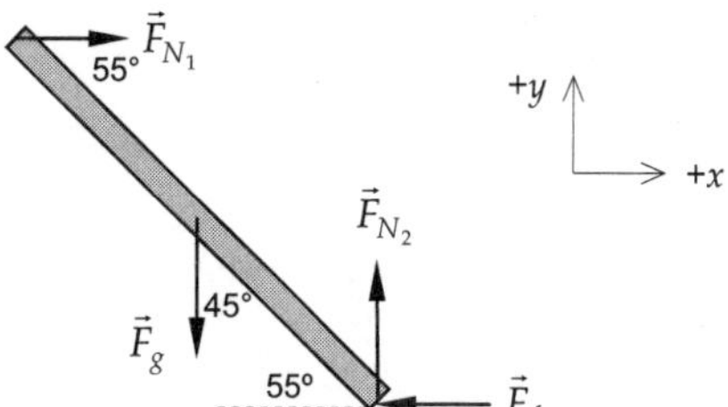

Figure 6-13: Free-body diagram of the ladder in Fig. 6-12

4. Find the angle between each force and the ladder, and calculate the torque generated by each force:

$$\tau_{\text{by }\vec{F}_{N_1}} = (2.0\,\text{m})F_{N_1}\sin 55° = (1.64\,\text{m})\cdot F_{N_1}\ \text{N·m, CW}$$

$$\tau_{\text{by }\vec{F}_g} = (1.0\,\text{m})(10\,\text{kg})(9.8\,\text{m/s}^2)\sin 45° = 69.3\ \text{N·m, CCW}$$

$$\tau_{\text{by }\vec{F}_{N_2}} = \tau_{\text{by }\vec{F}_f} = 0\ \text{N·m}$$

5. Set up Eq. (6.6) and solve for $\vec{F}_{N_1}$:

$$(1.64\,\text{m})\cdot F_{N_1} = 69.3\,\text{N·m} \Longrightarrow F_{N_1} = 42\ \text{N}$$

6. Apply the first condition of static equilibrium to find the other two forces. Basically the friction force is balanced by the normal force by the wall, and the normal force by the ground is balanced by the ladder's weight. Hence,

$$F_f = 42\ \text{N and } F_{N_2} = 98\ \text{N}$$

As you have seen from these static equilibrium problems, the solution procedure is relatively straightforward as long as you can draw accurate free-body diagrams and know how to find torque. As for the choice of the pivot point, place it where an unknown force is applied. The more forces acting at a certain spot, the better it is to put the pivot there. If you practice enough, you will know where to put the pivot without any difficulty.

PHYSICS 12 ANALYZED: STATIC EQUILIBRIUM

6-1. The First Condition of Static Equilibrium

- When an object at rest stays motionless, it is said to be in static equilibrium.
- The first condition of static equilibrium: "The forces acting on a body in static equilibrium must be balanced." (That is, $\vec{F}_{\text{net}} = 0$.)

6-2. Torque

- Torque is the product of the magnitude of a force and its lever arm: $\tau = F \underbrace{r \sin\theta}_{\text{lever arm}}$
- Torque is a vector. Its direction can be either clockwise or counterclockwise.
- Torque can be changed by varying
  1. the distance from the pivot to the force
  2. the magnitude of the force
  3. the angle between the body and the force

6-3. The Second Condition of Static Equilibrium

- A force can make a body accelerate linearly. Similarly, torque can make a body rotate faster or slower.
- The second condition of static equilibrium: "The torques applied to a body in static equilibrium must be balanced." (That is, $\tau_{\text{net}} = 0$.)

6-4. How to Solve Static Equilibrium Problems

1. Set up the coordinate system.
2. Draw the free-body diagram of the body.
3. Choose a point on the object to be the pivot.
4. Calculate the torque generated by each force in the diagram. (Don't forget the direction!)
5. Apply the second condition of static equilibrium.
6. Apply the first condition if necessary.

EXERCISES ON CHAPTER 6: STATIC EQUILIBRIUM

Exercise 6-1

Three forces are acting at a single point on an object in static equilibrium. If two of the forces act as shown right, determine the magnitude and direction of the third force.

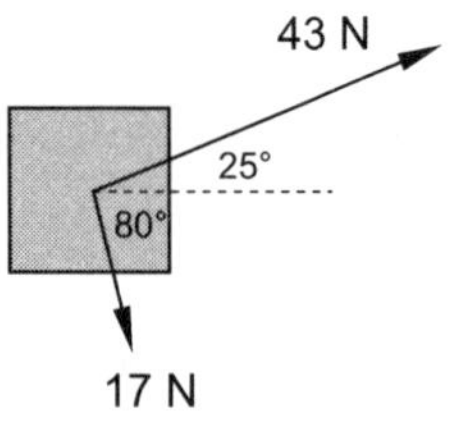

Exercise 6-2

A 7.5-kg mass suspended by a cable is kept in static equilibrium by a horizontal force of 26 N. Find the angle between the cable and the ceiling, θ.

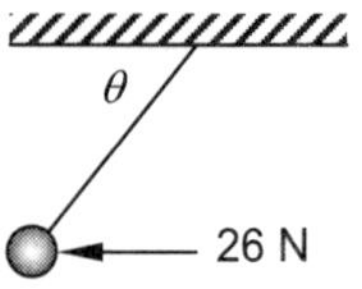

Exercise 6-3

An object is suspended from a ceiling and a wall, as shown. If the tension in the horizontal rope is 110 N, find the mass of the object.

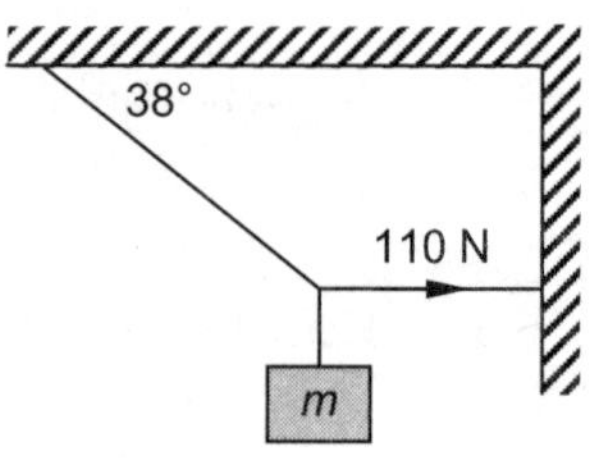

Exercise 6-4

A 28-kg traffic light is suspended by cables from two poles, as shown. Find the tension in each cable.

Exercise 6-5

A57-kg sphere is suspended by two cables as shown. Find the tension in each cable.

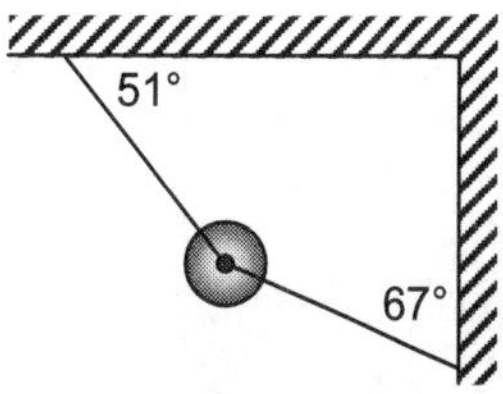

Exercise 6-6

A ball suspended by a string is held motionless by a 31-N force as shown. Find the mass of the ball.

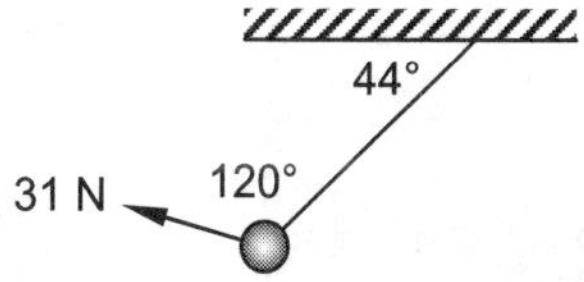

Exercise 6-7

A 1.2-kg picture frame is held at rest by a horizontal force, as shown. If the coefficient of friction between the wall and the frame is 0.28, what is the minimum magnitude of the horizontal force required to keep the frame in static equilibrium?

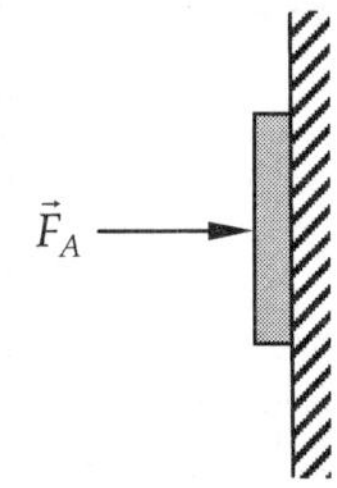

Exercise 6-8

A 220-g shirt is hanging from a ceiling as shown. Find the tension in each string.

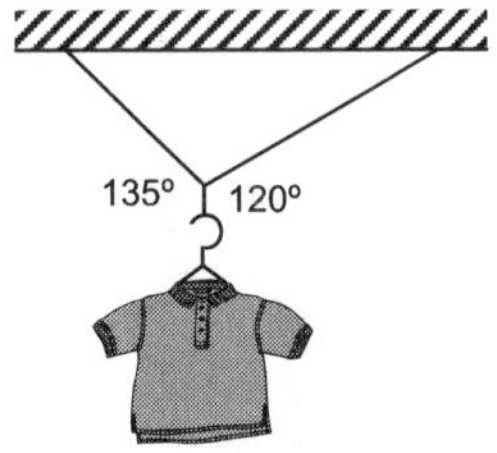

Exercise 6-9

A 48-kg lady is practicing tightrope walking with an 11-kg bar as shown. Find the tension in each part of the cable when she is at the position shown below.

Exercise 6-10

A 14-kg object suspended by a rope is pulled horizontally as shown. If the rope can withstand a maximum tension of 270 N, find the maximum angle θ to which the object can be pulled without breaking the rope.

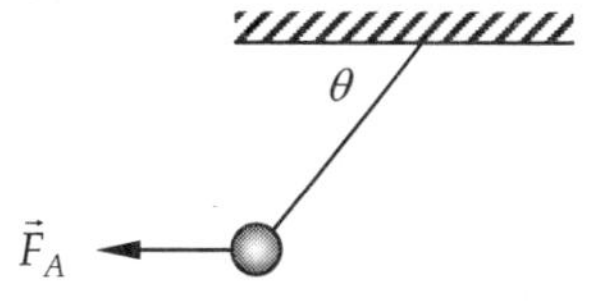

Exercise 6-11

A spherical object is suspended by two cords, one of which is connected to a 7.5-kg block resting on a frictionless incline, as shown. Find the mass of the spherical object.

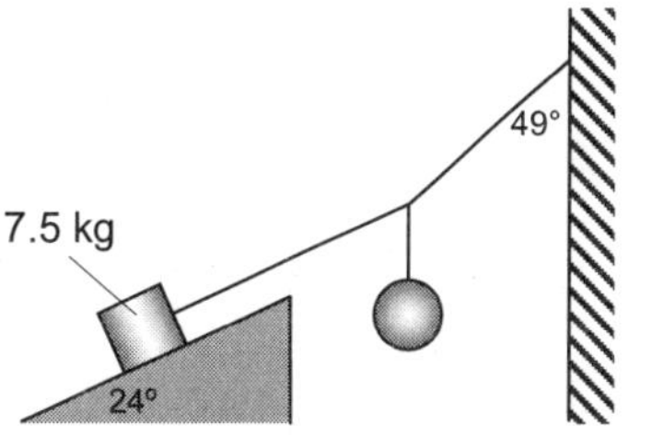

Exercise 6-12

Five forces are acting on a 3.2-m rod as shown below. Calculate the torque created by each force assuming the pivot is at (a) the left end, (b) the middle, and (c) the right end of the rod.

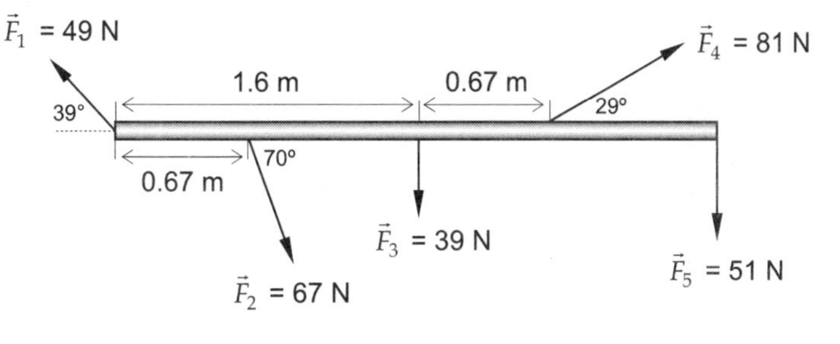

Exercise 6-13

A 42-kg detective is standing on a 3.7-m long beam, as shown. Find the mass of the beam.

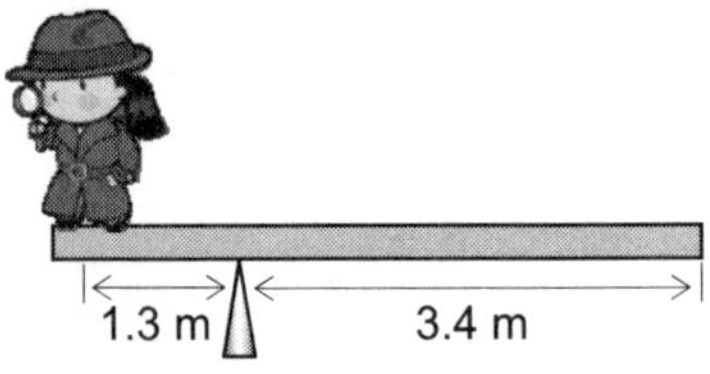

Exercise 6-14

A 16-kg beam is held stationary by a cord as shown. Find the tension in the cord and the force on the beam at the hinge.

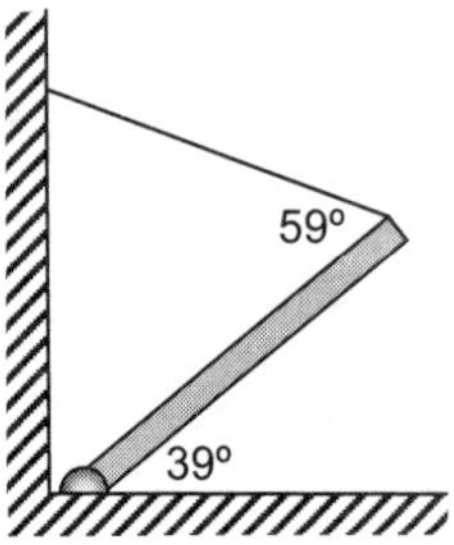

Exercise 6-15

A 1.7-m long beam of mass 7.8 kg is suspended by a cable as shown. Find the tension in the cable and the force on the beam at the hinge.

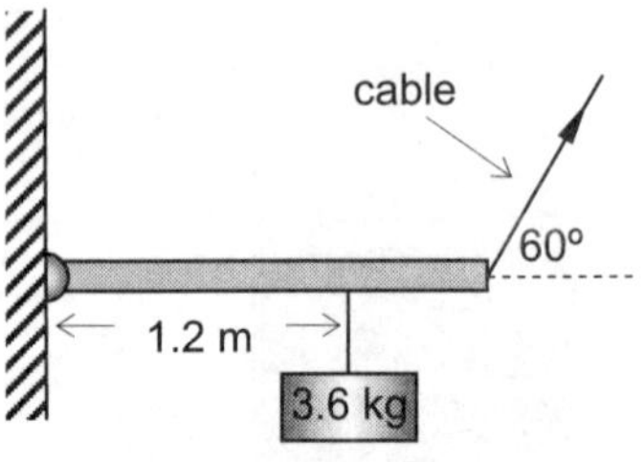

Exercise 6-16

The bicycle shown right has a wheel base of 1.3 m. If the bicycle's mass is 21 kg and the force exerted by the ground on the front wheel is 88 N, find how far the bicycle's center of mass is located from the front wheel.

Exercise 6-17

A 1.1-m long beam with a mass of 12 kg is kept stationary by a cable as shown. If the maximum tension the cable can withstand is 830 N, find the maximum mass of the load, *m*.

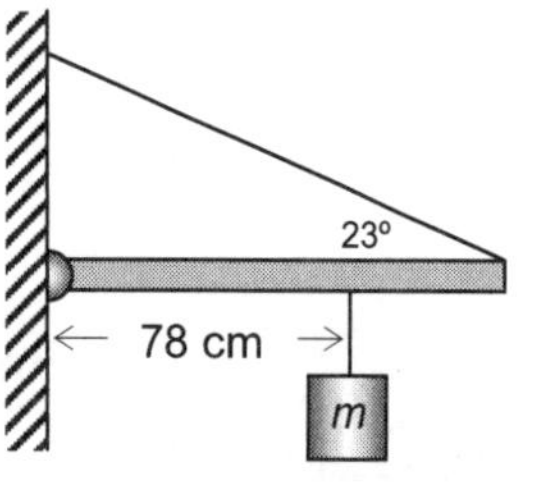

Exercise 6-18

An RCMP officer is walking on a 13-m long beam as shown. The officer and the beam are 53 kg and 27 kg, respectively.

a) Find the forces exerted by the supports when the officer is 2.0 m to the right of the left support.

b) If the officer keeps on walking, how far to the right of the right support can he walk before the beam tips over?

Exercise 6-19

A 63-cm long stick is held up by the force $\vec{F}$, and its bottom is about to slide to the left when it makes a 65° angle with the ground, as shown. If the stick's mass is 2.3 kg, what is the friction force acting on the stick?

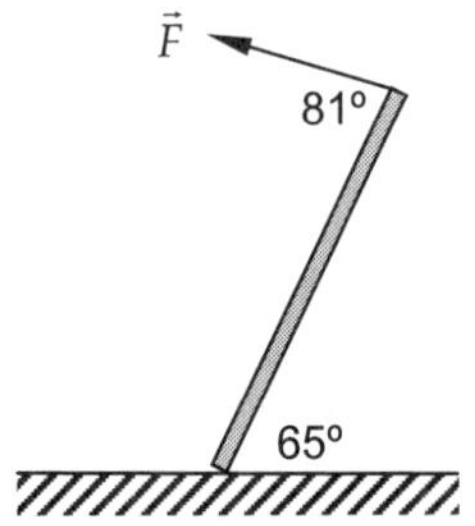

Exercise 6-20

A 4.7-m long beam with a mass of 13 kg is kept stationary by a cable as shown. If the cable makes a 48° angle with the ground, find the tension in the cable.

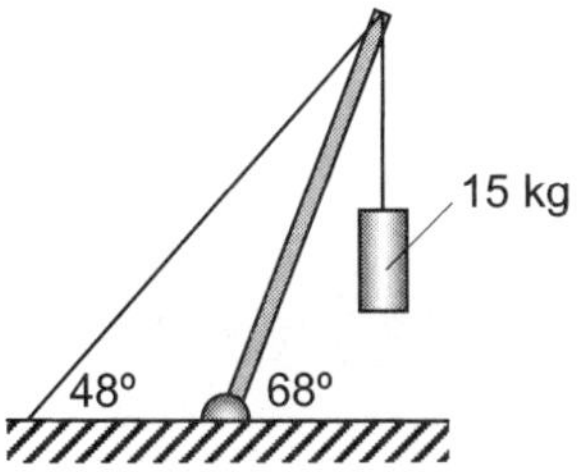

Exercise 6-21

A 78-kg man is fishing on a 3.5-m long board fixed to a pair of supports as shown. If the board's mass is 16 kg, find the force exerted by each support on the board.

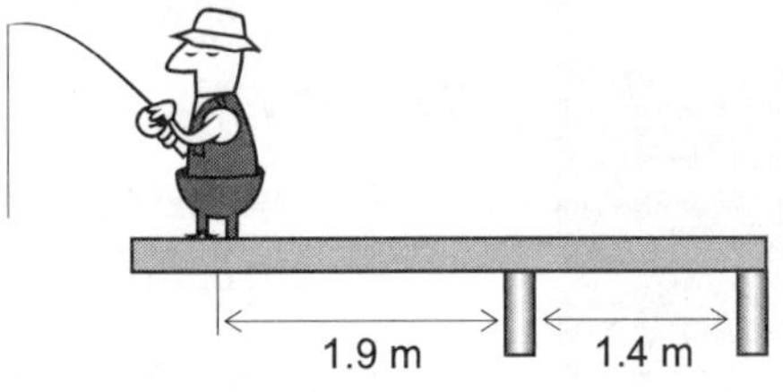

Exercise 6-22

A 3.8-m long beam of mass 9.0 kg is supported by a 2.6-kg rod as shown. Find (a) the force exerted by the rod on the beam and (b) the force exerted by the ground on the rod.

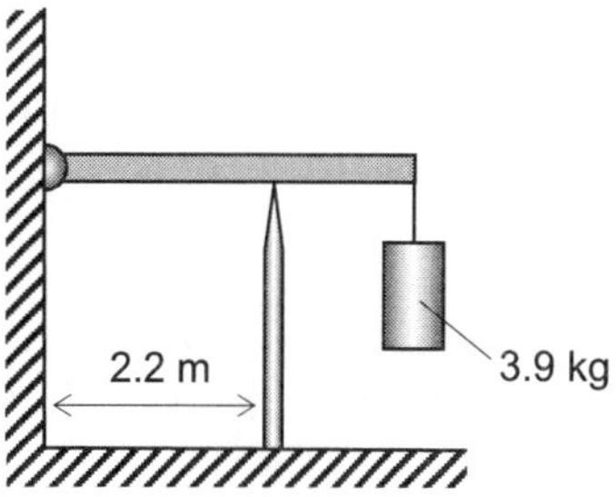

Exercise 6-23

A baseball bat is suspended horizontally by two strings as shown. If the tension in the left string is 5.9 N and the tension in the right string is 1.9 N, find (a) the mass of the bat and (b) how far the bat's center of mass is located from the left string.

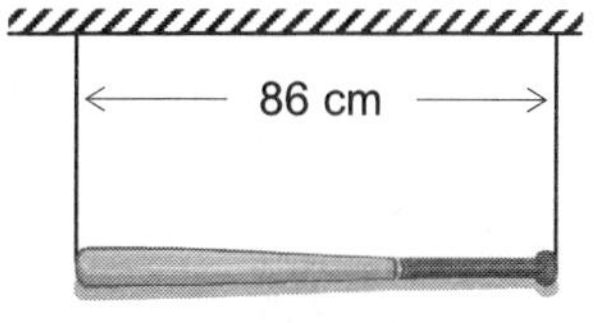

Exercise 6-24

A 1.5-m long rod with a mass of 7.3 kg is lying on the ground. If you want to lift one end of the rod off the ground by a cord as shown in the diagram, what is the minimum tension force required?

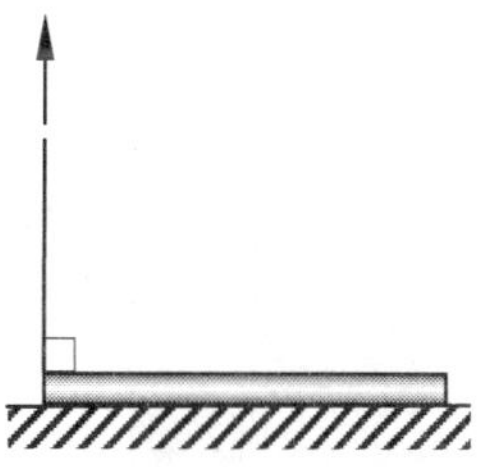

Exercise 6-25

A 1.8-m long beam with a mass of 3.5 kg is held horizontal by a string as shown. If the string can withstand a maximum tension of 500 N, find the minimum distance, *d*, the string can be attached without breaking.

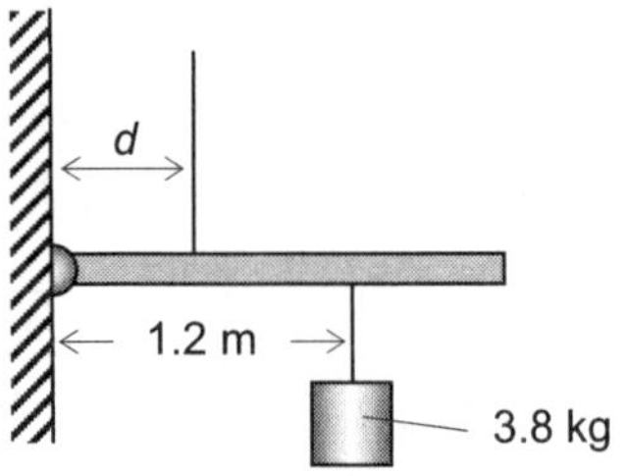

Exercise 6-26

A bear stands on a balance beam. When the bear moves forward (that is, to the left) 0.25 m from the position shown, the force exerted on the beam by the right support decreases from 3600 N to 3400 N. Find the bear's weight. (Ignore the beam's weight.)

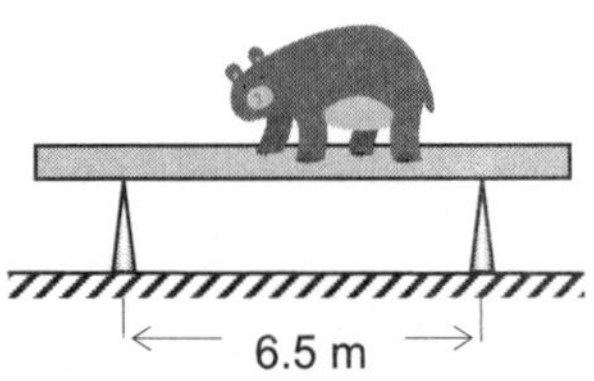

Exercise 6-27

An 8.5-kg beam, supported at its left end, is held stationary by a cord as shown. A 2.2-kg object is suspended from the beam's right end, and the force on the beam at the hinge is horizontal. Find the beam's length.

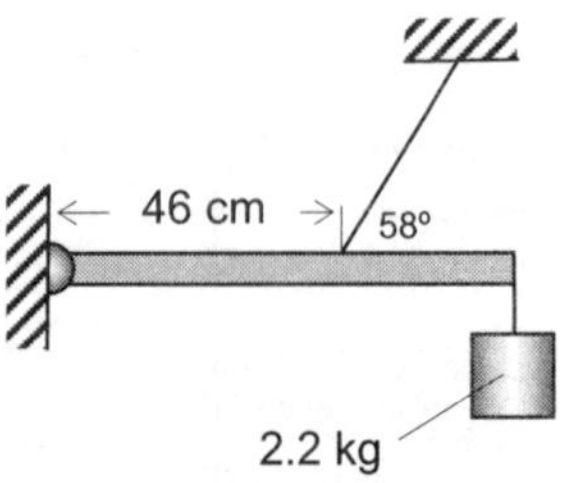

Exercise 6-28

A 6.2-m long beam of mass 35 kg is held stationary by a cable attached to the beam at three quarters of its length. A 52-kg object is suspended from the end of the beam. Find the tension in the cable and the force exerted on the beam at the hinge.

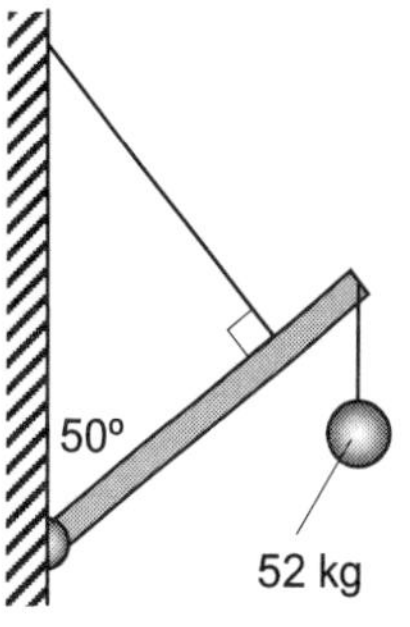

Exercise 6-29

A 49-kg woman climbs a ladder leaning against a smooth wall as shown. If the ladder is 3.5 m long and she is 2/7 of the way up, find the magnitude of each and every force acting on the ladder. (The ladder's mass is 8.4 kg.)

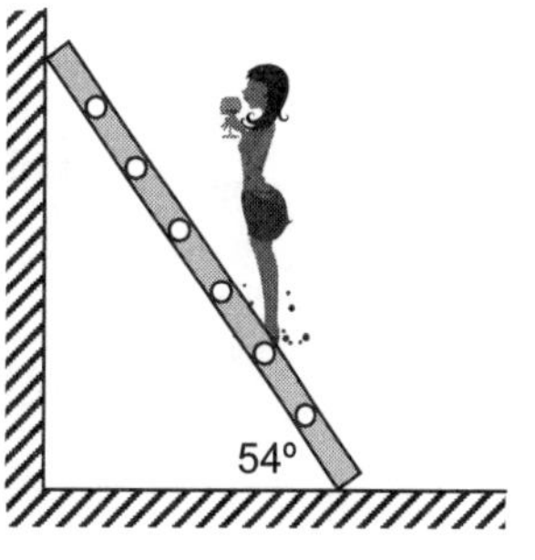

Exercise 6-30

An 80-cm long shelf with a mass of 1.5 kg is supported by a bracket as shown. Find the force exerted by the bracket on the shelf.

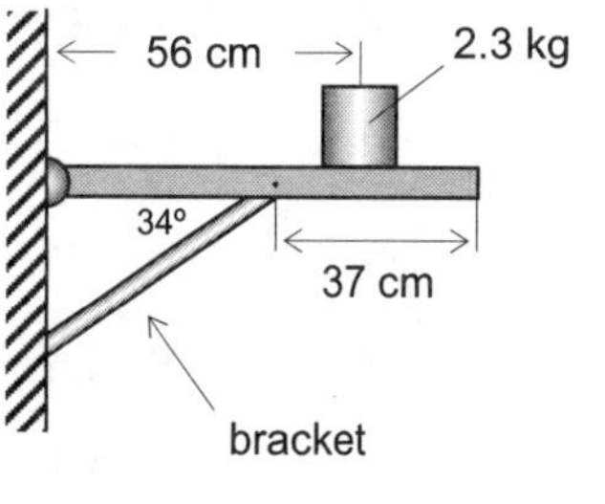

PROBLEMS ON CHAPTER 6: STATIC EQUILIBRIUM **(Answer Key on p. 370)**

1 – 5. Fill in the blanks

1. When the motion of an object does not change with time, the object is said to be in __________.

2. An object at rest is in __________ __________.

3. The modes of movement for objects modeled as lines are translation and ________.

4. A __________ __________ of a force is the perpendicular distance from the pivot to the line along which the force acts.

5. __________ is to rotation as force is to translation.

6 – 16. True/False questions

6. _____ A projectile is in dynamic equilibrium.

7. _____ If the net force on an object is zero, the object must be at rest.

8. _____ A positive net work must be done on an object to keep it in equilibrium.

9. _____ The Moon, as it revolves around the Earth, translates but does not rotate. (Hint: The observers on the Earth can see only one side of the Moon.)

10. _____ The standard unit of torque is the Joule [J].

11. _____ Adding torques to find the net torque requires consideration of their directions.

12. _____ One can increase the magnitude of torque solely by reducing the lever arm.

13. _____ Only the force component perpendicular to the body generates torque.

14. _____ A non-zero net torque changes the rate at which an object rotates.

15. _____ For an object in static equilibrium, one may choose any point on the object to be the pivot when calculating torques.

16. _____ If an object satisfies the second condition of static equilibrium, the first condition is automatically satisfied as well.

17 - 26. Multiple choice questions

17. A horizontal force is applied to keep a pendulum at rest as shown. As the angle θ is reduced, the magnitude of the horizontal force required to keep the pendulum at rest must

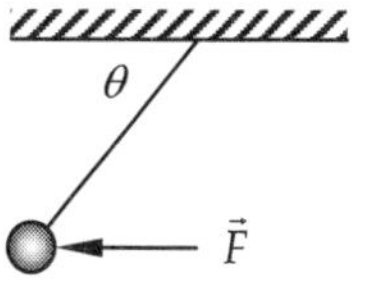

a) increase.
b) remain constant.
c) decrease.
d) increase at first but eventually decrease.

18. A spherical object is suspended by two cables from a ceiling and a wall. Is it possible for the cables to be perfectly straight as shown in the diagram?

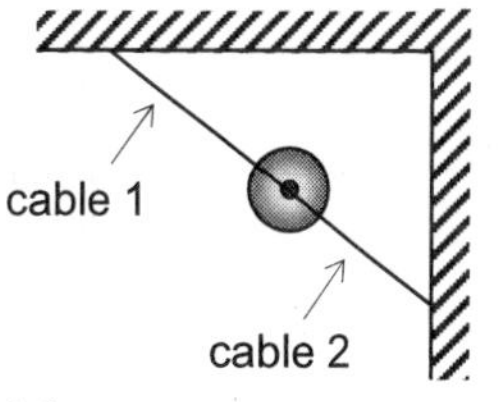

a) It is possible if the object is very light.
b) It is possible if the cables are made from the same material.
c) It is impossible because of the object's weight.
d) It is impossible because of the object's shape.

19. A mass is suspended by two cables as shown. Which of the following statements is correct? ($0° < \theta < 90°$)

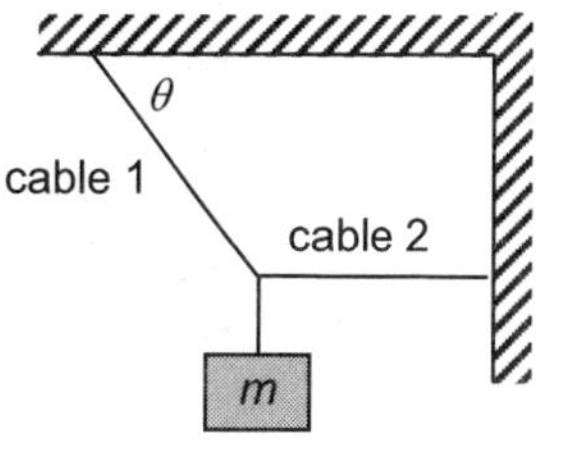

a) Cable 1 has greater tension than cable 2.
b) Cable 1 has the same tension as cable 2.
c) Cable 1 has less tension than cable 2.
d) Cable 1 has greater tension for small θ but less tension for large θ.

20. When unscrewing a bolt with a wrench, we grab the wrench as far away from the bolt as possible and apply the force perpendicular to the wrench in order to

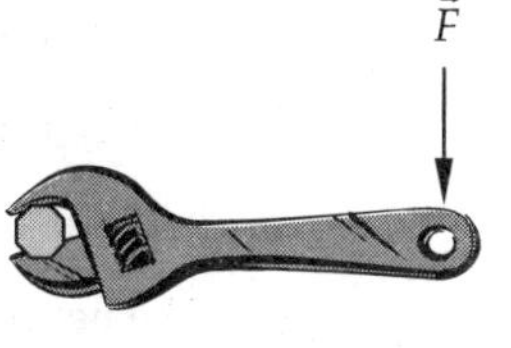

a) apply greater force.
b) maximize the lever arm.
c) avoid any slippage.
d) All of the above

21. For which of the following problems do you need to use torque?

a) 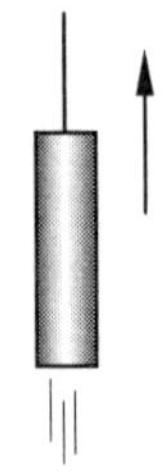

A long cylinder is accelerated upward by a string. Find the tension in the string.

b) 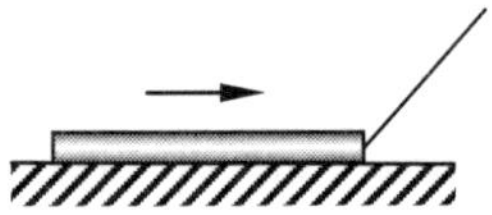

A long rod is pulled at constant velocity by a string. Find the friction force acting on the rod.

c) 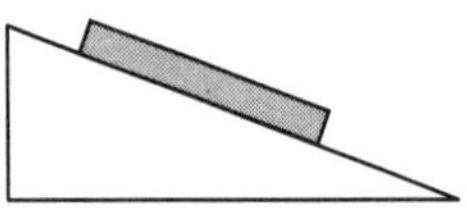

A long beam is at rest on an incline. Find the friction force acting on the beam.

d) 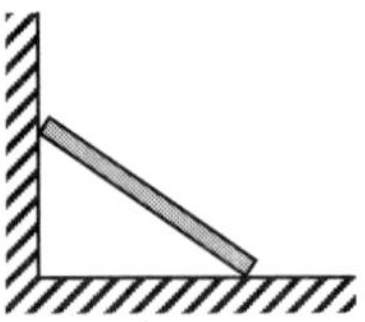

A long pole is leaning against a smooth wall. Find the friction force acting on the pole.

22. In which case the tension in the cable is the minimum? (The only difference is the direction of the cable.)

a) 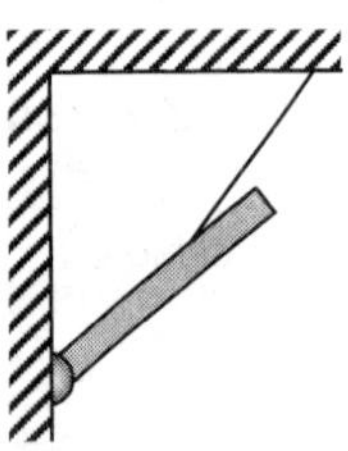
b) 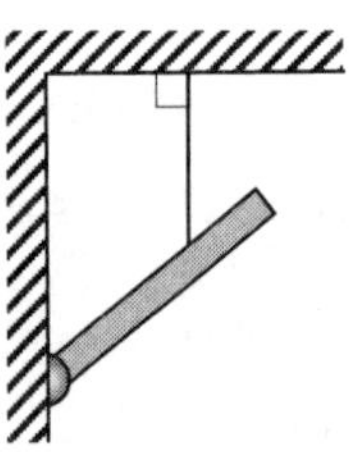
c) 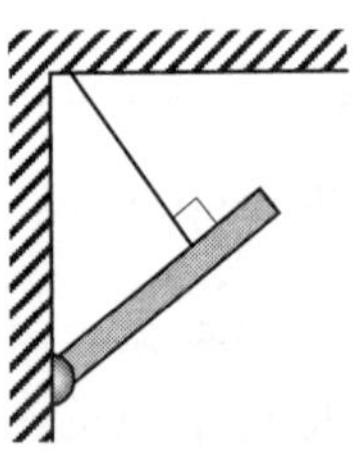
d) 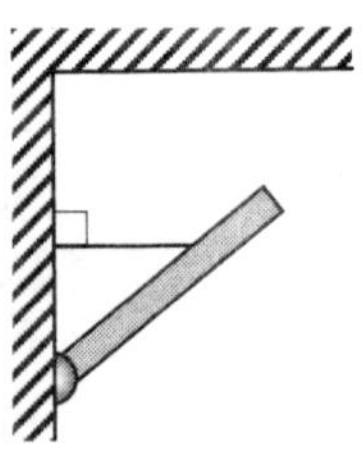

23. A beam hinged at one end to the ground is let go from the position shown. Which of the following statements is incorrect?

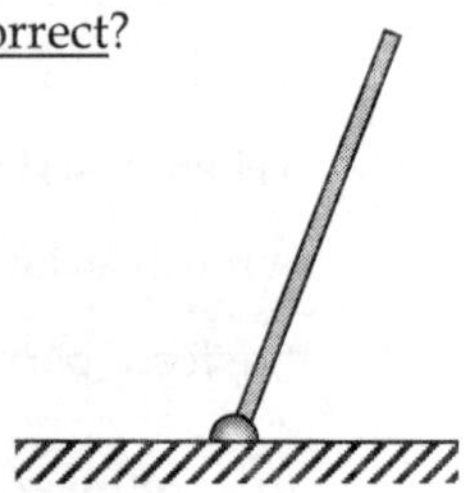

a) The beam is in dynamic equilibrium.
b) The beam rotates towards the ground because of its own weight.
c) The net torque on the beam is not zero.
d) The torque generated by the beam's weight increases as the beam rotates towards the ground.

24. A lady walks up a stationary ladder leaning against a smooth wall, as shown. What happens to the magnitude of the friction force exerted by ground on the ladder as she climbs the ladder?

a) It increases.

b) It remains the same.

c) It decreases.

d) We cannot determine it unless the coefficient of friction between the ladder and the ground is known.

25. A shelf of mass $2m$ is supported by a bracket. A block of mass m is on the shelf, and a cylinder of the same mass is suspended from the end of the shelf, as shown. About which point is the net torque on the shelf the greatest?

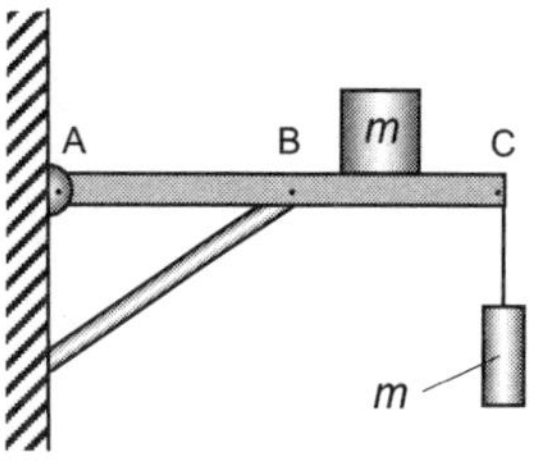

a) A b) B

c) C d) The net torque is the same.

26. A sign hangs from a beam as shown. Which of the following is the best way to reduce the tension in the cord?

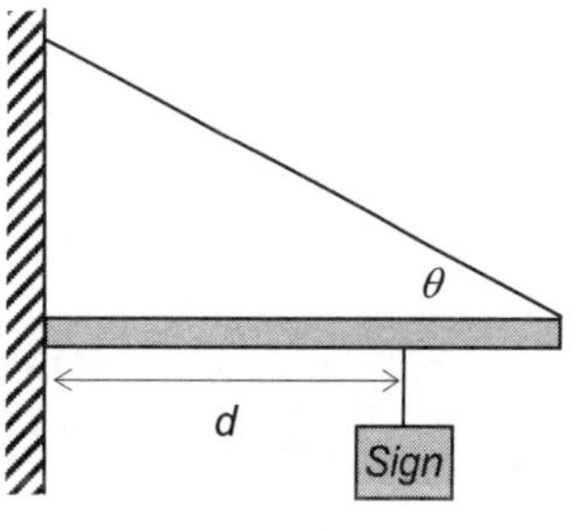

a) Decrease both θ and d.

b) Decrease θ and increase d.

c) Increase θ and decrease d.

d) Increase both θ and d.

27 – 35. Written-response questions

27. An object is kept in static equilibrium by a horizontal force whose magnitude is twice the object's weight. Find the angle θ.

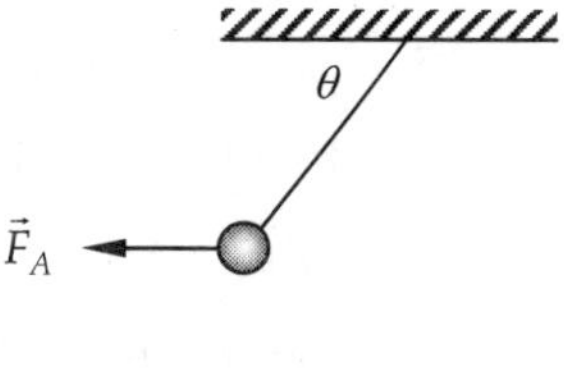

28. A block is suspended from a ceiling as shown. If the tension in Cable 1 is 170 N, find the tension in Cable 2 and the block's mass.

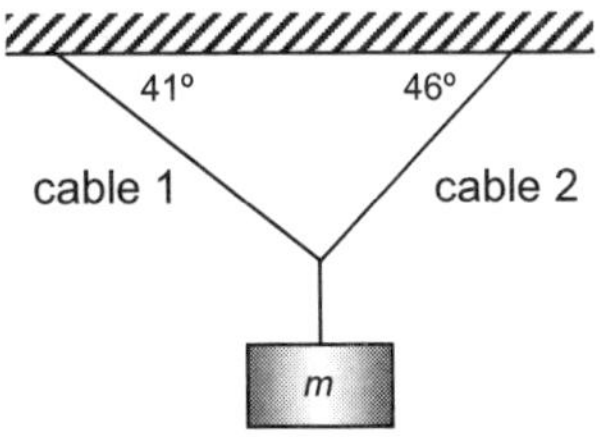

29. A 17-kg object is suspended by two ropes as shown. If the tension in the rope to the wall is 230 N, find the tension in the rope to the ceiling and the angle the rope makes with the ceiling.

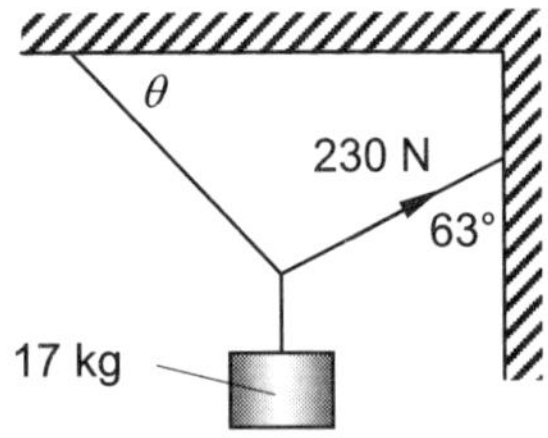

30. An object is suspended by two cords as shown. If each cable can withstand a maximum tension of 640 N, find the maximum mass of the object the cables can support without breaking.

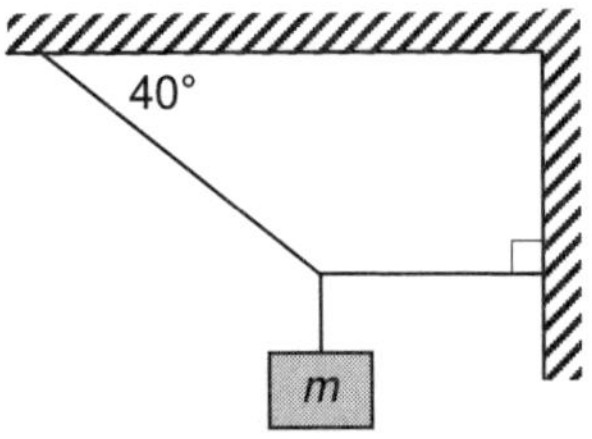

31. Two 44-kg gymnasts are practicing on a 4.5-m long beam as shown. The mass of the beam is 16 kg. Find the net torque on the beam.

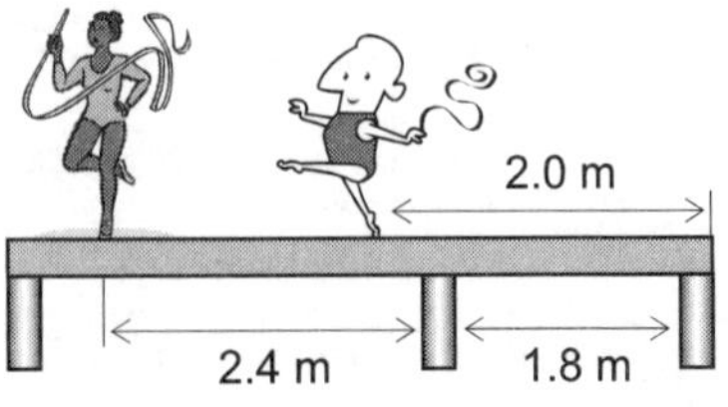

32. A 2.7-m long beam of mass 16 kg is held stationary by a horizontal cable as shown. The tension in the cable is 410 N. If the cable and the cord supporting the weight of the 24-kg mass are connected to the same point on the beam, find the length of the cable.

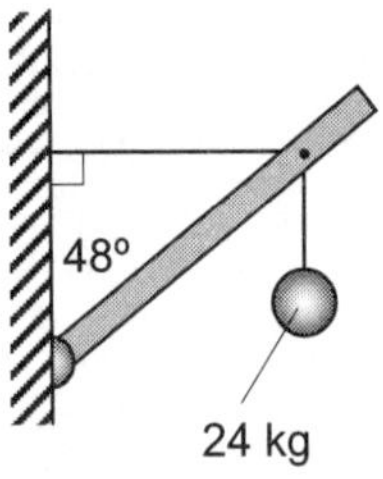

33. A 14-kg ladder is leaning against a wall as shown. Assuming no friction, find all the normal forces acting on the ladder.

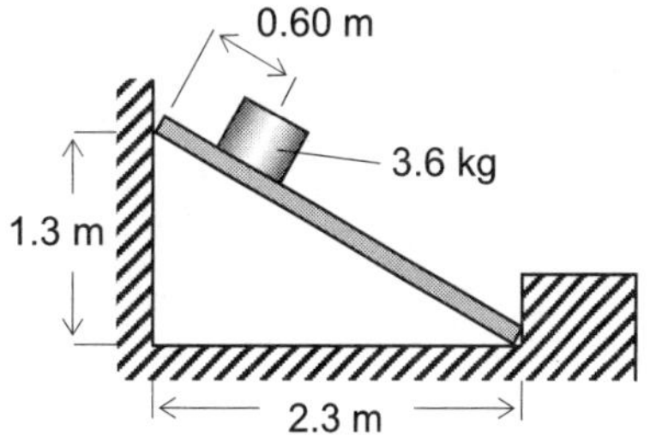

34. A 1.07-m long pole is fastened to the ground by two ropes as shown. The tension in the left rope is twice that in the right rope. Find the tension in each rope. (The mass of the pole is 0.85 kg.)

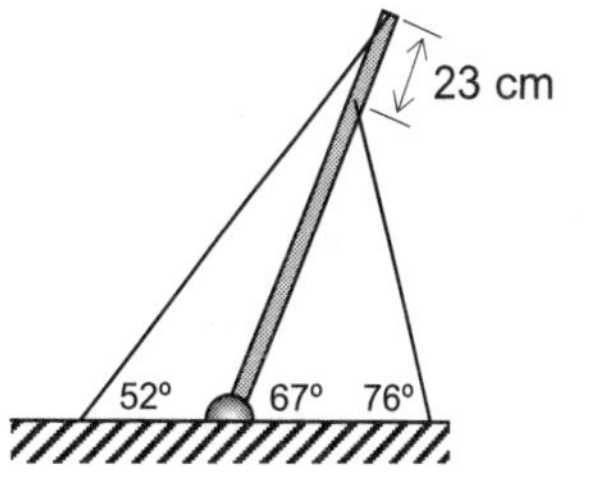

35.* A 3.1-kg object is suspended by two cables, one of which is connected to one end of a 1.2-kg meter stick as shown. The other end of the stick is hinged at the wall. Find the tension in each cable and the angle θ.

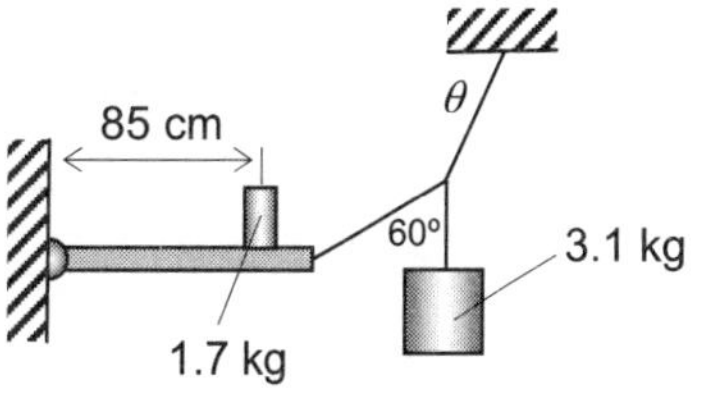

7

CHAPTER 7

CIRCULAR MOTION AND GRAVITATION

Human beings have looked up at the night skies for thousands of years and wondered about the motion of the planets and stars. In Europe, many people believed that the Earth was at the center of the universe, but after Copernicus and Galileo they learned that the planets in our solar system revolved around the Sun. They assumed that the planets moved at constant speed in circular orbits although they could not explain why.

In the late 17th century, Newton explained the movement of the planets using his three laws of motion (covered in Chapter 3) and the law of universal gravitation, which is one of the topics in this chapter. While the planets' orbits are in fact elliptical†, some orbits – such as the Earth's around the Sun and the Moon's around the Earth – are nearly circular, and this simplification allows us to explain the motion of the planets using the properties of circular motion and the law of gravitation.

In this chapter, you will:

- ✓ learn about uniform circular motion and centripetal force
- ✓ solve exercise problems on uniform circular motion
- ✓ learn about the law of universal gravitation and gravitational fields
- ✓ learn how to solve orbiting-object and launched-object problems
- ✓ solve exercise problems on gravitation
- ✓ test your knowledge of uniform circular motion and gravitation

† The discovery is credited to Kepler, who, in the early 17th century, found that the planets in our solar system have elliptic orbits and their speeds change as they move along their respective paths.

7-1. Uniform Circular Motion

Suppose an object is initially moving at constant speed to the right when a force is applied at an angle θ as shown in Fig. 7-1. If the angle is close to 0°, most of the force is used to change the object's speed, causing little change in its direction of travel. However, if applied at a greater angle, the force has less effect on the object's speed and instead causes the object to deviate more from the straight path it was initially following.

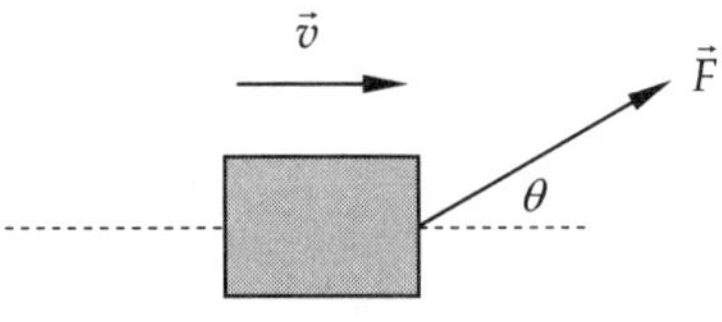

Figure 7-1: A force acting at an angle to the direction of motion

The above observation can be explained as follows: When the force vector is resolved into its components parallel and perpendicular to the direction of motion, the parallel component changes only the magnitude of the object's velocity, and the perpendicular component changes only the direction. As the angle θ in Fig. 7-1 increases from 0° to 90°, the perpendicular component increases while the parallel component decreases. That is why the object deviates more from that straight path with less change in speed as the force gets closer to being perpendicular to the direction of motion (compare Figs. 7-2 (a) and (c)). When the angle θ is greater than 90°, the object loses speed because the parallel component of the force is in the opposite direction to its initial velocity (see Fig. 7-2 (b)).

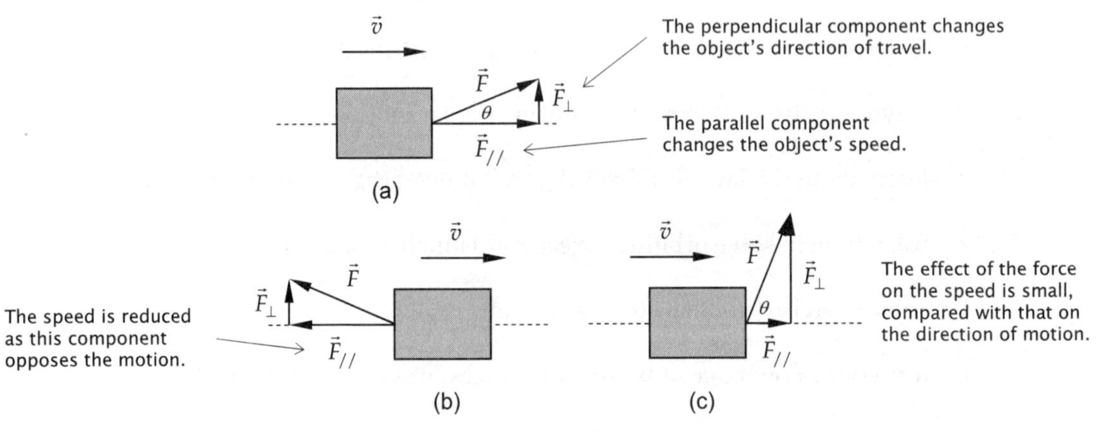

Figure 7-2: The effect of force's direction on a moving object's velocity

Now imagine that a constant force acts on a moving object so that it is always perpendicular to the object's motion. Then the object will follow a circular path as shown in Fig. 7-3, moving with constant speed. Such an object is said to be in **uniform circular motion**.

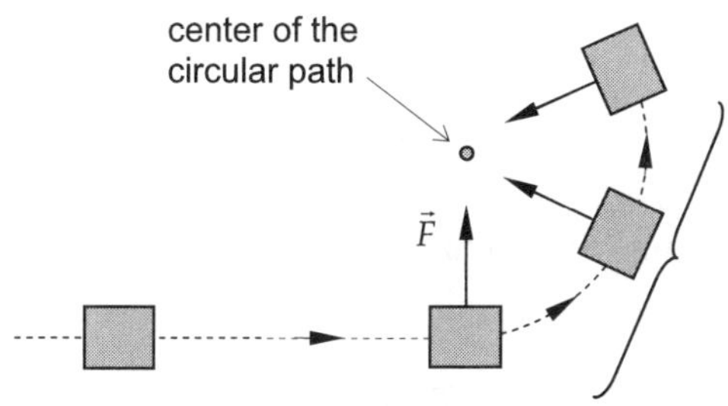

Figure 7-3: Uniform circular motion resulting from force applied perpendicular to the motion

Note that the net effect of all the forces acting on an object in uniform circular motion is to pull the object towards the center of the circular path. For this reason, the net force acting on an object in uniform circular motion is called the **centripetal force**. (The word 'centripetal' means 'center (*centri-*) seeking (*-petal*)'.) As shown in Fig. 7-4, it is always perpendicular to the velocity vector, which is tangent to the circular path of the object.

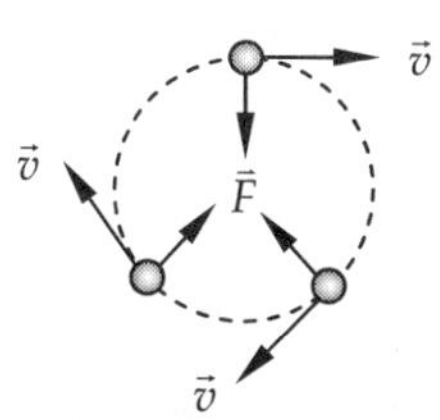

Figure 7-4: Velocity and centripetal force vectors in uniform circular motion

Traveling at constant speed, an object in uniform circular motion takes the same time to complete each revolution around the circular path. Hence, given the radius of the circular path (R) and the period (T), we can find the speed of the object as follows:

$$v = \frac{\text{distance}}{\text{time}} = \frac{\text{circumference}}{\text{period}} = \frac{2\pi R}{T} \quad \text{... (7.1)}$$

For instance, the Earth's speed around the Sun is

$$v_{\text{Earth}} = \frac{2\pi(\text{distance from Earth to the Sun})}{1\,\text{year}} = \frac{2\pi(1.496\times10^{8}\ \text{km})}{3.1536\times10^{7}\ \text{sec}} \cong 30\,\text{km/s}\,!$$

All objects obey Newton's second law of motion, including those in uniform circular motion. Since they all accelerate in the direction of the net force, the acceleration of an object undergoing uniform circular motion will be in the direction of the centripetal force, which is toward the center of the circular path. Such acceleration is called the **centripetal acceleration**, and its magnitude is related to the speed by

$$a_C = \frac{v^2}{R} \qquad \dots (7.2)$$

Then, for a body in uniform circular motion, we can express Newton's second law of motion as

after substituting Eq. (7.1)

$$F_C = m\,a_C = m\frac{v^2}{R} = m\frac{4\pi^2 R}{T^2} \qquad \dots (7.3)$$

Some examples of objects in uniform circular motion are shown in Fig. 7-5. While each object moves with constant speed, its velocity is not constant because of the continual change in its direction of motion. Each object constantly accelerates towards the center of the circular path that it follows. Since the centripetal force, or the net force, acting on each object is also towards the center, forces must act on it so that their vector sum points toward the center of the path. For instance, the sum of the tension and the weight acting on the ball is horizontal and towards the rod (Fig. 7-5 (a)); the sum of the weight and the normal force acting on the car at the position shown is vertically downward (Fig. 7-5 (b)); and the gravitational force, the only force acting on the moon, is towards the planet (Fig. 7-5 (c)). Keep in mind that a centripetal force is not another kind of force and should not appear in a free-body diagram because it is just the name given to a net force directed toward the center of a circle.

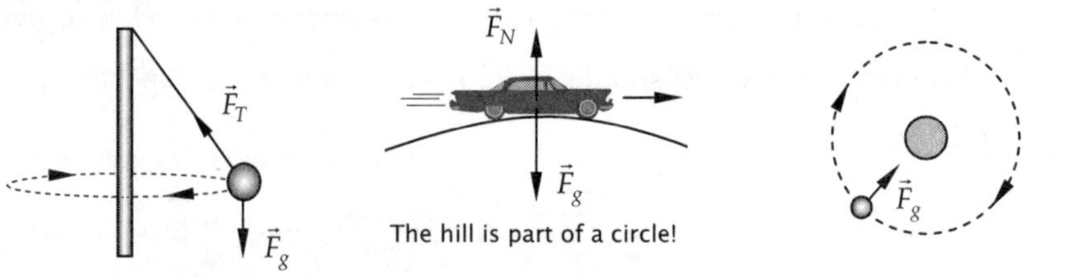

A ball is swung in a circle at constant speed

A car goes over a circular hill at constant speed

A moon orbits around a planet at constant speed

Figure 7-5: Examples of uniform circular motion

7-2. How to Solve Uniform Circular Motion Problems

Uniform circular motion problems require you to use Eq. (7.3). In each problem the unknown may be on the dynamic side or on the kinematic side, as shown below. In either case it is important to use the fact that the sum of the forces acting on a body must point towards the center of the circular path.

If a problem asks for one of the forces acting on a body, the unknown is in the dynamic side.

$$F_C = m a_C = m\frac{v^2}{R} = m\frac{4\pi^2 R}{T^2}$$

dynamic side | kinematic side

If a problem asks for mass, speed, radius of curvature, or period, the unknown is in the kinematic side.

The solution procedure is straightforward and summarized as follows:

1. Read the problem carefully and draw a free-body diagram of the object, showing all the forces acting on it.
2. Identify the side of the unknown. If it is the dynamic side, go to Step 3a. If it is the kinematic side, go to Step 3b.
3. a. Calculate the right-hand side of Eq. (7.3) to obtain the magnitude of the centripetal force. Knowing both the magnitude and direction of the centripetal force, you can proceed to find the unknown force with Eqs. (3.6) and (3.7) (on p. 72)

   b. Add all the forces in the free-body diagram to find the magnitude of the centripetal force. Apply Eq. (7.3) to solve for the unknown.
4. Check whether the answer is reasonable as you think about the characteristics of uniform circular motion.

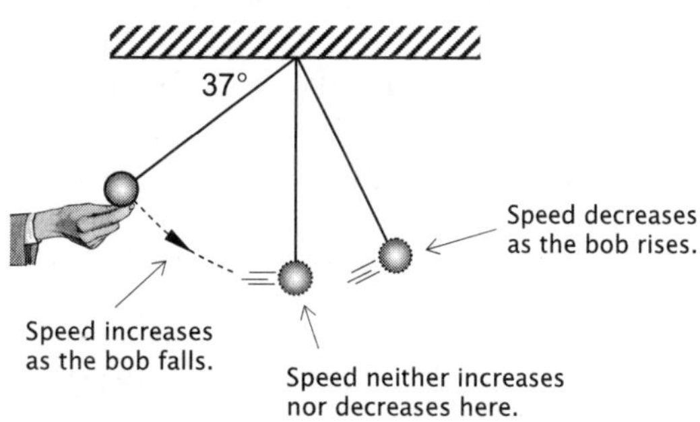

Figure 7-6: Treating the motion of pendulum as uniform circular motion

Let's apply this solution procedure to several examples. In the first example, a 3.5-kg pendulum bob is let go from the position shown in Fig. 7-6. The bob initially gains speed as its potential energy is converted to kinetic energy. Past the vertical position, however, the bob loses speed as it gains altitude. Overall, this is not uniform circular motion because the bob's speed changes with position. Yet if we focus on the moment when the pendulum becomes vertical, we can treat the bob's motion as uniform circular motion because it is neither gaining nor losing speed. This allows us to apply Eq. (7.3) to analyze the bob's motion at that instant.

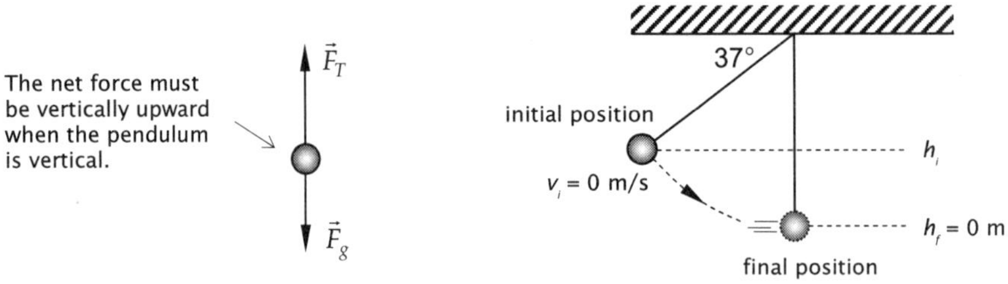

(a) free-body diagram (b) application of work-energy theorem

Figure 7-7: Analysis of the pendulum bob's motion in Fig. 7-6

Assuming the pendulum is 1.3 m long, let's find the tension force when the pendulum is vertical. The free-body diagram of the pendulum bob is shown in Fig. 7-7 (a), and we know that the magnitude of the tension force must be greater than that of the bob's weight because the net force must be directed vertically upward. Since the unknown, the tension force, is on the dynamic side of Eq. (7.3), we need to evaluate its right-hand side. Unfortunately, the speed of the bob is not given, and we need to find it by applying the work-energy principle (Eq. (4.17) on p. 106). The initial and final positions of the bob are shown in Fig. 7-7 (b). The work done on the bob by the tension force is zero (because it is always perpendicular to the direction of the bob's motion), so $W_{NC} = 0\,\text{J}$. Thus, Eq. (4.17) becomes

$$0 = \frac{1}{2}(3.5\,\text{kg})\left[v_f^2 - (0\,\text{m/s})^2\right] + (3.5\,\text{kg})(9.8\,\text{m/s}^2)\left[0\,\text{m} - 1.3\,\text{m}\cdot(1 - \sin 37°)\right]$$

Solve the above equation for the final speed, v_f, and we get $v_f = 3.2\,\text{m/s}$. Now we can obtain the magnitude of the centripetal force:

$$F_C = m\frac{v^2}{R} = (3.5\,\text{kg})\frac{(3.2\,\text{m/s})^2}{1.3\,\text{m}} = 27.3\,\text{N}$$

Therefore, the magnitude of the tension is

$$F_T = 27.3\,\text{N} + (3.5\,\text{kg})(9.8\,\text{m/s}^2) \cong 62\,\text{N}$$

In the next example, a car is traveling at constant speed over a hill whose radius of curvature is 75 m, as shown in Fig. 7-8 (a), and we want to find out the maximum speed at which the car can travel at the top of the hill without losing contact with the surface. The free-body diagram of the car at the top of the hill is shown in Fig. 7-8 (b). (There may be friction acting on the car, but if there is, it has to be balanced by the thrust provided by the car's engine. Otherwise, the net force cannot be directed towards the center of the circular path.)

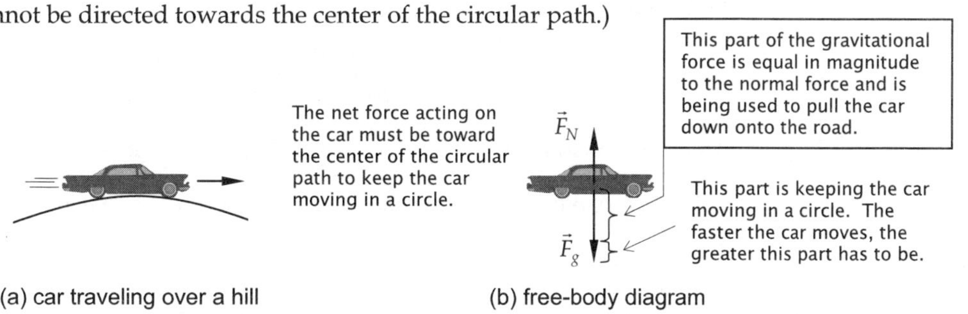

Figure 7-8: Analysis of car's motion over a hill

The magnitude of the centripetal force can be found from the free-body diagram:

$$F_C = F_g - F_N = m\left(9.8\,\mathrm{m/s^2}\right) - F_N$$

Thus, Eq. (7.3) becomes

$$F_C = m\left(9.8\,\mathrm{m/s^2}\right) - F_N = m\left(\frac{v^2}{75\,\mathrm{m}}\right)$$

We can see from the above equation that, as the car's speed (v) increases, the centripetal force on the car has to increase as well. This means that more of the gravitational force is needed to keep the car moving in a circle, which results in a decrease in the magnitude of the normal force as the car is pulled down onto the road less and less. (Note that the magnitude of the gravitational force is fixed, but the normal force, being an accommodating force, can change under different conditions.) When all of the gravitational force is acting as a centripetal force, the normal force disappears, and the car loses contact with the road. The magnitude of the centripetal force is maximized at that point, and so is the car's speed. Therefore,

$$m\left(9.8\,\mathrm{kg}\right) - 0\,\mathrm{N} = m\left(\frac{v_{\mathrm{max}}^2}{75\,\mathrm{m}}\right)$$

Cancel out the mass of the car and solve for v_{max}; we find that the car's maximum speed without it losing contact with the ground is 27 m/s, or about 98 km/h. Any speed above this will cause the car to be airborne past the top of the hill.

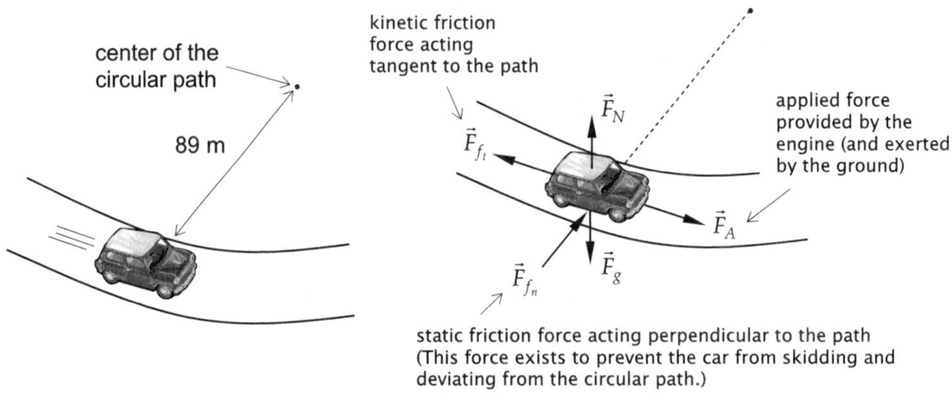

(a) a car turning a corner (b) free-body diagram of the car

Figure 7-9: A car undergoing uniform circular motion while turning a corner

In the last example, a car is turning a corner on level ground as shown in Fig. 7-9 (a), and we'd like to find the coefficient of static friction if the maximum speed at which the car can turn the corner without skidding is 25 m/s. Since the sum of the forces shown in the free-body diagram of the car (Fig. 7-9 (b)) is toward the center of the circular path, the vertical forces, $\vec{F}_g$ and $\vec{F}_N$, should add to zero, and so should the forces tangent to the path, $\vec{F}_A$ and $\vec{F}_{f_t}$. The only remaining force is $\vec{F}_{f_n}$, which pushes the car toward the center of the path. Hence Eq. (7.3) can be written as

$$F_C = F_{f_n} = m\frac{v_{\max}^2}{R}$$

Since $F_f = \mu F_N = \mu m g$ in this example, the above equation becomes

$$F_{f_n} = \mu \cancel{m}(9.8\,\text{kg}) = \cancel{m}\frac{(25\,\text{m/s})^2}{89\,\text{m}} \Longrightarrow \mu = 0.72$$

As you can see, the coefficient of static friction is independent of the car's mass.

PHYSICS 12 ANALYZED: UNIFORM CIRCULAR MOTION

7-1. Uniform Circular Motion

- An object moving in a circular path with constant speed is said to be in uniform circular motion.
- The net force acting on a body in uniform circular motion is called the centripetal force.
- Centripetal force and centripetal acceleration always point towards the center of a circular path and are perpendicular to the velocity.
- The equation of motion for a body undergoing uniform circular motion is

$$F_C = m\,a_C = m\frac{v^2}{R} = m\frac{4\pi^2 R}{T^2} \qquad \ldots (1)$$

7-2. How to Solve Uniform Circular Motion Problems

1. Draw a free-body diagram of the object.
2. Identify the side of the unknown. If it is the dynamic side, go to Step 3a. If it is the kinematic side, go to Step 3b.
3. a. Calculate the right-hand side of Eq. (1) to obtain the magnitude of the centripetal force. Knowing both the magnitude and direction of the centripetal force, you can proceed to find the unknown force with

$$\vec{F}_{\text{net}_x} = m\,\vec{a}_x \text{ and } \vec{F}_{\text{net}_y} = m\,\vec{a}_y$$

   b. Add all the forces in the free-body diagram to find the magnitude of the centripetal force. Apply Eq. (1) to solve for the unknown.

EXERCISES ON CHAPTER 7: UNIFORM CIRCULAR MOTION

Exercise 7-1

A 4.8-kg object on the end of a string is hurled in a vertical circle of radius 0.77 m. If the object's speed at the top of the circle is 2.8 m/s, find the tension at that position.

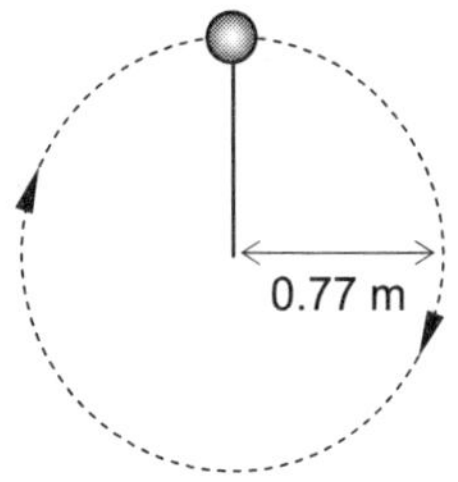

Exercise 7-2

A 76-kg man travels in a circular path of radius 2.2 m on a horizontal platform, as shown. It takes 3.4 s for him to complete one revolution around the pole. If the friction acting on him is 170 N towards the pole, find the tension force in the rope.

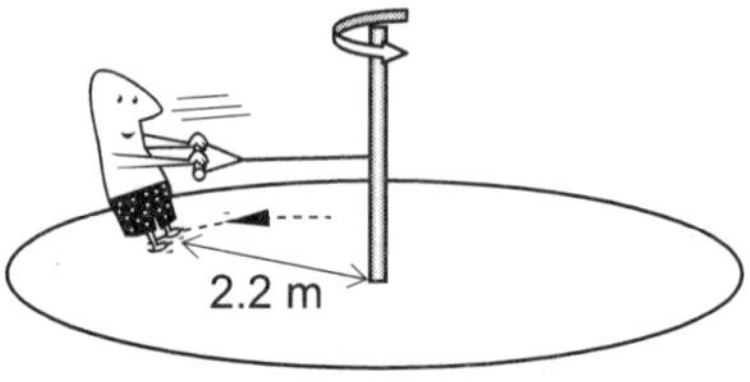

Exercise 7-3

A 58-kg snowboarder slides down a half-pipe 1.8 m from the bottom, as shown. The radius of the half-pipe is 1.9 m. When the snowboarder is at the bottom, find (a) his speed and (b) the normal force acting on him. (Ignore friction.)

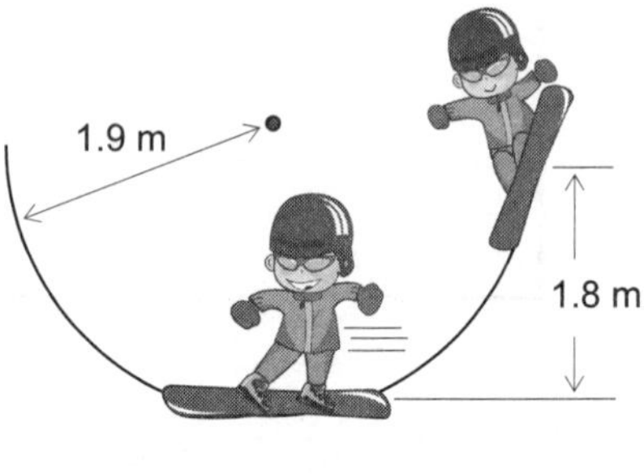

Exercise 7-4

Low gravity in outer space has negative long-term effects on human body. One way to alleviate this problem is create artificial gravity by spinning the space station, as shown. If the radius of the station is 140 m, at how many revolutions per second does the station need to rotate in order for the astronauts to feel the same weight as on the Earth?

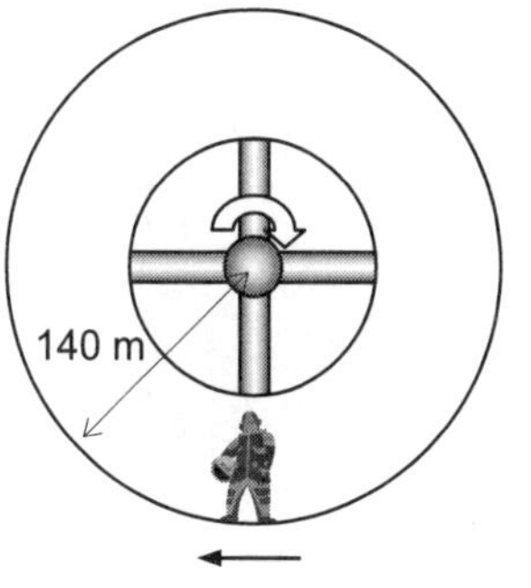

Exercise 7-5

A 950-kg car is traveling on a horizontal circular track of radius 80 m. If the coefficient of friction between the tires and the track is 0.67, find the minimum time for the car to complete one lap around the track.

Exercise 7-6

A 3.8-kg object, suspended from the ceiling by a 1.4-m cable, travels in a horizontal circle, as shown. If the angle between the cable and the ceiling is 53°, find (a) the tension in the cable and (b) the speed of the object.

Exercise 7-7

A 1300-kg car is traveling at 18 m/s on a hill of radius 85 m. Find the force exerted by the hill on the car when the car is at the top of the hill.

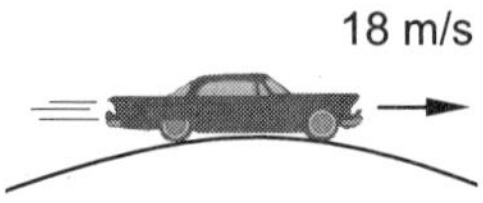

Exercise 7-8

An 81-kg office worker holds on to a horizontal cable of length 1.2 m, as shown. If he jumps off the ground, find the tension in the cable when he is at the lowest point of the circular path.

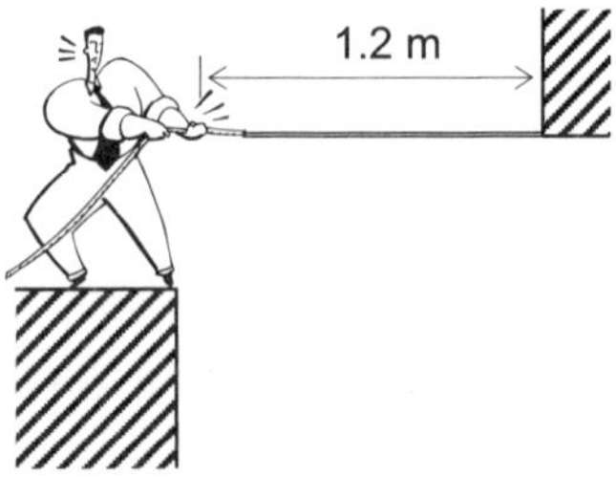

Exercise 7-9

A group of young adults are riding "washing machine" at an amusement park, as shown. The machine is designed so that, when it rotates at high speed, the occupants will become stuck to the wall and not fall. If the machine is 6.2 m in diameter, and the coefficient of friction is 0.33, find the minimum speed needed to prevent people from falling.

Exercise 7-10

A 45-kg girl fastened to a seat moves in a vertical circle of radius 1.2 m at constant speed. If she completes one rotation every 6.1 s, find the force the seat exerts on the girl at the top of the circle.

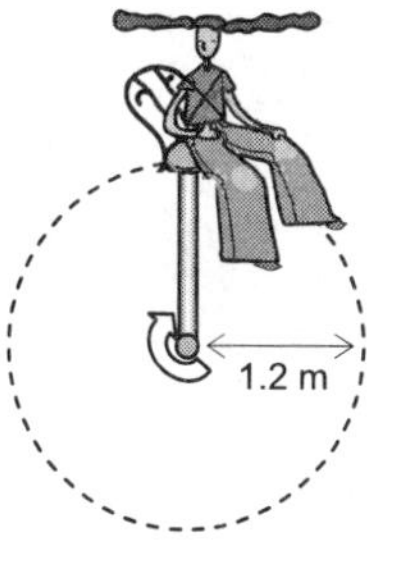

Exercise 7-11

A deliver boy traveling at 7.3 m/s goes around a loop as shown. If his speed at the bottom of the loop is just enough for him to safely go around the loop, find the diameter of the loop, *h*.

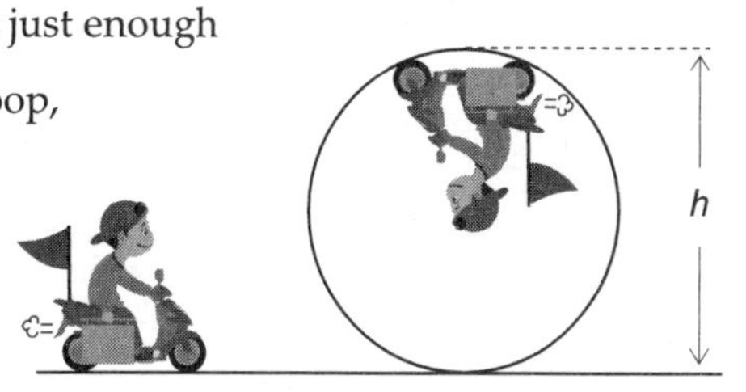

Exercise 7-12

An 1100-kg car turns a corner along a horizontal circular path whose radius of curvature is 73 m. What minimum friction force is required for the car to travel at the speed of 23 m/s without slipping?

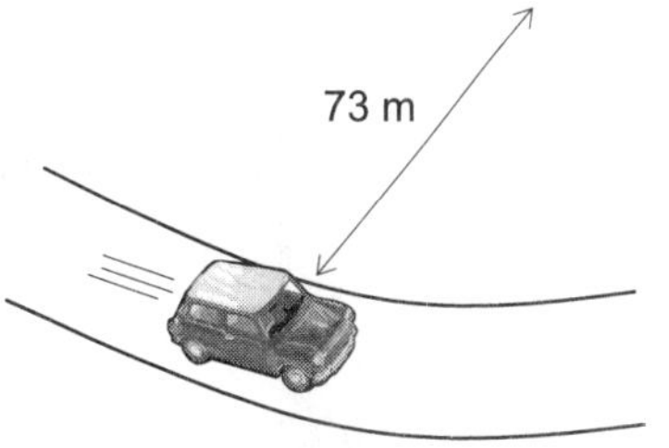

Exercise 7-13

During an air show a fighter flies in a vertical circular path of radius 1.3 km, as shown. At the top, the 74-kg pilot feels total downward force of four times his weight. Find the speed of the fighter at the top. (The pilot is fastened to his seat.)

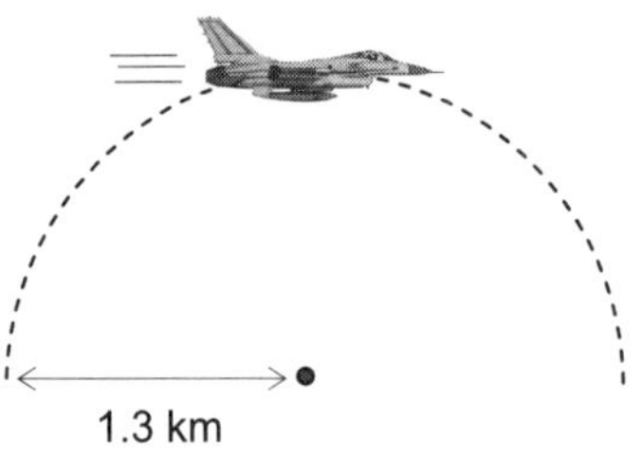

Exercise 7-14

A 2.7-kg object is on a horizontal rotating platform, 0.84 m from the pole as shown. If the object completes one revolution every 1.8 s, find the minimum friction force required to keep the object from sliding.

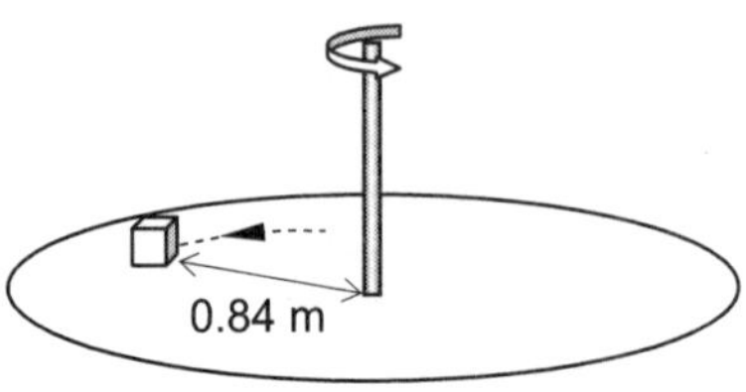

Exercise 7-15

A 12-kg object slides down a frictionless slope and rotates around the loop of diameter 5.4 m, as shown. If the object is released 8.8 m above the ground, find the normal force on the object at the top of the loop.

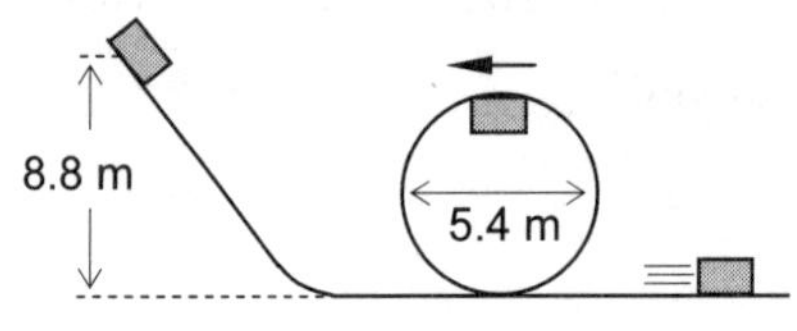

7-3. Newton's Law of Universal Gravitation

One of Newton's greatest contributions to science was his formula for **the law of universal gravitation**. It states that every object in the universe exerts a force on every other object, and the magnitude of this force is determined by

$$F_g = G\frac{mM}{r^2} \quad \ldots (7.4)$$

where m and M are the masses of the objects involved, r is the distance between the objects' centers, and G is a universal constant equal to $6.67\times10^{-11}\,\mathrm{N{\cdot}m^2/kg^2}$. For example, if a 60-kg boy is standing 2.0 m away from a 40-kg girl, the magnitude of the gravitational force they exert on each other is

It's a very small force!

$$F_g = \left(6.67\times10^{-11}\,\mathrm{N{\cdot}m^2/kg^2}\right)\frac{(60\,\mathrm{kg})(40\,\mathrm{kg})}{(2.0\,\mathrm{m})^2} = 4.0\times10^{-8}\,\mathrm{N}$$

To find the direction of the gravitational force, use the fact that it is always attractive. Therefore, the boy and the girl in the above example pull each other with 4.0×10^{-8} N force, as shown in Fig. 7-10.

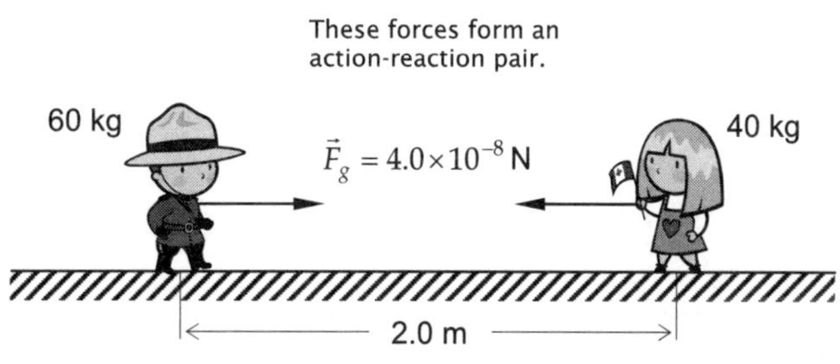

Figure 7-10: Gravitational force between a 60-kg boy and a 40-kg girl

This gravitational force between the boy and the girl is too small in magnitude to influence the motion of either person and may be ignored. In order for a gravitational force to be influential, at least one of the objects has to have substantial mass, such as the Earth. For instance, the magnitude of the gravitational force between the 60.0-kg boy and the Earth is

Earth's mass

$$F_g = \left(6.67\times10^{-11}\,\mathrm{N{\cdot}m^2/kg^2}\right)\frac{(60.0\,\mathrm{kg})(5.98\times10^{24}\,\mathrm{kg})}{(6.38\times10^{6}\,\mathrm{m})^2} = 5.88\times10^{2}\,\mathrm{N}$$

Earth's radius

While this force is too small in magnitude to influence the motion of the Earth, it is large enough to influence the boy's motion and must be considered when analyzing his motion. Figure 7-11 shows the gravitational forces between the boy and the Earth.

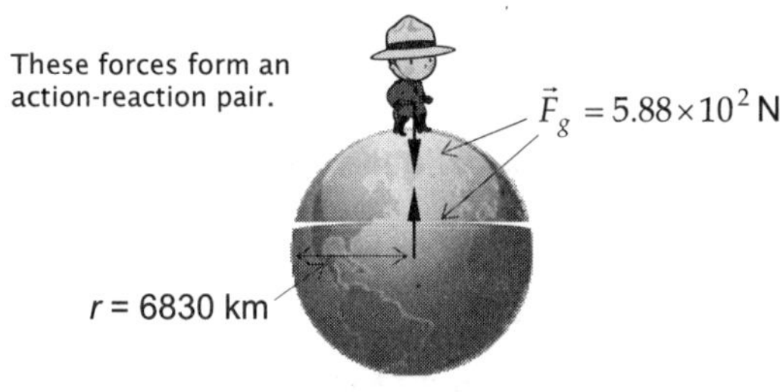

Figure 7-11: Gravitational force between a 60-kg boy and the Earth

When there are more than two objects involved, calculating the gravitational force acting on each object requires vector addition. Suppose we have three objects A, B, and C as shown in Fig. 7-12 (a). The three objects exert gravitational force on one another, as shown in Fig. 7-12 (b). To find the net gravitational force on B, for instance, we need first to find $F_{\text{on B by A}}$ and $F_{\text{on B by C}}$ using Eq. (7.4):

$$F_{\text{on B by A}} = G\frac{m_A m_B}{r_{AB}^2} = \left(6.67\times10^{-11}\ \text{N}\cdot\text{m}^2/\text{kg}^2\right)\frac{\left(64\times10^6\ \text{kg}\right)\left(27\times10^6\ \text{kg}\right)}{(8.4\,\text{m})^2} = 1630\ \text{N}$$

$$F_{\text{on B by C}} = G\frac{m_B m_C}{r_{BC}^2} = \left(6.67\times10^{-11}\ \text{N}\cdot\text{m}^2/\text{kg}^2\right)\frac{\left(27\times10^6\ \text{kg}\right)\left(8.0\times10^6\ \text{kg}\right)}{(3.9\,\text{m})^2} = 947\ \text{N}$$

Then the magnitude and direction of the net gravitational fore on B are

$$\left(F_{\text{net}}\right)_{\text{on B}} = \sqrt{(1630\,\text{N})^2 + (947\,\text{N})^2} \cong 1900\ \text{N}$$

$$\theta = \tan^{-1}\left(\frac{947\,\text{N}}{1630\,\text{N}}\right) \cong 30°$$

Thus, the net gravitational force on B is 1900 N, 30° below the line joining A and B.

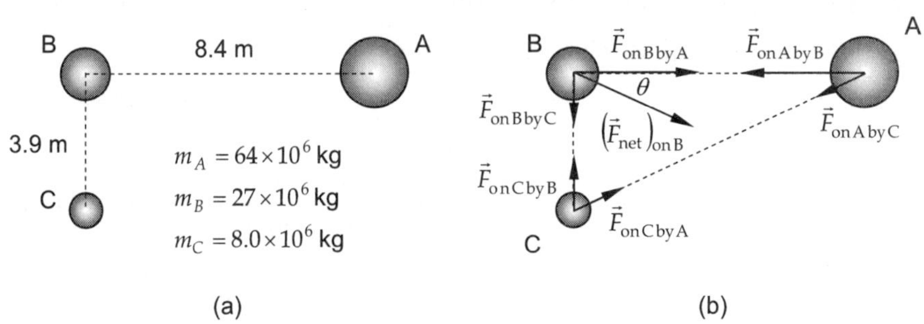

Figure 7-12: Finding net gravitational force on a body

7-4. Gravitational Field

When an object is on or near the surface of the Earth, we can find the magnitude of its weight using Eq. (7.4) as follows:

Object's mass → m; Earth's mass → M_E; Earth's radius → r_E

$$F_g = G\frac{m M_E}{r_E^2} = m\underbrace{\left(G\frac{M_E}{r_E^2}\right)}_{\text{Earth's gravitational field on its surface}} \quad \ldots (7.5)$$

Since the gravitational constant (G) and the Earth's mass (M_E) and radius (r_E) are constant, we may write Eq. (7.5) as

$$\vec{F}_g = m\vec{g} \quad \ldots (7.6)$$

Earth's gravitational field ($\vec{g}$); This object (m) enters Earth's gravitational field and experiences the force.

where

$$g = G\frac{M_E}{r_E^2} = \left(6.67\times10^{-11}\,\text{N}\cdot\text{m}^2/\text{kg}^2\right)\frac{5.98\times10^{24}\,\text{kg}}{\left(6.38\times10^6\,\text{m}\right)^2} \cong 9.8\,\text{N/kg} \quad \ldots (7.7)$$

$\vec{g}$ in Eq. (7.6) is known as the **gravitational field**, and its magnitude, or strength, on or near the Earth's surface is 9.8 N/kg. ($\vec{g}$ is also called the gravitational acceleration because the unit N/kg is equivalent to that of acceleration, m/s^2.)

Every object, no matter how small its mass, creates a gravitational field around itself, whose magnitude is given by

$$g = G\frac{M}{r^2} \quad \ldots (7.8)$$

This mass (M) produces the gravitational field, g.

where M is the object's mass and r is the distance from the object's center to a point where g is to be found. Hence, the gravitational field strength 100 km away from the surface of the Earth is

$$g = G\frac{M_E}{\underbrace{(r_E + 100000\,\text{m})^2}_{\text{Earth's radius + 100 km}}} = \left(6.67\times10^{-11}\,\text{N}\cdot\text{m}^2/\text{kg}^2\right)\frac{5.98\times10^{24}\,\text{kg}}{\left(6.48\times10^6\,\text{m}\right)^2} \cong 9.5\,\text{m/s}^2\ [\text{or N/kg}]$$

and the gravitational field strength on the surface of the Moon is

$$g = G\frac{M_{moon}}{\underbrace{r_{moon}^2}_{\text{Moon's radius}}} = \left(6.67\times10^{-11}\,\text{N}\cdot\text{m}^2/\text{kg}^2\right)\frac{7.35\times10^{22}\,\text{kg}}{\left(1.74\times10^6\,\text{m}\right)^2} \cong 1.6\,\text{m/s}^2$$

which is about one-sixth the gravitational field strength on the Earth's surface.

The gravitational field is a vector, and the direction of the gravitational field created by an object is always towards its center, as shown in Fig. 7-13. Note that the field strength decreases as you move away from the object (and r increases).

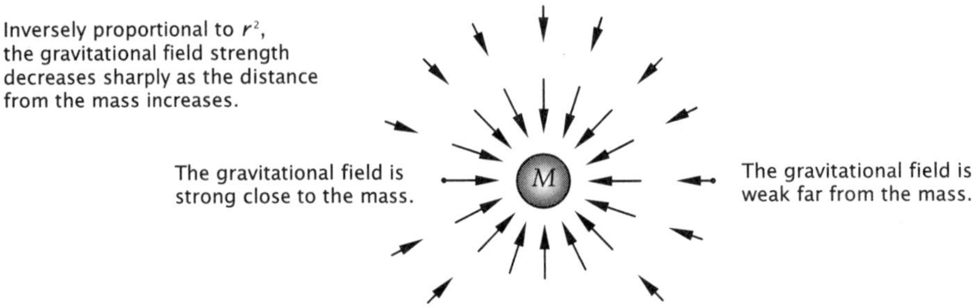

Figure 7-13: Gravitational field around a mass

When two or more bodies are located in the vicinity of a point, the gravitational field at that point is determined by adding the gravitational field vectors created by each body. For example, assume we have two objects A and C as shown in Fig. 7-14 (a). Each object generates a gravitational field towards itself, and, at point P, the field created by A points to the right, and that by C points downward, as shown in Fig. 7-14 (b). These field vectors are

$$\vec{g}_{\text{by A}} = G\frac{m_A}{r_A^2} = \left(6.67\times10^{-11}\,\text{N}\cdot\text{m}^2/\text{kg}^2\right)\frac{64\times10^6\,\text{kg}}{(8.4\,\text{m})^2} = 6.05\times10^{-5}\,\text{N/kg, right}$$

$$\vec{g}_{\text{by C}} = G\frac{m_C}{r_C^2} = \left(6.67\times10^{-11}\,\text{N}\cdot\text{m}^2/\text{kg}^2\right)\frac{8.0\times10^6\,\text{kg}}{(3.9\,\text{m})^2} = 3.51\times10^{-5}\,\text{N/kg, down}$$

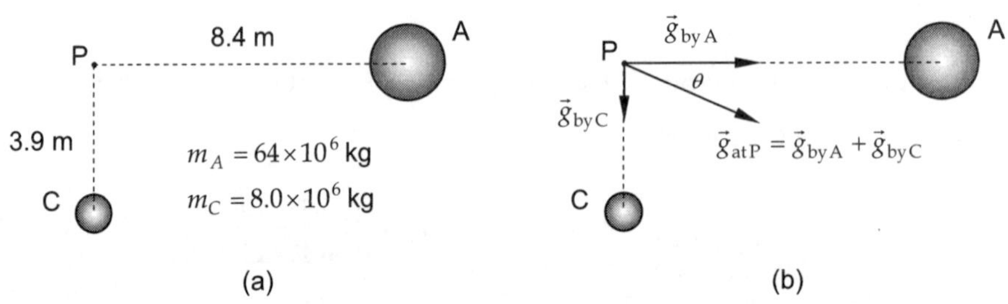

Figure 7-14: Finding net gravitational field at a point

The gravitational field at point P is the vector sum of those two field vectors:

$$g_{\text{at P}} = \sqrt{\left(6.05\times10^{-5}\,\text{N/kg}\right)^2 + \left(3.51\times10^{-5}\,\text{N/kg}\right)^2} \cong 7.0\times10^{-5}\,\text{N/kg}$$

$$\theta = \tan^{-1}\left(\frac{3.51\times10^{-5}\,\text{N/kg}}{6.05\times10^{-5}\,\text{N/kg}}\right) \cong 30°$$

Therefore, the gravitational field at P is 7.0×10^{-5} N/kg, 30° below the line joining P and A.

Using Eq. (7.6), we can express the gravitational field around mass M also as

Gravitational field is a vector. Its direction is the same as that of the gravitational force on m. →

$$\vec{g} = \frac{\vec{F}_g}{m} \quad \text{[N/kg]} \qquad \dots (7.9)$$

Equation (7.9) gives us another way to find the gravitational field at a particular point in space: Measure the gravitational force on an object at that point and divide that force by its mass. For example, to find the gravitational field at point P in Fig. 7-14 (a), we first locate an object at that point and measure the gravitational force on it. If that object's mass is 27×10^{6} kg, we have the situation shown in Fig. 7-12 (a), and the net gravitational force on B has already been calculated in the previous section to be 1900 N, 30° below the line joining A and B. Now we divide that force by the mass of B to find the gravitational field at point P:

$$\vec{g} = \frac{1900\,\text{N}}{27\times10^{6}\,\text{kg}} = 7.0\times10^{-5}\ \text{N/kg, 30° below the line joining A and P (or B)}$$

Note that it is the same as the gravitational field calculated using Eq. (7.8) in the previous page.

7-5. Absolute Gravitational Potential Energy

When two objects accelerate towards each other because of the gravitational force between them, the gravitational potential energy decreases as it is converted into kinetic energy, as shown in Fig. 7-15. Hence, the closer the objects are to each other, the less the gravitational potential energy between them.

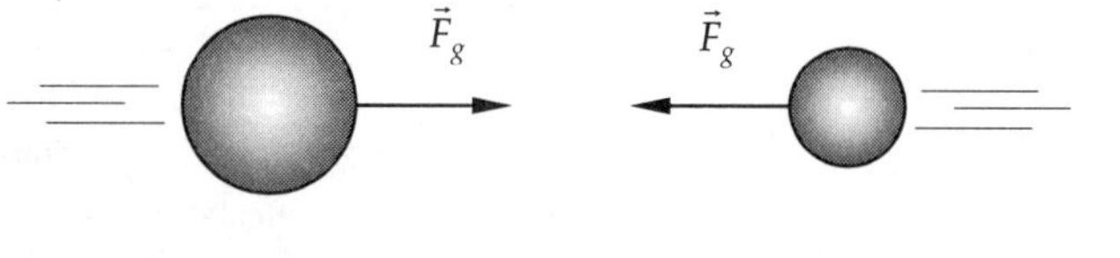

Figure 7-15: Conversion of gravitational potential energy into kinetic energy

Recall that we have been using the following definition for gravitational potential energy:

$$E_P = m\,g\,h \qquad \dots (7.10)$$

(h: distance above an arbitrary baseline)

This definition is valid only when an object is on or near the surface of the Earth, where the value of g is 9.8 N/kg. A more general definition is

$$E_P = -G\frac{mM}{r} \qquad \ldots (7.11)^{\dagger}$$

The negative sign in the equation ensures that gravitational potential energy (E_P) decreases as the distance between two objects (r) decreases, as shown in a plot of E_P vs. r in Fig. 7-16. Defined as in Eq. (7.11), the gravitational potential energy is always negative in value, which should not be a concern because what counts is the change in energy, not the actual value.

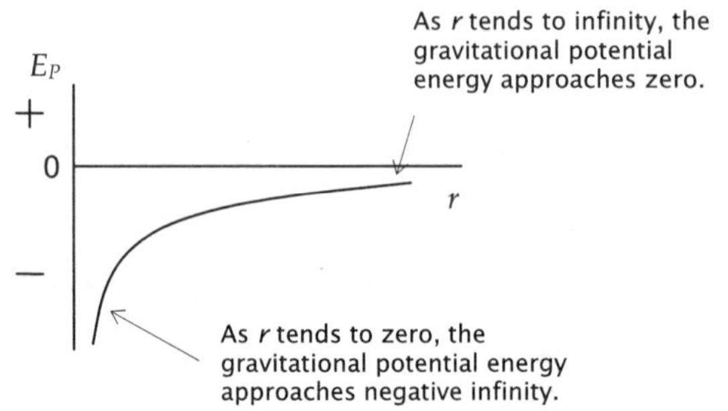

Figure 7-16: A plot of gravitational potential energy vs. the distance between two objects

Using Eq. (7.11), we can calculate the change in gravitational potential energy of a 500-kg object launched from the surface of the Earth to an altitude of 100 km. Initially, the distance between the object and the Earth's center is the Earth's radius, 6.38×10^6 m. When the object is 100 km above the surface of the Earth, the distance increases to $6.38\times10^6\text{ m}+1.00\times10^5\text{ m}=6.48\times10^6\text{ m}$. Thus,

$$\Delta E_P = -G\frac{mM}{r_f} - \left(-G\frac{mM}{r_i}\right) = -GmM\left(\frac{1}{r_f}-\frac{1}{r_i}\right)$$

$$= -\left(6.67\times10^{-11}\text{ N}\cdot\text{m}^2/\text{kg}^2\right)\left(500\text{ kg}\right)\left(5.98\times10^{24}\text{ kg}\right)\left(\frac{1}{6.48\times10^6\text{ m}}-\frac{1}{6.38\times10^6\text{ m}}\right) = 4.82\times10^8\text{ J}$$

(object's mass: 500 kg; Earth's mass: 5.98×10^{24} kg; final distance: 6.48×10^6 m; initial distance: 6.38×10^6 m)

Had we used Eq. (7.10) to find the change in gravitational potential energy, the change in gravitational potential energy would have been

$$\Delta E_P = mgh_f - mgh_i = \left(500\text{ kg}\right)\left(9.8\text{ m/s}^2\right)\left(100\times10^3\text{ m}\right) = 4.90\times10^8\text{ J}$$

† The derivation involves 'integration' of Eq. (7.4), which is beyond the scope of Physics 12 and thus not shown.

The error is about 8×10^6 J, which results from assuming a constant gravitational field strength of 9.8 m/s^2 throughout the motion of the object, and the error increases as the object moves farther from the Earth.

7-6. Uniform Circular Motion in Space

In our solar system, a number of planets have one or more moons orbiting them, and the planets themselves orbit the Sun. For those bodies whose orbits are nearly circular, we can simplify their motion as uniform circular motion. For example, the Earth and the Moon each follow a nearly circular orbit at an almost uniform speed. Such motion requires a centripetal force to pull the orbiting body towards the center of the orbit. In space, the gravitational force between the central body and the orbiting body is that centripetal force, as shown in Fig. 7-17.

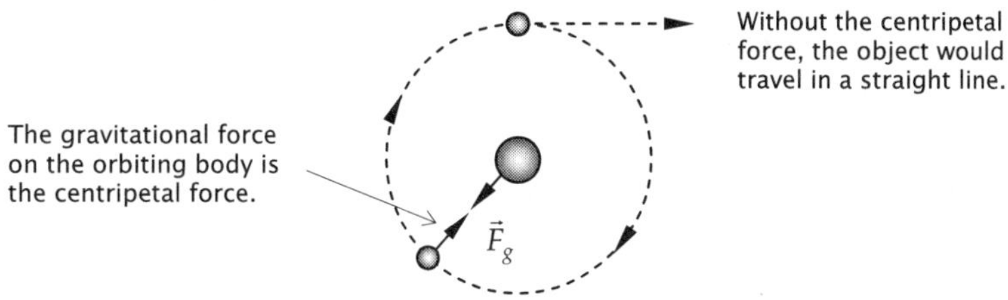

Figure 7-17: Centripetal force on an orbiting object in space

Thus, for an orbiting object in space, the equation of motion in Eq. (7.3) (on p. 190) can be written as

Substitute Eq. (7.4). →

$$\underbrace{G\frac{mM}{r^2}}_{\text{centripetal force on } m} = m\frac{v^2}{r} \quad \text{... (7.12)}$$

Then the **orbital speed**, v, becomes

The orbital speed is independent of the orbiting body's mass, m.

$$v = \sqrt{G\frac{M}{r}} \quad \text{... (7.13)}$$

Hence, an orbiting object must move at the speed determined by Eq. (7.13) in order to sustain its periodic motion. If it travels any faster, it will spiral away from the central body; if it travels any slower, it will spiral towards and collide with the central body.

For instance, if an artificial satellite is to orbit 200 km above the surface of the Earth, its speed has to be

$$v = \sqrt{\left(6.67\times10^{-11}\ \mathrm{N\cdot m^2/kg^2}\right)\frac{\left(5.98\times10^{24}\ \mathrm{kg}\right)}{\left(6.38\times10^{6}\ \mathrm{m}+2.00\times10^{5}\ \mathrm{m}\right)}} \cong 7.8\ \mathrm{km/s}$$

A satellite whose orbital radius is 500 km travels with the speed equal to

$$v = \sqrt{\left(6.67\times10^{-11}\ \mathrm{N\cdot m^2/kg^2}\right)\frac{\left(5.98\times10^{24}\ \mathrm{kg}\right)}{\left(6.38\times10^{6}\ \mathrm{m}+5.00\times10^{5}\ \mathrm{m}\right)}} \cong 7.6\ \mathrm{km/s}$$

As the above calculations show, an object orbiting at a larger distance has to move at slower speed because of the weaker gravitational pull by the central body.

Multiplying Eq. (7.12) by r, we obtain

$$G\frac{mM}{r} = mv^2 \qquad \ldots (7.14)$$

Using the definitions of the absolute gravitational potential energy and kinetic energy, we can re-write Eq. (7.14) as

$$-E_P = 2E_K \qquad \ldots (7.15)$$

Thus, the total mechanical energy of an orbiting object is

$$E_T = E_P + E_K = E_P + \frac{-E_P}{2} = \frac{E_P}{2} \qquad \ldots (7.16)$$

For example, a 1600-kg satellite orbiting at 500 km above the surface of the Earth has the gravitational potential energy equal to

$$E_P = -G\frac{mM}{r} = -\left(6.67\times10^{-11}\ \mathrm{N\cdot m^2/kg^2}\right)\frac{\left(1600\ \mathrm{kg}\right)\left(5.98\times10^{24}\ \mathrm{kg}\right)}{\left(6.38\times10^{6}\ \mathrm{m}+5.00\times10^{5}\ \mathrm{m}\right)} = -9.3\times10^{10}\ \mathrm{J}$$

Hence, its kinetic energy and total mechanical energy, respectively, are

$$E_K = -\frac{E_P}{2} = -\frac{-9.3\times10^{10}\ \mathrm{J}}{2} \cong 4.6\times10^{10}\ \mathrm{J}$$

$$E_T = \frac{E_P}{2} = \frac{-9.3\times10^{10}\ \mathrm{J}}{2} \cong -4.6\times10^{10}\ \mathrm{J}$$

Notice that Eqs. (7.13), (7.15), and (7.16) have all been derived from the equation of motion in Eq. (7.12). Thus, if you can set up Eq. (7.12) and work from it, you do not have to memorize all the derived relationships. For that reason, it is advised that you make it a habit to always start with Eq. (7.12).

7-7. Weightlessness

When you sit in a chair, you experience two forces: the force of gravity exerted by the Earth and the normal force by the chair. The gravitational force is not a contact force, meaning a contact is not necessary for the force to exist. You cannot feel such non-contact forces, but you can feel contact forces. So, as you sit in a chair, it is the normal force that provides you with the sensation of weight. If there were no normal force acting on your body, you would have no sensation of your own weight.

Weightlessness is simply a sensation experienced by someone when there are no forces acting on his or her body except the gravitational force. You usually feel weightless when you become a projectile yourself. If you are aboard a satellite orbiting the Earth, you feel weightless because you are in a state of free-fall! Astronauts inside a space station may seem to be floating in the air, but they are, in fact, falling at the same rate as the space station. The fact that you feel weightless does not mean that you have lost weight. It is just that there is no force to make you feel your own weight.

7-8. How to Solve Gravitation Problems

Most of the gravitation problems can be classified into two groups: one involving orbiting objects and the other involving non-orbiting objects, as shown in Fig. 7-18. As they require different approaches, you need to examine whether the object is orbiting in space or not. If it is, you may assume it is in uniform circular motion, with the gravitational force acting as the centripetal force. Thus, follow the solution procedure outlined in Section 7-2 (p. 191), using Eq. (7.12) as the equation of motion.

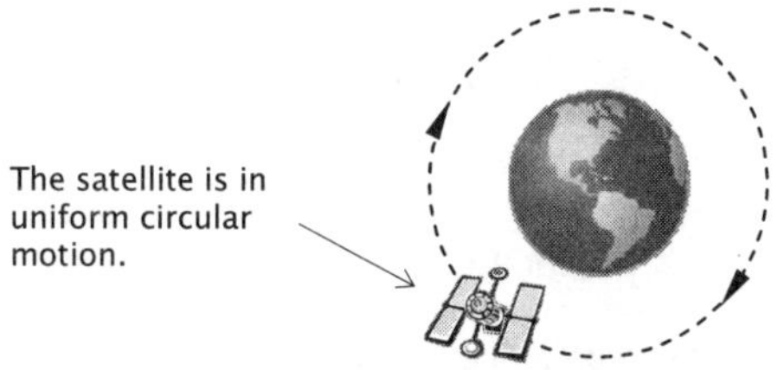

(a) orbiting satellite

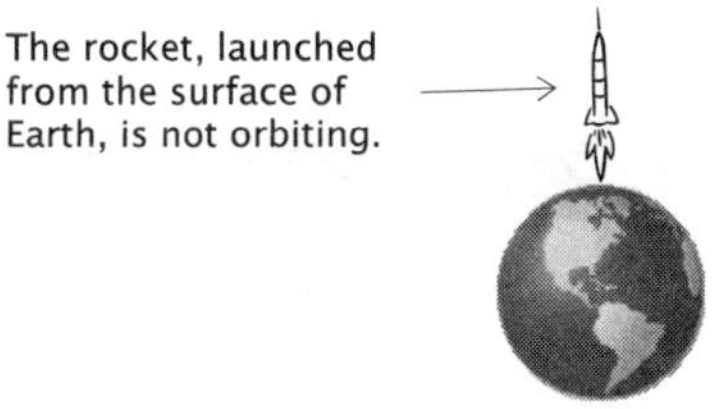

(b) launched rocket

Figure 7-18: Two types of gravitation problems

For example, suppose we need to find the speed of a geosynchronous satellite, which completes one revolution around the equator of the Earth in 24 hours. We begin by setting up Eq. (7.12). Solving the equation for the orbital speed v, we obtain Eq. (7.13):

$$v = \sqrt{G\frac{M}{r}} = \sqrt{\left(6.67\times10^{-11}\,\mathrm{N\cdot m^2/kg^2}\right)\frac{5.98\times10^{24}\,\mathrm{kg}}{r}} = \frac{2.0\times10^{7}\,\mathrm{m^{1.5}/s}}{\sqrt{r}}$$

where M is the Earth's mass, and r is the orbital radius, for which we do not have the value. Instead, we know the satellite's period of revolution, which is 24 hours, or 86400 seconds. Using Eq. (7.1) (on p. 189), we can find another relationship between v and r:

$$v = \frac{2\pi r}{T} = \frac{6.28\,r}{86400\,\mathrm{s}} = \left(7.3\times10^{-5}\,\mathrm{s^{-1}}\right)\cdot r$$

Set the above two equations equal to each other and solve for r:

$$v = \frac{2.0\times10^{7}\,\mathrm{m^{1.5}/s}}{\sqrt{r}} = \left(7.3\times10^{-5}\,\mathrm{s^{-1}}\right)\cdot r \quad\Longrightarrow\quad r = 4.2\times10^{7}\,\mathrm{m}$$

Hence, the orbital speed is

$$v = \left(7.3\times10^{-5}\,\mathrm{s^{-1}}\right)\left(4.2\times10^{7}\,\mathrm{m}\right) = 3.1\,\mathrm{km/s}$$

Whether an object is orbiting or not, it is a projectile as long as it is traveling in air or in space. When it moves near the surface of the Earth, we may assume a constant gravitational acceleration of 9.8 m/s^2 and use the kinematic equations to analyze it. However, we cannot make the same assumption for objects that travel a substantial distance away from the Earth, and, therefore, we may not apply the kinematic equations for those objects.

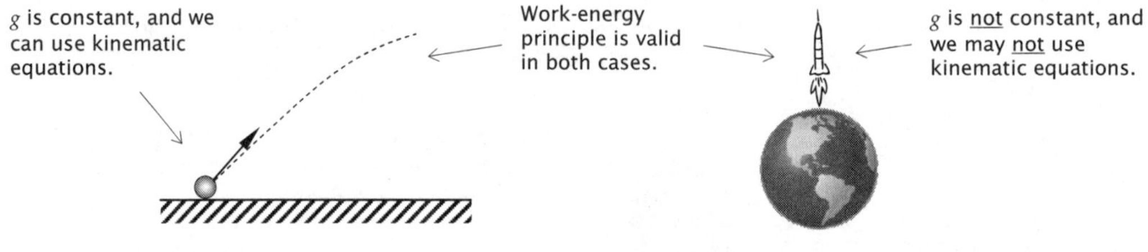

Figure 7-19: Comparison of two projectile motions

Fortunately, the work-energy principle (we learned in Chapter 4) is valid whether the acceleration is constant or not. Thus, we can apply it to non-orbiting objects in the same way we did to projectiles in Chapter 4, but with one change: Use Eq. (7.11) for gravitational potential energy to accommodate changing gravitational field strength. The work-energy principle is then written as

$$W_{NC} = \Delta E_K + \Delta E_P = \frac{1}{2}m\left(v_f^2 - v_i^2\right) - G\,m\,M\left(\frac{1}{r_f} - \frac{1}{r_i}\right) \qquad \text{... (7.17)}$$

For instance, let's find the amount of work required to put a 1500-kg geosynchronous satellite into the proper orbit. The satellite is initially at rest on the surface of the Earth and, at the final state, is orbiting at 3.1 km/s with the orbital radius of 4.2×10^{7} m (as we found from the previous example). Hence,

$$v_i = 0\ \text{m/s} \qquad v_f = 3100\ \text{m/s}$$

$$r_i = 6.38\times10^{6}\ \text{m} \qquad r_f = 4.2\times10^{7}\ \text{m}$$

Substitute all known quantities into Eq. (7.17), and solve for W_{NC}:

$$\begin{aligned} W_{NC} &= \frac{1}{2}(1500\,\text{kg})\left[(3100\,\text{m/s})^2 - (0\,\text{m/s})^2\right] \\ &\quad - \left(6.67\times10^{-11}\,\text{N}\cdot\text{m}^2/\text{kg}^2\right)(1500\,\text{kg})\left(5.98\times10^{24}\,\text{kg}\right)\left(\frac{1}{4.23\times10^{7}\,\text{m}} - \frac{1}{6.38\times10^{6}\,\text{m}}\right) \\ &= 8.7\times10^{10}\,\text{J} \end{aligned}$$

We can also use the work-energy principle to find an object's **escape velocity**, the minimum speed an object needs on the surface of a planet or a moon in order to escape from its gravitational pull and never return. You can imagine that the escape velocity on the Earth has to be quite high; otherwise, objects will eventually fall back towards the Earth.

Assume an object is moving away from the surface of the Earth at the escape velocity. As the object continues its journey, it keeps losing speed, and when it is an infinite distance away from the Earth, its speed is practically zero. Thus,

Initially, the object is on Earth's surface and moving at escape velocity.

$$v_i = v_e\ \text{m/s} \implies E_{K_i} = \frac{1}{2}m v_e^2$$

$$r_i = 6.38\times10^{6}\ \text{m} \implies E_{P_i} = -G\frac{Mm}{r_i}$$

Finally, the object is at an infinite distance from Earth, and its speed is practically zero.

$$v_f \approx 0\ \text{m/s} \implies E_{K_f} \approx 0$$

$$r_f \approx \infty \implies E_{P_f} \approx 0$$

Since there is no non-conservative force acting on the object once it leaves the ground, W_{NC} is zero as well. Equation (7.17) can then be simplified to

$$0 = \frac{1}{2}mv_e^2 - G\frac{Mm}{r_i}$$

Solve the above equation for v_e, and we find the following expression for the escape velocity:

$$v_e = \sqrt{\frac{2GM}{R}} \qquad \text{... (7.18)}$$

where M and R are the mass and the radius of the body being escaped from, respectively. Using the Earth's mass and radius, we can obtain the escape velocity on the surface of the Earth to be about 11 km/s. To find the escape velocity on the Moon instead, we need to use the Moon's mass and radius as shown below.

Moon's mass

$$v_e = \sqrt{\frac{2\left(6.67\times10^{-11}\ \text{N}\cdot\text{m}^2/\text{kg}^2\right)\left(7.35\times10^{22}\ \text{kg}\right)}{1.74\times10^{6}\ \text{m}}} = 2.4\ \text{km/s}$$

Moon's radius

PHYSICS 12 ANALYZED: GRAVITATION

7-3. Newton's Law of Universal Gravitation

- Newton's law of universal gravitation: "Every object in the universe attracts every other object with a force equal to

$$F_g = G\frac{mM}{r^2}$$ ← Gravitational force is always attractive.

- When a body experiences gravitational forces exerted by multiple objects, the net gravitational force on it is the vector sum of the individual forces.

7-4. Gravitational Field

- Every object creates a gravitational field whose magnitude is given by $g = G\frac{M}{r^2}$.

  The direction of the gravitational field is always towards the center of the object.

- The gravitational field can also be expressed as $\vec{g} = \frac{\vec{F}_g}{m}$.

- When there are multiple objects near a point, the gravitational field at that point is the vector sum of the individual gravitational field generated by each object.

7-5. Absolute Gravitational Potential Energy

- Gravitational potential energy is defined as $E_P = -G\frac{mM}{r}$.

- As the distance between two masses increases, the gravitational potential energy increases as well and approaches the value of zero.

- As the distance between two masses decreases, the gravitational potential energy decreases and approaches negative infinity.

7-6. Uniform Circular Motion in Space

- For an orbiting object, the equation of motion is

$$G\frac{mM}{r^2} = m\frac{v^2}{r}$$

- The orbital speed is given by $v = \sqrt{G\frac{M}{r}}$.

- The gravitational potential energy of an orbiting body is -2 times its kinetic energy: $E_P = -2E_K$. The orbiting body's total mechanical energy is then half its gravitational potential energy: $E_T = \frac{E_P}{2}$

7-7. Weightlessness

- Usually, the normal force on your body is what you regard as your own weight.
- Weightlessness is a feeling experienced by someone when there are no forces acting on his or her body except the force of gravity.

7-8. How to Solve Gravitation Problems

- For orbiting objects, use the equation of motion for uniform circular motion:

$$G\frac{mM}{r^2} = m\frac{v^2}{r} = m\frac{4\pi^2 r}{T^2}$$

- For non-orbiting objects, use the work-energy principle:

$$W_{NC} = \Delta E_K + \Delta E_P = \frac{1}{2}m\left(v_f^2 - v_i^2\right) - GmM\left(\frac{1}{r_f} - \frac{1}{r_i}\right)$$

- The escape velocity on the surface of a planet with mass M and radius R is

$$v_e = \sqrt{\frac{2GM}{R}}$$

EXERCISES ON CHAPTER 7: GRAVITATION (Data given in Appendix B, p. 360)

Exercise 7-16

Find the gravitational field strength at the surface of a planet of mass 8.9×10^{25} kg and radius 8.3×10^{7} m.

Exercise 7-17

a) Find the orbital speed of the Earth as it revolves around the Sun.

b) Find the centripetal acceleration of the Moon as it orbits the Earth.

Exercise 7-18

An asteroid is traveling in an orbit of radius 5.3×10^{11} m around the Sun. If the gravitational force between the asteroid and the Sun is 8.6×10^{16} N, find the mass of the asteroid.

Exercise 7-19

A 1300-kg satellite travels around Mars at an orbital radius of 6.4×10^{6} m. Find the total energy of this satellite relative to infinity. (The mass of Mars = 6.42×10^{23} kg)

Exercise 7-20

An astronaut standing on the surface of a spherical asteroid of radius 6.2×10^5 m drops a 7.3-kg object and watches it accelerate at 0.23 m/s^2. Find this asteroid's mass.

Exercise 7-21

A 1400-kg spacecraft rests on the surface of the Moon. If 2.5×10^9 J of work is done on the spacecraft, how high above the surface can the spacecraft be lifted?

Exercise 7-22

A 2800-kg spacecraft circles a planet once every 2.3×10^5 s with an orbital radius of 3.7×10^6 m. Find the mass of this planet.

Exercise 7-23

A stationary 52-kg object is released from 3500 km above the surface of the Earth. Find the speed of the object just before impact. (Ignore air resistance.)

Exercise 7-24

A satellite orbits the Moon at a speed of 1.6 km/s. Find the orbital radius of this satellite.

Exercise 7-25

Find the escape velocity for an object on the surface of a 8.7×10^{26} kg planet of radius 2.7×10^{8} m.

Exercise 7-26

A satellite is traveling in an orbit two Earth radii above the surface of the Earth, as shown. Find the satellite's orbital period.

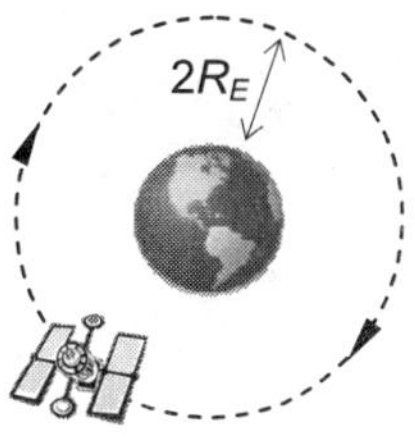

Exercise 7-27

An 83-kg astronaut, standing on the surface of a spherical asteroid of radius 1.3 km, jumps with the speed of 1.8 m/s and reaches a maximum height of 120 m above the surface. Find the mass of the asteroid.

Exercise 7-28

A 2.5×10^{5} kg rocket is to be launched from the surface of the Earth. Find the amount of work needed to lift the rocket to a maximum altitude of 2.6×10^{6} m.

Exercise 7-29

A 1150-kg satellite orbits the Earth at an altitude where the gravitational field strength is 8.9 N/kg. Find this satellite's orbital speed and total mechanical energy.

Exercise 7-30

A 3.8×10^{27} kg planet has an escape velocity of 7.4×10^{4} m/s. Find the planet's radius.

Exercise 7-31

A 3100-kg satellite orbits a 6.5×10^{27} kg planet. If this satellite's orbital period and radius are 9.6×10^{3} s and twice the planet's radius, respectively, find (a) the planet's radius and (b) the satellite's gravitational potential energy.

Exercise 7-32

A 4.8×10^5 kg rocket is launched from the surface of the Earth with an initial speed of 4.8 km/s. Find the rocket's speed when it is 600 km above the surface of the Earth.

Exercise 7-33

A space shuttle is orbiting the Earth at half the orbital radius of a geosynchronous satellite. Find the orbital period of this space shuttle. (A geosynchronous satellite has an orbital period of 24 hours.)

Exercise 7-34

Find the minimum energy required to put a 2200-kg satellite on the surface of the Earth into an orbit of radius 8.6×10^6 m.

Exercise 7-35

Find the minimum kinetic energy a space vehicle of mass 8500 kg needs at the surface of the Earth in order to escape from the Earth's gravitational pull and travel to infinity.

PROBLEMS ON CHAPTER 7: CIRCULAR MOTION & GRAVITATION

1 – 5. Fill in the blanks (Answer Key on p. 371)

1. An object following a circular path at constant speed is said to be in _________ _________ motion.

2. The net force acting on the object in Problem 1 is always directed towards the center of the circular path and is called the _________ force.

3. Newton's law of _______ _________ states that every object in the universe exerts a gravitational pull on every other object.

4. The _________ _________ at a particular point in space can be calculated by measuring the gravitational force on an object at that point and dividing that force by its mass.

5. _________ is a sensation experienced by someone when there are no forces acting on his or her body except the gravitational force.

6 – 15. True/False questions

6. _____ A body in uniform circular motion has a constant period.

7. _____ The momentum of a body in uniform circular motion is constant.

8. _____ The forces acting on a body in uniform circular motion are not balanced.

9. _____ The work done by the centripetal force on a body in uniform circular motion is always zero.

10. _____ The gravitational force the Earth exerts on us is stronger than the gravitational force we exert on it.

11. _____ When the distance between two masses is tripled, the gravitational force between the two decreases by a factor of 1/3.

12. _____ An orbiting object's centripetal acceleration is the gravitational field at its position in the orbit.

13. _____ A satellite with a larger orbital radius travels faster than one with a smaller orbital radius.

14. _____ The gravitational potential energy of an object launched from the surface of a planet decreases as it moves away from the planet.

15. _____ You feel weightless only when the force of gravity on your body becomes very weak.

16 – 28. Multiple choice questions

16. You are driving a car at a constant speed over a mountainous road with many ups-and-downs. Between the top of a hill and the dip between two hills, where would you weigh more? Where would you 'feel' heavier?

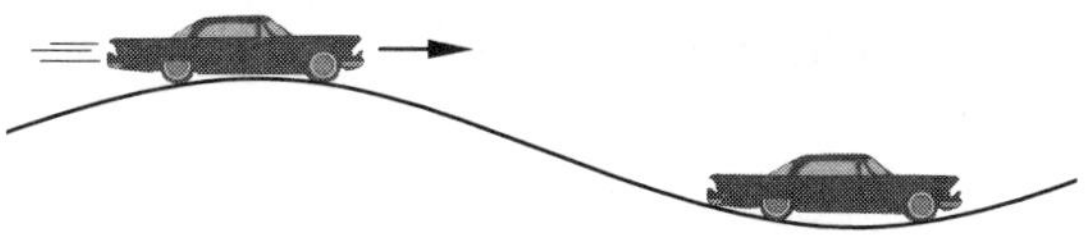

a) You weigh more and feel heavier at the top.
b) You weigh more and feel heavier at the dip.
c) You weigh the same but feel heavier at the top.
d) You weigh the same but feel heavier at the dip.

17. A pendulum swings from one side to the other, as shown. Which of the following graphs best represents the tension in the string?

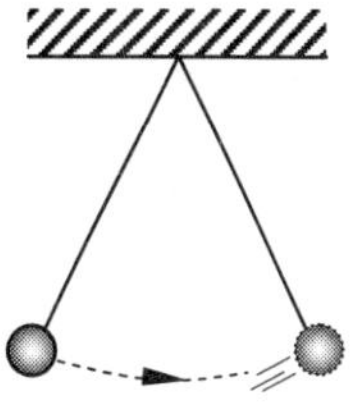

a) 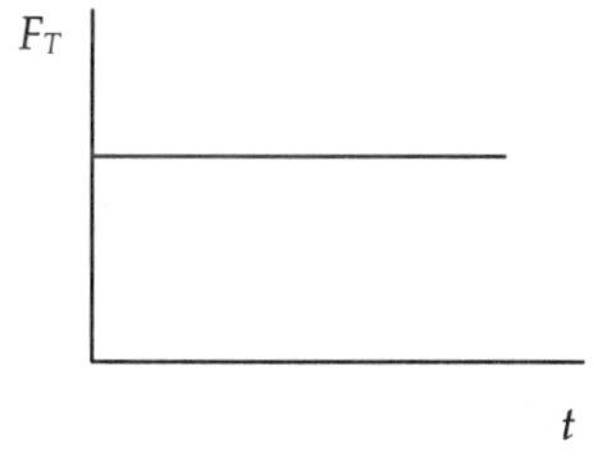

b) 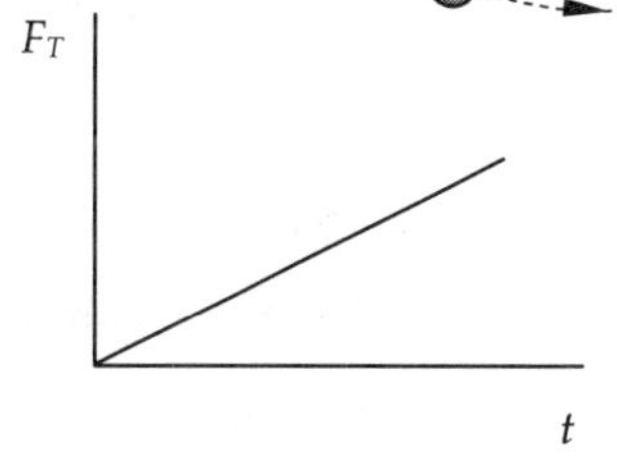

c) 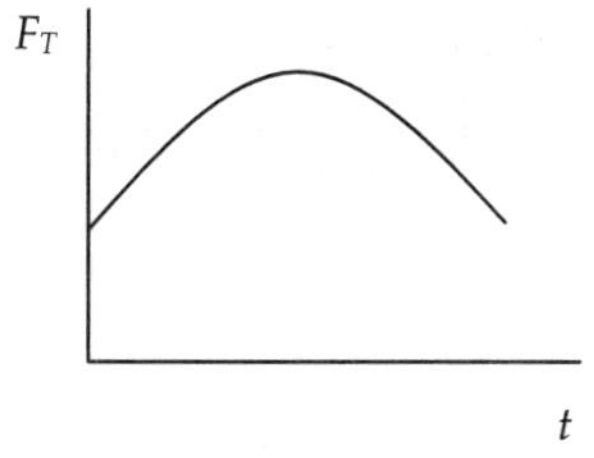

d) 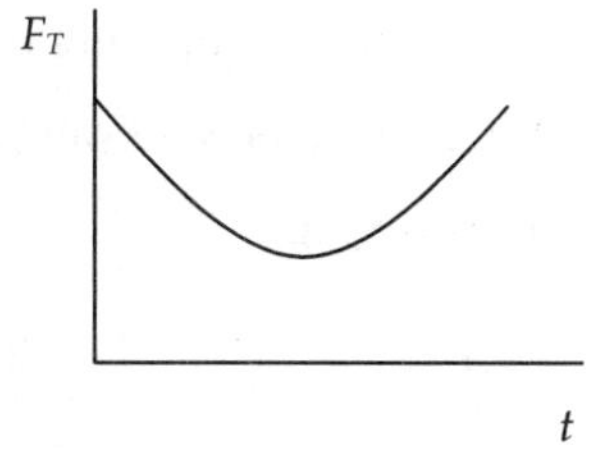

18. When riding in a car turning a sharp corner, you feel as if your body is pulled outwards. Which of the following best explains this phenomenon?

    a) The air inside the car is momentarily sucked out of the car.
    b) The centripetal force on your body is too weak.
    c) There is a force called centrifugal force pulling your body outward.
    d) There is no force; your body wants to continue moving straight ahead because of inertia while the car deviates from that straight path.

19. An object attached to a cable moves in a horizontal circle at constant speed. Which is the correct free-body diagram for the object at the position shown?

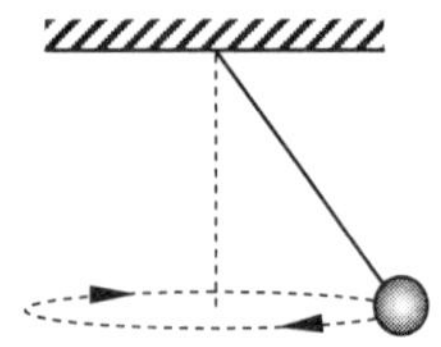

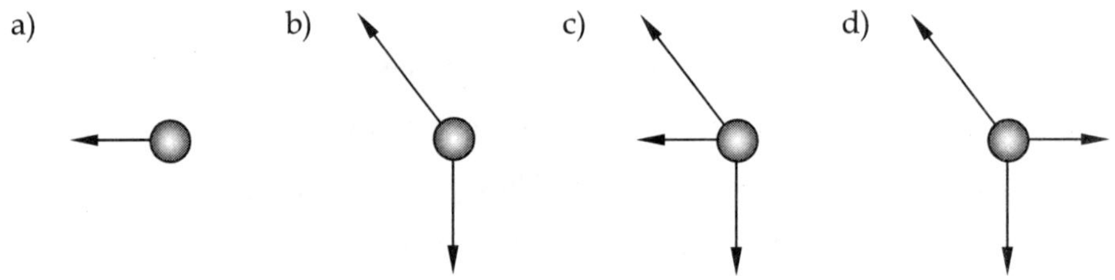

20. Two satellites are orbiting the Earth, as shown. Satellite A is in a higher orbit than satellite B. When satellite A completes ten revolutions, how many revolutions has satellite B completed?

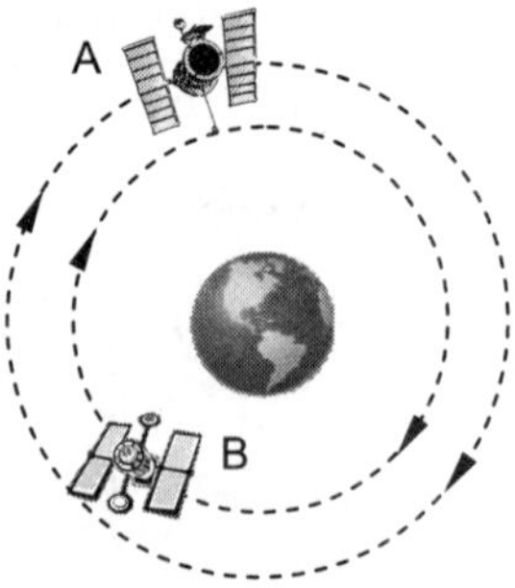

    a) Less than ten revolutions
    b) Exactly ten revolutions
    c) More than ten revolutions
    d) We cannot determine unless we know their orbital radii.

21. Consider the satellites in Problem 20. If satellite A's mass and orbital radius are both 1.5 times those of satellite B, what is the ratio of the centripetal force on satellite A to B?

    a) 1:1   b) 2:3   c) 1:2   d) 1:3

22. A space probe is orbiting a star when the star gradually collapses to one-fifth its initial radius. If there was no change in the star's mass, what happens to the space probe's orbital radius?

a) It increases.　　b) It decreases.

c) It stays the same.

d) We cannot determine unless we know the probe's initial orbital radius.

For Problems 23 and 24, refer to the quantities below:

I. Planet's mass	II. Planet's radius
III. Space shuttle's mass	IV. Orbital radius

23. On which of the above quantities does an orbiting space shuttle's speed depend?

a) I and II only　　b) I and IV only

c) II and III only　　d) III and IV only

24. A space shuttle is launched from the surface of a planet. On which of the above quantities does the escape velocity on the planet depend?

a) I and II only　　b) II and III only

c) I, II, and III　　d) None of the statements is true.

25. Which of the following graphs best represents the gravitational field strength (g) as a function of the distance from a planet (r)? (R is the planet's radius.)

a)
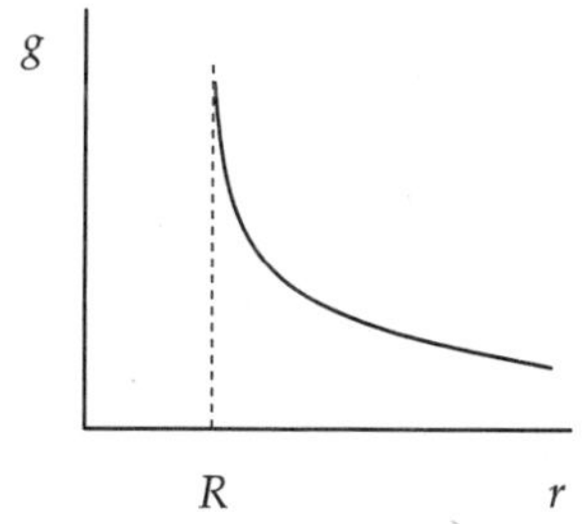

b)
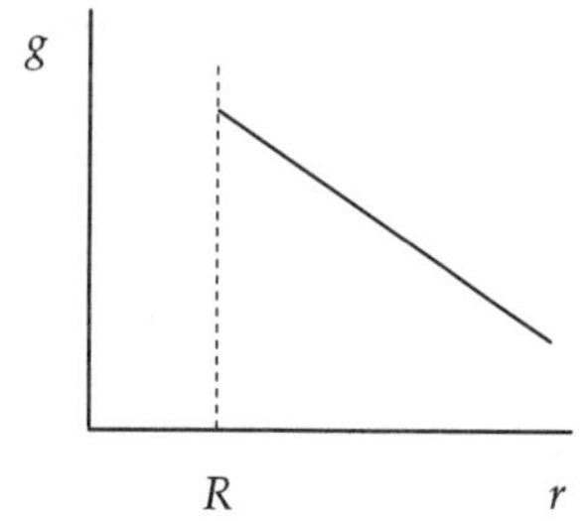

c)
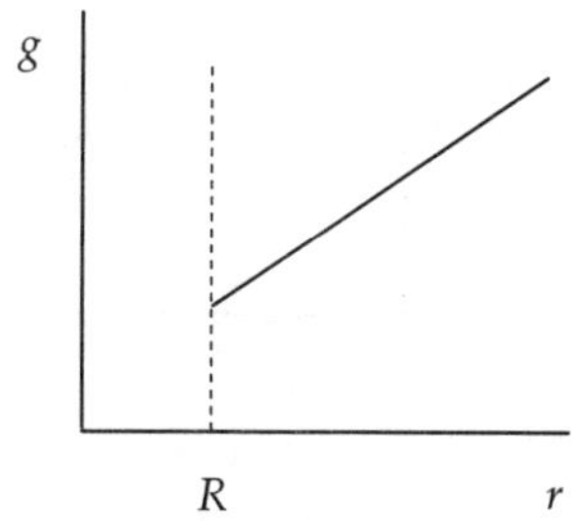

d)
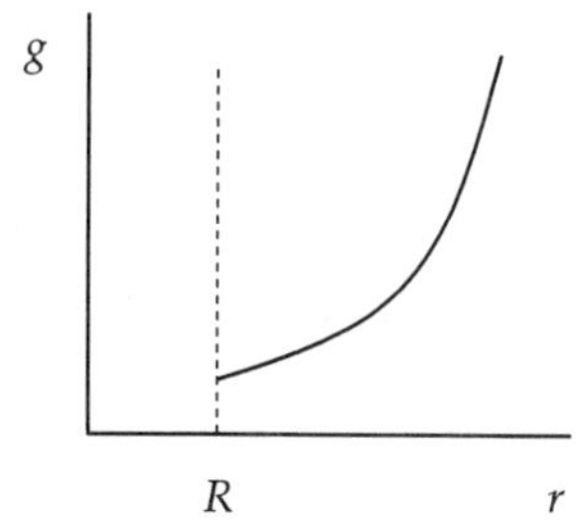

26. A satellite of mass m is orbiting the Earth with a constant speed, v. The distance between the satellite and the Earth is r. If the gravitational potential energy and the kinetic energy are defined as shown below, which of the following is a possible value for the total mechanical energy of the satellite?

$$E_P = -G\frac{mM}{r} \quad \text{and} \quad E_K = \frac{1}{2}mv^2$$

a) -5123 J   b) 0 J   c) 0.05123 J   d) 5123 J

27. The graph on the right shows the magnitude of the gravitational force exerted by the Earth (F_g) versus the distance from the center of the Earth (r). Which of the following does the shaded area under the graph represent?

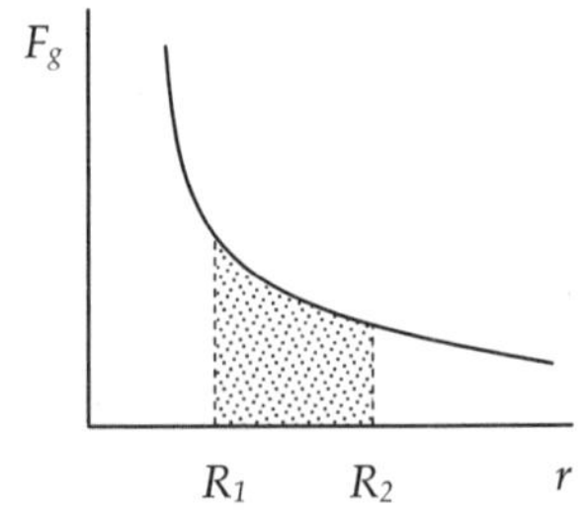

a) The kinetic energy of an object
b) The gravitational potential energy of an object
c) The total mechanical energy of an object
d) The work required to move an object in the Earth's gravitational field

28. A comet travels around the Sun as shown. Which graph best represents the kinetic energy of the comet as it moves from a to b?

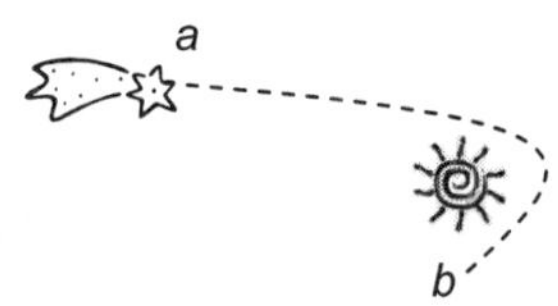

a)
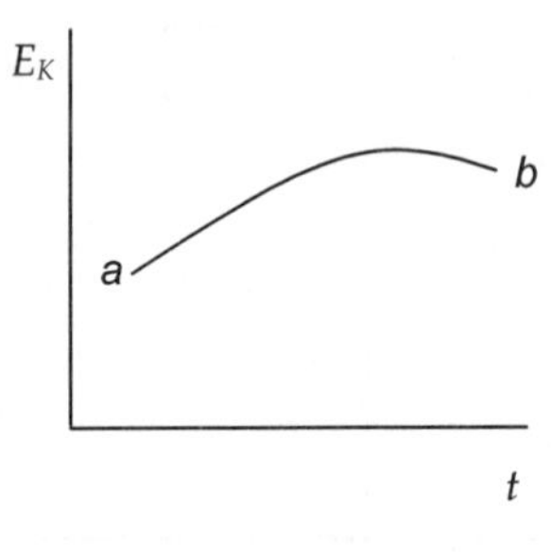

b)
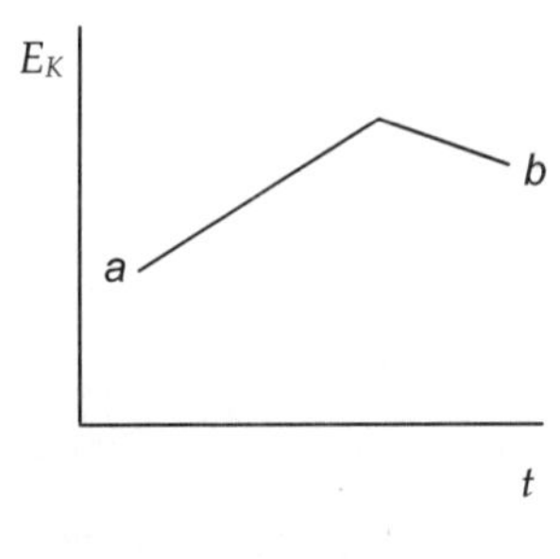

c)
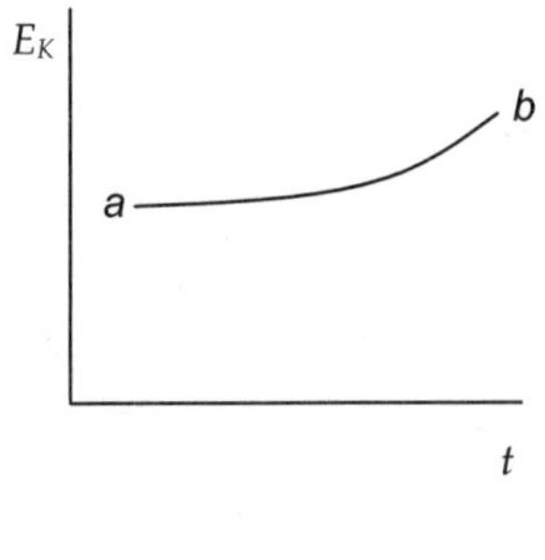

d)
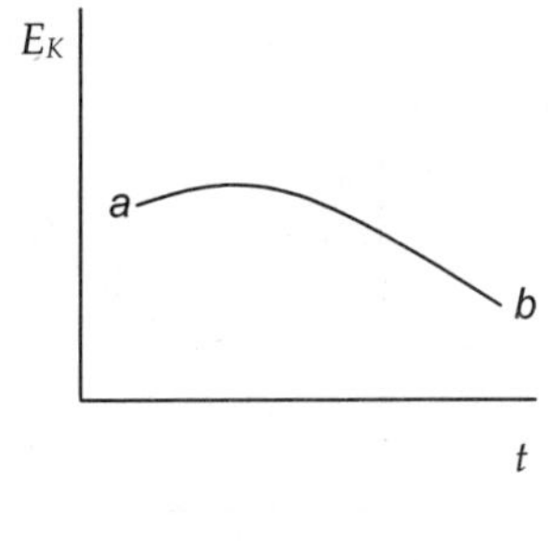

29 – 50. Written-response questions

29. A racing car is moving around a circular track at constant speed. The magnitude of its centripetal acceleration is a when it races at normal speed. Find the magnitude of its centripetal acceleration when the car travels at one-third its normal speed. (Answer in terms of a.)

30. A 6.0-kg object, suspended from the ceiling by a 2.1-m cable, travels in a horizontal circle, as shown. If the angle between the cable and the ceiling is 71°, find (a) the magnitude of the centripetal force on the object and (b) the period of revolution of the object.

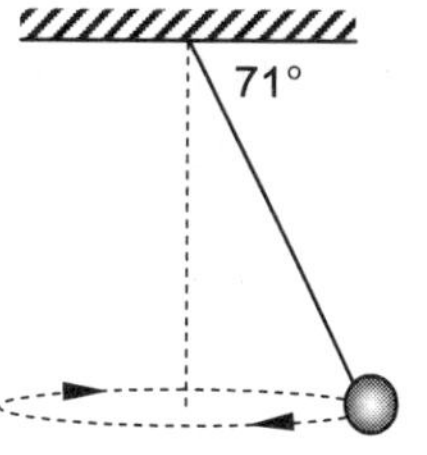

31. A car turns a corner along a horizontal circular path whose radius of curvature is 86 m. On a dry day, the car can travel at the maximum speed of 21 m/s without slipping. On snowy days, the coefficient of friction is reduced to one quarter of its value on a dry day. Find the maximum speed the car can travel on a snowy day.

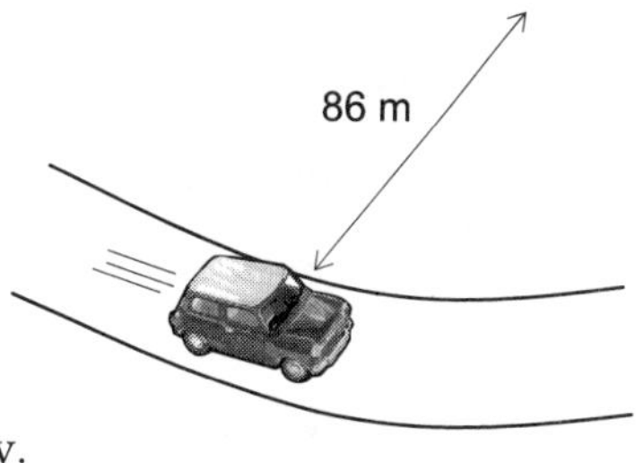

32. An object is swung in a vertical circle of radius 0.98 m. Find the minimum speed of the object at the top if it is to continue moving in a circular path.

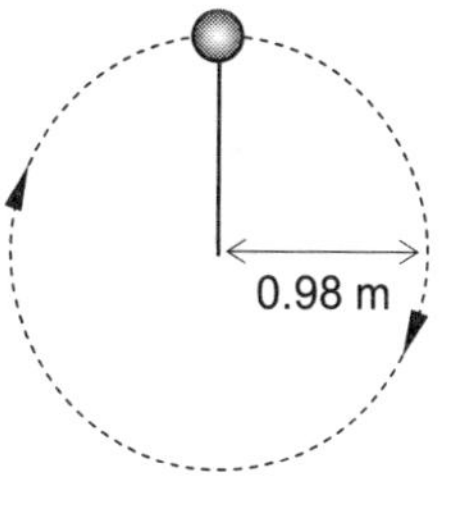

33. A 66-kg man travels in a circular path of radius 1.8 m on a horizontal platform, as shown. If the rope can withstand the maximum tension of 3500 N before breaking, find the maximum speed with which the man can move safely on the platform. (Ignore friction.)

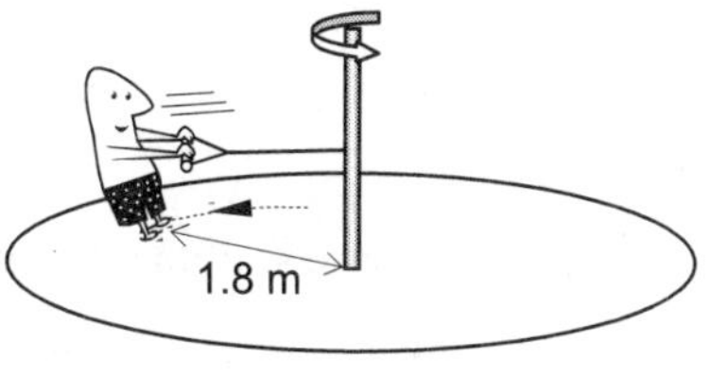

34. A 0.26-kg object, attached to a 0.77-kg mass by a cable through a hole in the table, travels in a circular path as shown. Assuming no friction exists between the object and the table, find the object's speed when the object is 0.35 m from the hole.

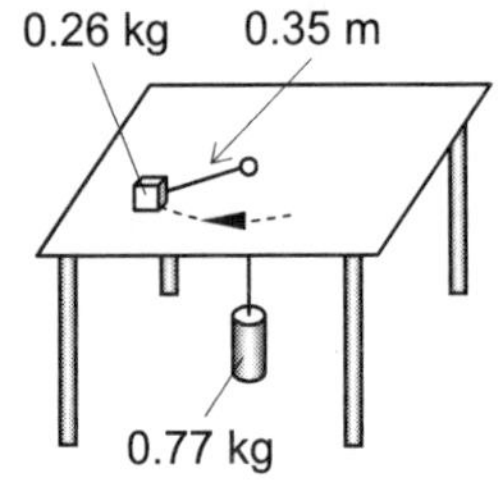

35. An object slides down a frictionless slope and rotates around the loop of diameter 6.5 m, as shown. Find the minimum height, h, the object needs to be released from so that it does not fall off the loop at the top.

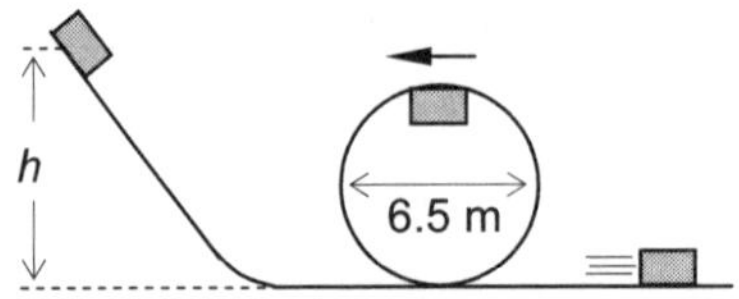

36. An 'artificial' gravitational field is to be created by rotating a circular space station of radius 140 m. If the field strength of 7.8 m/s^2 is needed at this radius, what should be the period of rotation of the space station?

37. A lunar explorer of mass 570 kg takes off from the surface of the Moon. In order to accelerate the spacecraft at 1.7 m/s^2, how much thrust force must its rocket engine exert?

38. A racing car is traveling on a level circular track of radius 40 m. On Earth, it can race at a maximum speed of 15 m/s without skidding. This car and the track are then transported to another planet, where the maximum speed without skidding is found to be only 8.4 m/s. What is the gravitational field strength on the surface of this planet?

39. A 1600-kg satellite circles the Earth at an altitude of 480 km above its surface. Find (a) the orbital speed and (b) the total energy of this satellite.

40. A space shuttle orbiting the Earth completes one revolution every 1.5×10^4 s. Find the gravitational force on a 75-kg astronaut inside the shuttle.

41. A planet of radius 3.7×10^6 m has a gravitational field strength of 86 N/kg at its surface. If a satellite is to orbit this planet at 7.4×10^6 m above the planet's surface, what will be the orbital period of this satellite?

42. A 2.5-kg object dropped 15 m above the surface of an unknown planet hits the ground 4.0 s later. If the planet's radius is 8.3×10^5 m, find the mass of the planet.

43. A satellite travels around a planet at 11 km/s with an orbital radius of 4.7×10^6 m. If an identical satellite orbits the same planet with half the kinetic energy, find its orbital radius.

44. A 1600-kg satellite orbits a planet in a circle of radius 4.9×10^6 m. If this satellite's kinetic energy is 4.6×10^{11} J, find the mass of the planet.

45. A 2700-kg satellite, having reached the end of its useful life, is falling towards the Earth. If its speed is 4.3 km/s at a distance of 620 km above the Earth's surface, find (a) the total energy of this satellite and (b) the impact speed at the surface.

46. A 19000-kg space lab, initially 5.7×10^8 m from the center of a 1.5×10^{25} kg planet, falls to the planet's surface and gains 4.5×10^{11} J of kinetic energy on the way. Find the radius of the planet.

47. A spacecraft leaves the surface of the Earth with a speed of 1.8×10^4 m/s. Find the speed of the spacecraft when it is infinitely far from the Earth.

48. Venus completes one revolution around the Sun every 225 Earth-days. Find the orbital radius of Venus in AU. (1 AU is the distance from the Earth to the Sun, or about 150 million kilometers.)

49. A 4500-kg space vehicle is circling the Earth with an orbital radius of 3.9×10^7 m. Find the minimum energy this vehicle needs in order to escape the Earth's gravitational pull and reach a position infinitely far from the Earth.

50. A rocket is launched from the surface of the Earth with a speed of 8.0 km/s. Will it be able to travel infinitely far from the Earth? If not, how far away from the Earth can it travel? (Ignore air resistance.)

8

CHAPTER 8

ELECTROSTATICS

Beginning in this chapter, we will study the forces between electric charges and their motion. While the size and the mass of electric charges are in many cases exceptionally small in comparison with the objects we have considered so far, all the physical laws and principles introduced in this course are still valid. Although the forces exerted on them are usually very small in magnitude, electric charges have little inertia, and can therefore accelerate at an incredible rate and move at very high speeds.

In this chapter, you will be introduced to electric force, field, and potential energy, whose equations look very similar to the ones for gravitational force, field, and potential energy. In fact, being non-contact forces, the gravitational and electric forces share many characteristics, and it is advised that you compare and contrast the electric quantities with the gravitational ones, discussed in Chapter 7. Another electric quantity to be introduced here is electric potential, which is useful in analyzing one-dimensional motions of charges, such as the motion between two charges or charges moving between two oppositely charged parallel plates.

In this chapter, you will:

- ✓ learn about electric force and electric field
- ✓ solve exercise problems on electric force and field
- ✓ learn about electric potential energy, electric potential, charges' motion between two parallel plates and in a cathode-ray tube
- ✓ solve exercise problems on electric potential energy, electric potential, and cathode-ray tubes
- ✓ test your knowledge of electrostatics

8-1. Electric Charges

Electric charge is one of the fundamental attributes of the particles of which matter is made. Among the three basic sub-atomic particles, only protons and electrons carry electric charge; neutrons are electrically neutral. The charge carried by a single proton is known as the **elementary charge, *e***, where

The SI unit for electric charge is the Coulomb. ⟶ $e = 1.60 \times 10^{-19}\ \text{C}$ [Coulomb] ... (8.1)

Thus, the sum of the charges, or the net charge, on five protons is 5*e*, or 8.00×10^{-19} C. An electron has the same charge as a proton except that it is negative. Hence, the net charge on six electrons is -6*e*, or -9.60×10^{-19} C.

Then the net charge on an object composed of as many electrons as protons is zero; the object is electrically neutral. If the net charge on an object is positive, it means that there are more protons than electrons, and we can find the number of excess protons by dividing the net charge by the elementary charge. For instance, suppose an object has the net charge of 3.00 μC. The number of excess protons is

$$\frac{3.00 \times 10^{-6}\ \text{C}}{1.60 \times 10^{-19}\ \text{C}} = 1.875 \times 10^{13} \text{ more protons than electrons}$$

Note that the net charge must be a whole number multiple of the elementary charge because an object cannot gain or lose a fraction of an electron. (Only electrons are transferred from one object to another. Protons never move because they are tightly bound in the nucleus.) Therefore, the net charge of +1*e*, +2*e*, or +3*e* may be possible, but it can never be +1.2*e*, -2.5*e*, and so forth.

8-2. Coulomb's Law

The electric force is one of the fundamental forces in nature, and, as you all know, a pair of charged particles exert an electric force on each other just as two masses exert gravitational force on each other. Indeed, the electric force shares many similarities with the gravitational force. For example, the electric force is a non-contact, conservative force, just like the gravitational force. In addition, the magnitude of the electric force is given by

$$F_e = k\frac{|qQ|}{r^2} \quad \text{... (8.2)}$$

← The absolute value sign is needed to keep the magnitude from becoming negative.

where q and Q are the charges involved, r is the distance between the charges, and k is a proportionality constant equal to $9.0 \times 10^9\ \mathrm{N \cdot m^2/C^2}$. Notice that the equation is very similar in form to Newton's law of universal gravitation in Eq. (7.4) (on p. 201). Equation (8.2) is known as **Coulomb's law**, and it is to electric force what Newton's law of gravitation is to gravitational force. It basically describes the magnitude of the electric force between two charges when they are at rest.

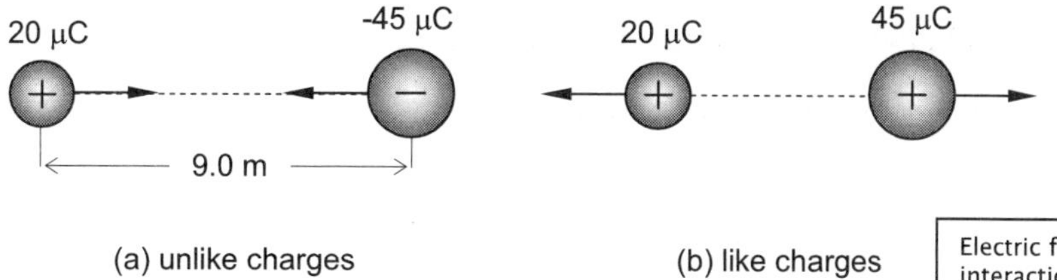

Figure 8-1: Electric force between two charges

Electric force is an interaction between two charges.

Consider the two charged objects shown in Fig. 8-1 (a). According to Eq. (8.2), the magnitude of the electric force between the objects is

$$F_e = 9.0 \times 10^9\ \mathrm{N \cdot m^2/C^2} \frac{\left|(20 \times 10^{-6}\ \mathrm{C})(-45 \times 10^{-6}\ \mathrm{C})\right|}{(9.0\ \mathrm{m})^2} = 0.10\ \mathrm{N}$$

Since the objects have unlike charges, the force is attractive. Thus, the objects attract each other with 0.10 N forces. If the charges on the objects had been of the same polarity, the objects would have repelled each other, as shown in Fig. 8-1 (b). In fact, one critical difference between the electric force and the gravitational force is that the former can be repulsive as well as attractive. Therefore, when determining the direction of an electric force, we need to know whether we have like charges or unlike charges in addition to a diagram showing the location of the charges.

The process of finding the net electric force on a charge surrounded by multiple charges is practically the same as that used in finding the net gravitational force on a mass surrounded by multiple bodies. That is, first calculate the electric force exerted by each surrounding charge, and then find the vector sum of these forces.

Suppose three charged objects A, B, and C are arranged as shown in Fig. 8-2 (a). The objects, located at the vertices of an equilateral triangle, are an equal distance away from each other. Each object exerts electric forces on the others and vice versa, as shown in Fig. 8-2 (b). Note that some forces are attractive and some are repulsive. Then to find the net electric force on B, we need first to find $F_{\text{on B by A}}$ and $F_{\text{on B by C}}$ using Eq. (8.2):

$$F_{\text{on B by A}} = k\frac{|q_A q_B|}{r_{AB}^2} = (9.0\times10^9\ \text{N}\cdot\text{m}^2/\text{C}^2)\frac{|(-2.0\times10^{-6}\ \text{C})(3.0\times10^{-6}\ \text{C})|}{(4.0\,\text{m})^2} = 3.38\times10^{-3}\ \text{N}$$

$$F_{\text{on B by C}} = k\frac{|q_B q_C|}{r_{BC}^2} = (9.0\times10^9\ \text{N}\cdot\text{m}^2/\text{C}^2)\frac{|(3.0\times10^{-6}\ \text{C})(5.0\times10^{-6}\ \text{C})|}{(4.0\,\text{m})^2} = 8.44\times10^{-3}\ \text{N}$$

Adding these two vectors by components, we find the net electric force on B to be 7.4×10^{-3} N, 23° above the $-x$-axis.

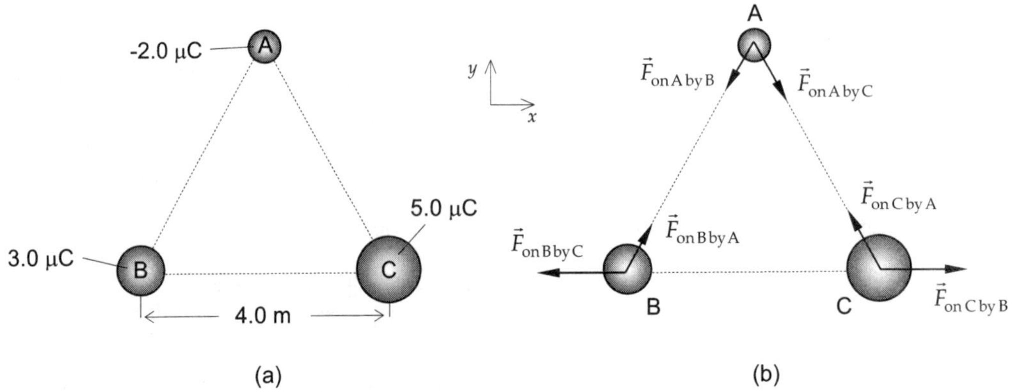

Figure 8-2: Finding the net electric force on a charge

8-3. Electric Field

Just as each and every mass creates a gravitational field around itself, each and every charge generates an **electric field**, denoted by $\vec{E}$. It is a vector quantity whose magnitude is defined as

> The electric field is similar to the gravitational field in many ways, so it is a good idea to try to understand the former by comparing it with the latter.

Electric field created by Q ⟶ $$E = k\frac{|Q|}{r^2} \quad [\text{N/C}] \qquad \ldots (8.3)$$

Compare it with Eq. (7.8)

Therefore, the electric field strength 12 m away from a 35-μC charge is

$$E = (9.0\times10^9\ \text{N}\cdot\text{m}^2/\text{C}^2)\frac{|35\times10^{-6}\ \text{C}|}{(12\,\text{m})^2} = 2.2\times10^3\ \text{N/C}$$

To find the direction of the electric field at a certain point, we can put a positive test charge there and observe the direction in which it accelerates. Figure 8-3, for example, shows a test charge placed at point A being repelled by a positive charge. Then the direction of the electric field at point A is 'away from the positive charge.'

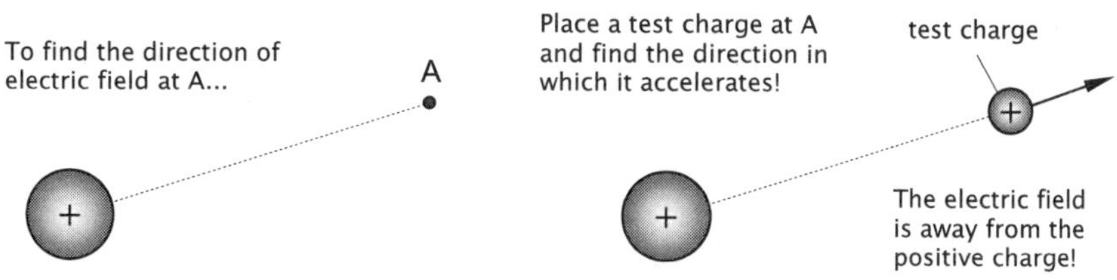

Figure 8-3: Finding the direction of electric field using a positive test charge

Using the above method, we can make the following conclusions: The electric field created by a positive charge points away from the charge, and the one generated by a negative charge points towards the charge, as shown in Fig. 8-4.

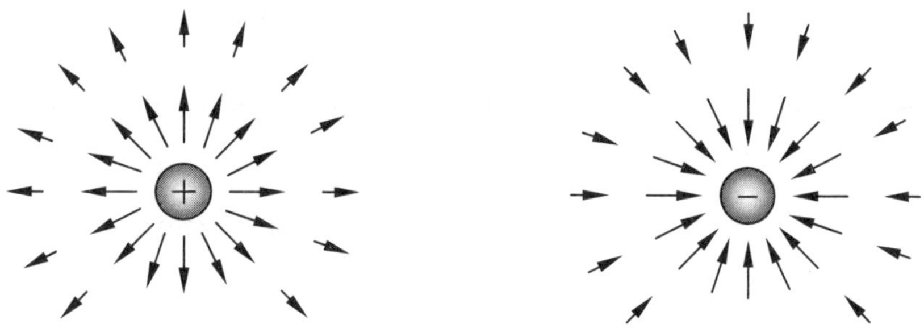

(a) positive charge

(b) negative charge

Figure 8-4: Electric fields around electric charges

Alternatively, electric field can be defined as electric force per unit charge q:

Electric field created by Q → $\vec{E} = \frac{\vec{F}_e}{q}$ [N/C] ← electric force on q ... (8.4)

The electric force is then the product of the charge q and the electric field of Q:

$$\vec{F}_e = q\vec{E} \quad \text{... (8.5)}$$

Note that Eqs. (8.4) and (8.5) are similar in form to Eqs. (7.9) (on p. 205) and (7.6) (on p. 203), respectively. However, the charge (q) may be either positive or negative whereas the mass (m) is always positive. This implies that, depending on the sign of the charge, the electric force on it may be in the same direction as the electric field or in the opposite direction, as illustrated in Fig. 8-5.

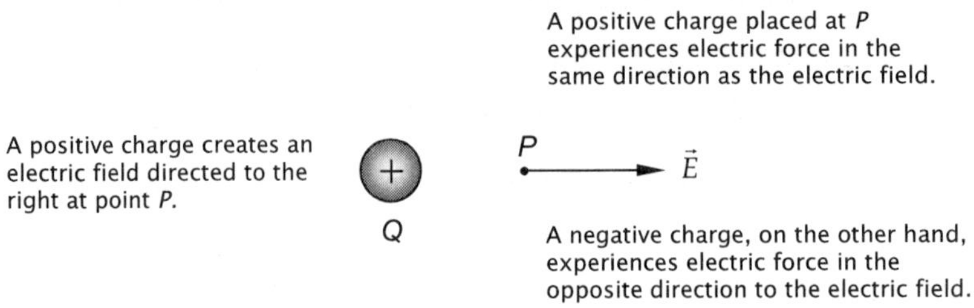

Figure 8-5: The directions of electric force and electric field

For instance, if a 3.0-C charge is located at a point where the electric field is 20 N/C, west, the electric force on the charge, according to Eq. (8.5) will be

$$\vec{F}_e = (3.0\,\text{C})(20\,\text{N/C}) = 60\,\text{N, west}$$

The direction is same as the electric field because the charge is positive.

On the other hand, if a -4.5-μC charge experiences 9.0-N electric force, north, the electric field is

$$\vec{E} = \frac{9.0\,\text{N}}{-4.5\times10^{-6}\,\text{C}} = -2.0\times10^{6}\,\text{N/C, north, which is } 2.0\times10^{6}\,\text{N/C, south}$$

Note that the direction of the electric field (south) is opposite to that of the electric force (north) because the charge is negative.

As in finding the gravitational field, when several charges are located in the vicinity of a point, the electric field at that point is determined by the vector sum of the individual fields created by the charges. Let's revisit the example of three charged objects in Section 8-2 (p. 232) and find the electric field at point P, the midpoint of the line segment joining objects B and C (see Fig. 8-6 (a)).

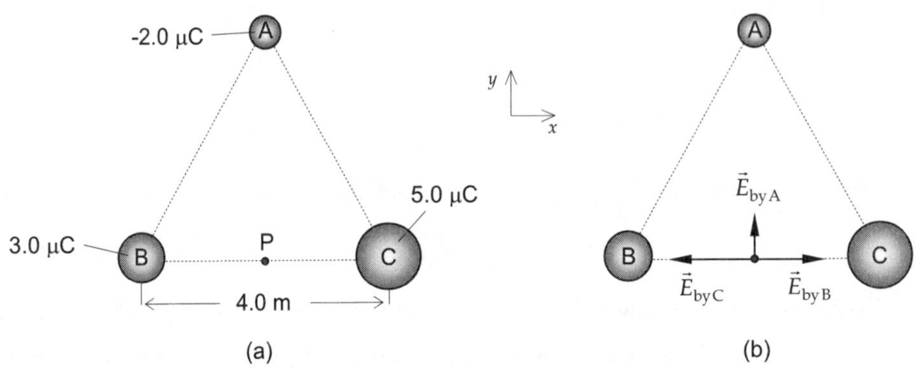

Figure 8-6: Finding the net electric field at a point

The electric field created at P by each object is

Since the charge on A is negative, the electric field is toward A. → $\vec{E}_{\text{by A}} = k\dfrac{|q_A|}{r_{AP}^2} = (9.0\times10^9\ \text{N}\cdot\text{m}^2/\text{C}^2)\dfrac{|-2.0\times10^{-6}\ \text{C}|}{(4.0\,\text{m}\cdot\sin 60°)^2} = 1.5\times10^3\ \text{N/C, up}$

Since each charge is positive, the electric field is away from it. →

$$\vec{E}_{\text{by B}} = k\frac{|q_B|}{r_{BP}^2} = (9.0\times10^9\ \text{N}\cdot\text{m}^2/\text{C}^2)\frac{|3.0\times10^{-6}\ \text{C}|}{(2.0\,\text{m})^2} = 6.75\times10^3\ \text{N/C, right}$$

$$\vec{E}_{\text{by C}} = k\frac{|q_C|}{r_{CP}^2} = (9.0\times10^9\ \text{N}\cdot\text{m}^2/\text{C}^2)\frac{|5.0\times10^{-6}\ \text{C}|}{(2.0\,\text{m})^2} = 11.25\times10^3\ \text{N/C, left}$$

According to the coordinate system shown in Fig. 8-6,

$$(\vec{E}_{\text{net}})_x = \vec{E}_{\text{by B}} + \vec{E}_{\text{by C}} = 6.75\times10^3\ \text{N/C} - 11.25\times10^3\ \text{N/C} = -4.5\times10^3\ \text{N/C}$$

$$(\vec{E}_{\text{net}})_y = \vec{E}_{\text{by A}} = 1.5\times10^3\ \text{N/C}$$

The net electric field at P is then

$$\vec{E}_{\text{net}} = \sqrt{(4.5\times10^3\ \text{N/C})^2 + (1.5\times10^3\ \text{N/C})^2} = 4.7\times10^3\ \text{N/C, } 18°\ \text{above } -x\text{-axis}$$

Using this method, we can find electric fields at various points near a group of charges. For instance, the electric field around a pair of unlike charges of equal magnitude is shown in Fig. 8-7. Another way to visualize the field around a charge or a group of charges is to draw **electric field lines**. They are drawn so that the field vectors become tangent to the lines.

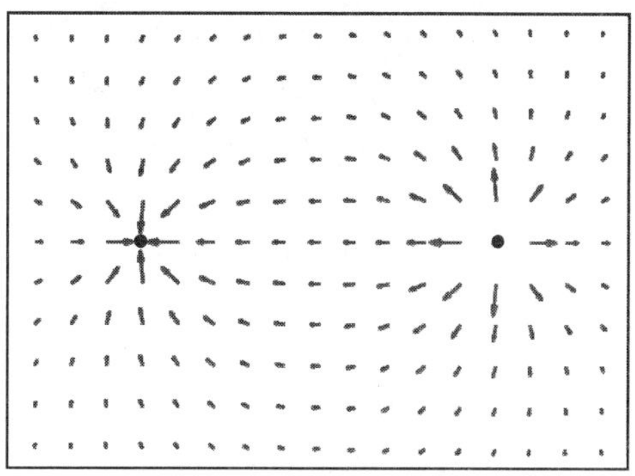

Fig. 8-7: Electric field vectors

[source: http://en.wikipedia.org/wiki/Electric_field]

For example, the electric field lines for the pair of charges in Fig. 8-7 are drawn in solid lines in Fig. 8-8 (a). Note that the electric field lines start from the positive charge and end at the negative charge. The electric field lines for a pair of positive charges are shown in Fig. 8-8 (b), and those for a positive charge in Fig. 8-8 (c). Between two oppositely charged parallel plates, all electric field lines are perpendicular to the surface of the plates and equally spaced, as shown in Fig. 8-8 (d). Again, the lines start from the positively charged plate and terminate at the negatively charged plate. Since the lines are equally spaced, the electric field strength is the same everywhere between the plates.

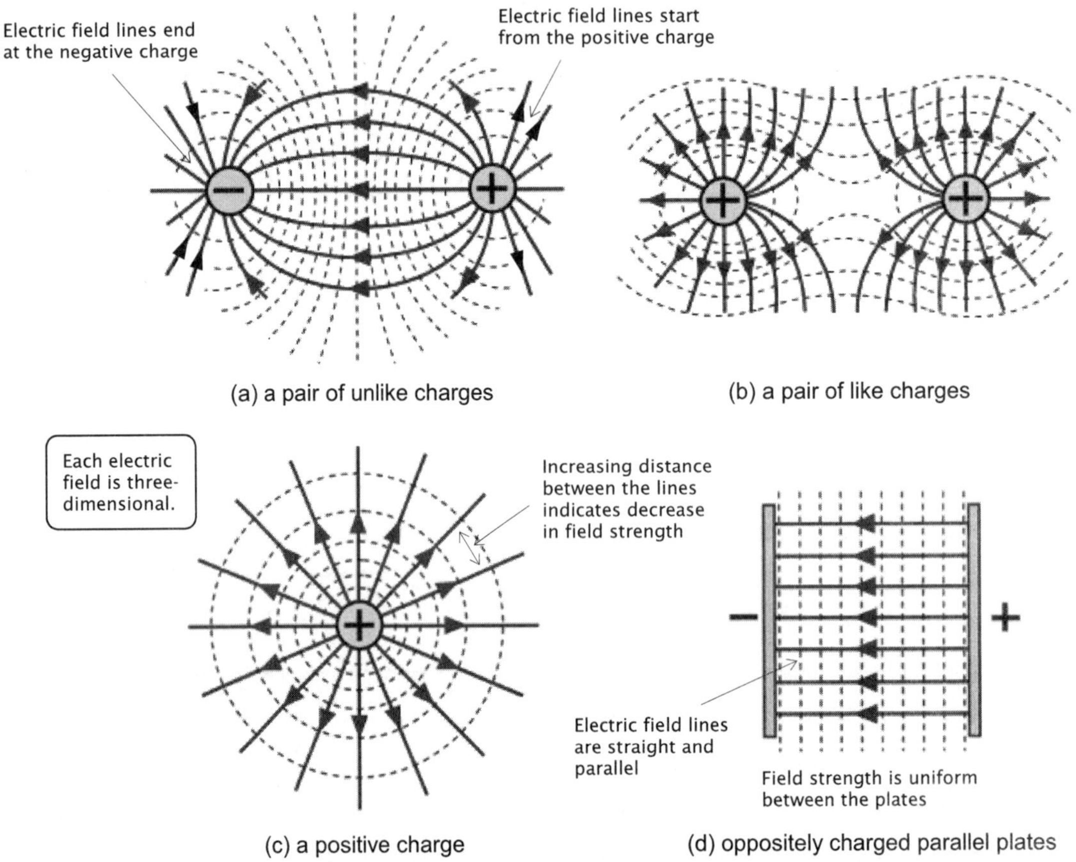

Figure 8-8: Electric field lines

The characteristics of electric field lines are summarized as follows:

1. You can always find the direction of the electric field (which is the same as the direction of the electric force on a positive test charge) from the field lines. It is the direction tangent to the field lines at any given point.
2. The strength of the electric field is proportional to the number of lines crossing a unit area perpendicular to the lines. In other words, the closer the lines, the stronger the field.
3. Electric field lines always start from positive charges and terminate at negative charges. The number of lines starting or ending is proportional to the magnitude of the charge.

PHYSICS 12 ANALYZED: ELECTRIC FORCE & ELECTRIC FIELD

8-1. Electric Charges

- Electric charge is a scalar quantity and is measured in Coulombs [C].
- A proton and an electron carry electric charge of the same magnitude but of opposite signs. The charge on a single proton is called the elementary charge, e, where $e = 1.60 \times 10^{-19}$ C.
- The net charge of an object must be a whole number multiple of the elementary charge. A neutral object has the net charge of zero and contains as many electrons as protons.

8-2. Coulomb's Law

- Coulomb's law: $F_e = k\frac{|qQ|}{r^2}$ ← Electric force can be repulsive as well as attractive!
- When a charge experiences electric forces exerted by multiple charges, the net electric force on it is the vector sum of the individual forces.

8-3. Electric Field

- The strength of the electric field created by charge Q at distance r away from it is given by $E = k\frac{|Q|}{r^2}$ [N/C].
- The electric field created by a positive charge points away from the charge; that generated by a negative charge points toward the charge.
- Electric field is also the electric force per unit charge: $\vec{E} = \frac{\vec{F}_e}{q}$ [N/C]
- The electric field at a point in space is the vector sum of the individual fields created by the surrounding charges.
- Characteristics of electric field lines:
  1. Electric field vectors are tangent to the electric field lines.
  2. The closer the lines, the stronger the field.
  3. Electric field lines start from positive charges and end at negative charges.
- Electric field strength between two oppositely charged parallel plates is constant.

EXERCISES ON CHAPTER 8: ELECTRIC FORCE & ELECTRIC FIELD

Exercise 8-1

A 7.4-μC charge exerts an attractive force of magnitude 0.13 N on an unknown charge. If the distance between the charges is 0.42 m, find the magnitude and polarity of the unknown charge.

Exercise 8-2

An alpha particle experiences an electric force of 8.3×10^{-10} N at a distance of 1.7×10^{-9} m from an unknown charge. If the alpha particle is brought closer, to a distance of 5.7×10^{-10} m from the unknown charge, find the magnitude of the electric force on the alpha particle.

Exercise 8-3

The electric field strength 3.1 m from a negative charge is 240 N/C. Find the magnitude and direction of the electric field 8.6 m from the charge.

Exercise 8-4

An electron is placed in a uniform, horizontal electric field, as shown. If the electric field strength is 3400 N/C, find the magnitude and direction of (a) the electric force on the electron and (b) the electron's acceleration.

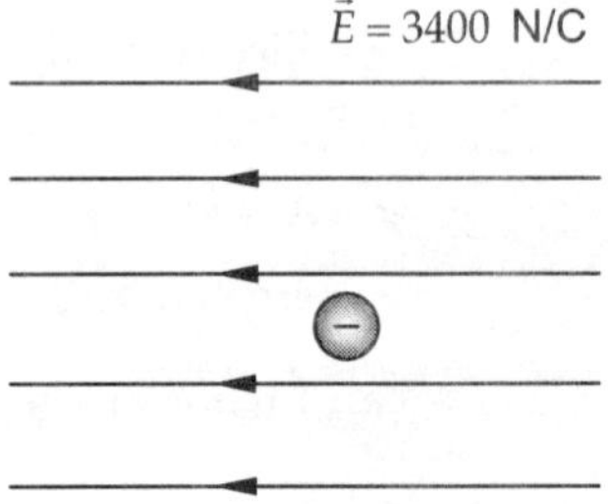

Exercise 8-5

In the diagram shown right, the net electric force acting on the 2.2-μC charge is 0.38 N, left. Find the magnitude and polarity of the unknown charge Q.

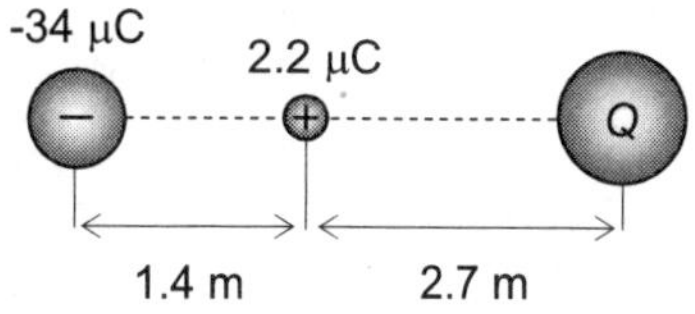

Exercise 8-6

A pair of unlike charges are placed 16.8 m apart as shown. Find the magnitude and direction of the electric field at point P.

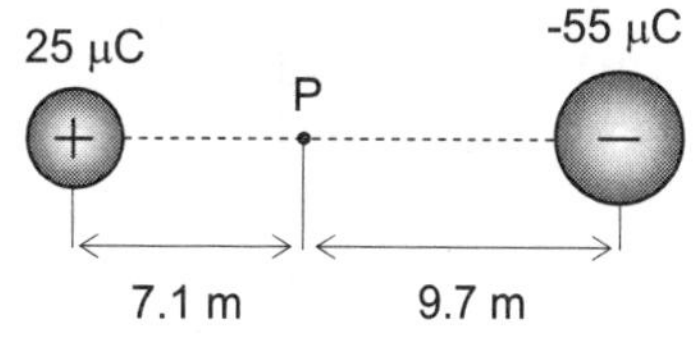

Exercise 8-7

Consider Exercise 8-6. If a 15-μC charge is placed at point P, find the magnitude and direction of the electric force on (a) the 15-μC charge and (b) the -55-μC charge.

Exercise 8-8

A proton is placed in a uniform, vertical electric field. The weight of the proton is supported by the electric force on the proton, as shown. Find the magnitude and direction of the electric field.

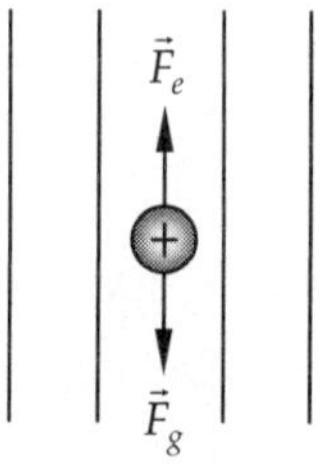

Exercise 8-9

Four charged objects are located at the vertices of a square, as shown right. It is known that the charges on B and C are each +1.4 μC and that the charges on A and D are both negative. If the net force on D is zero, what is the magnitude of the charge on A?

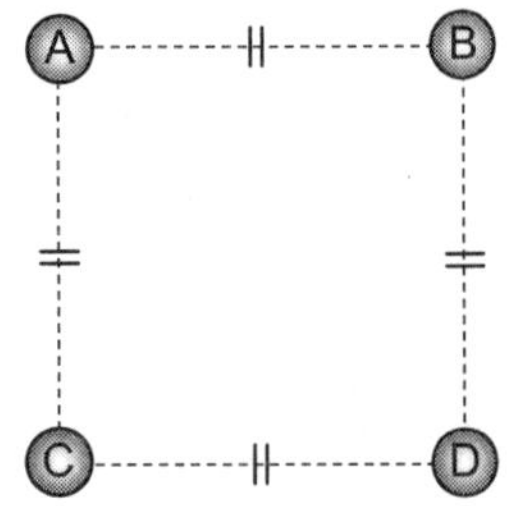

Exercise 8-10

Find the magnitude and direction of the electric field at point P due to the two fixed charges shown right.

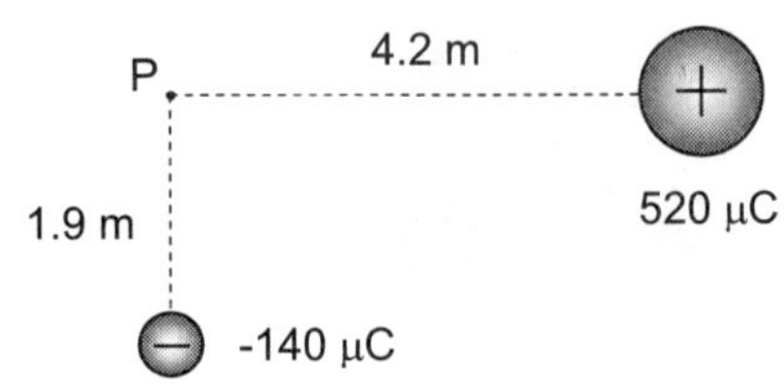

Exercise 8-11

A proton is fired into a uniform electric field as shown and enters the field at 27 km/s. Find the proton's acceleration while it is within the electric field. (Find both magnitude and direction.)

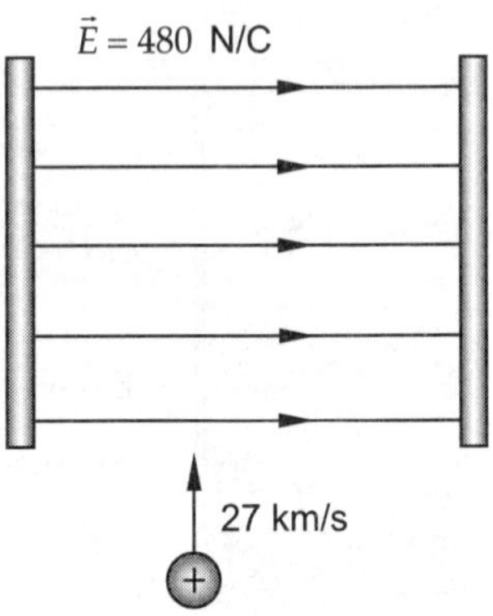

Exercise 8-12

In the diagram shown right, the magnitude of the electric field at point P is 2.3×10^4 N/C. Find the magnitude of the unknown charge Q.

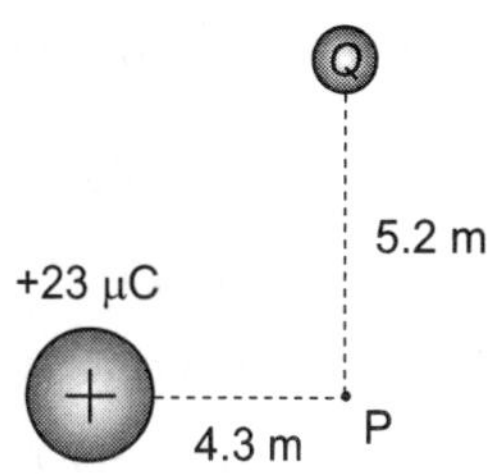

Exercise 8-13

In a hydrogen atom, an electron orbits a proton at a distance of 5.3×10^{-11} m. Find the orbital speed of the electron.

Exercise 8-14

The nucleus of a Carbon-14 atom consists of six protons and eight neutrons. If an electron is orbiting the nucleus at a distance of 2.1×10^{-10} m, find (a) the magnitude of the centripetal force on the electron, (b) the orbital speed, and (c) the orbital period of the electron.

Exercise 8-15

A 28-g mass is suspended by a string in a uniform electric field as shown. If the strength of the electric field is 1300 N/C, find the magnitude and polarity of the net charge on the mass.

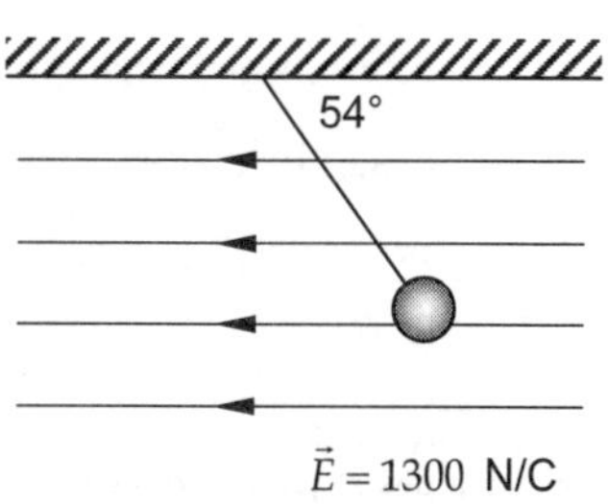

8-4. Electric Potential Energy

A charge moves in an electric field as the result of an electric force doing work on the charge. Since electric force is a conservative force, the work causes a change in the potential energy of the charge, as shown in Eq. (4.14) (on p. 105). The potential energy associated with electric force is the **electric potential energy**, defined as

Similarly, an object falls in a gravitational field as the result of the gravitational force doing work on the object.

Although potential energy can be positive or negative, it is NOT a vector. →

$$E_P = k\frac{qQ}{r} \quad \text{... (8.6)}$$

← There is no absolute value sign!

Thus, the electric potential energy between a 12-μC charge and a 13-μC charge separated by a distance of 2.0 m is

$$E_P = k\frac{qQ}{r} = \left(9.0\times10^9\ \mathrm{N\cdot m^2/C^2}\right)\frac{\left(12\times10^{-6}\ \mathrm{C}\right)\left(13\times10^{-6}\ \mathrm{C}\right)}{2.0\,\mathrm{m}} = 0.70\ \mathrm{J}$$

If let go, the two like charges move away from each other, and as the distance between them increases, the electric potential energy decreases. For instance, when the distance between them is 5.0 m, the potential energy is

$$E_P = k\frac{qQ}{r} = \left(9.0\times10^9\ \mathrm{N\cdot m^2/C^2}\right)\frac{\left(12\times10^{-6}\ \mathrm{C}\right)\left(13\times10^{-6}\ \mathrm{C}\right)}{5.0\,\mathrm{m}} = 0.28\ \mathrm{J}$$

As these charges move away from each other, the potential energy gets converted into their kinetic energy, as shown in Fig. 8-9.

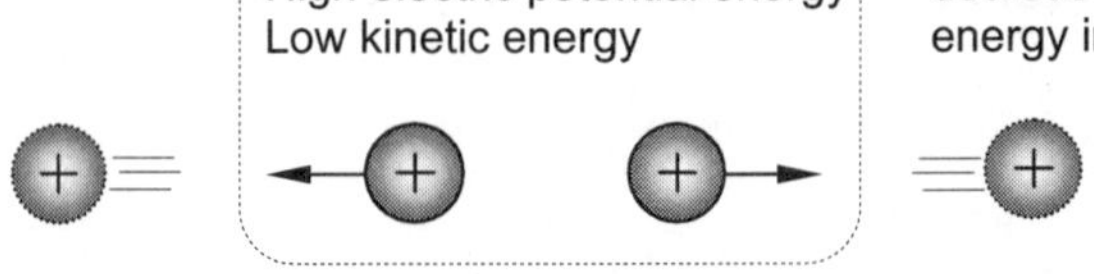

Figure 8-9: Electric potential energy between two like charges

Note that electric potential energy is negative for a pair of unlike charges. In this case, the electric potential energy increases when the distance between the charges increases, as shown in Fig. 8-10.

Work must be done to bring like charges together, so their potential energy increases as the charges are brought closer together.

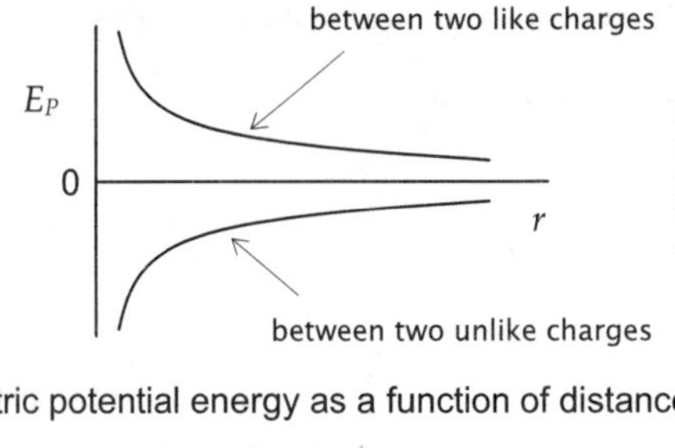

Work must be done to separate unlike charges, so their potential energy increases as the charges are moved farther away from each other.

Figure 8-10: Electric potential energy as a function of distance between two charges

In summary, electric potential energy is given to an electric charge when it is separated from another charge of opposite sign, or when it is pushed closer to another charge with the same sign. In either case, the work done on the charge is transformed into the electric potential energy of the charges.

For instance, suppose we want to calculate the work required to bring two electrons initially 0.40 m apart to within 0.010 m. Initially, the potential energy between the two electrons is

$$(E_P)_i = k\frac{qQ}{r} = (9.0\times10^9\ \mathrm{N\cdot m^2/C^2})\frac{(-1.6\times10^{-19}\ \mathrm{C})(-1.6\times10^{-19}\ \mathrm{C})}{0.40\ \mathrm{m}} = 5.76\times10^{-28}\ \mathrm{J}$$

At 0.010 m from each other, the potential energy between them increases to

$$(E_P)_f = k\frac{qQ}{r} = (9.0\times10^9\ \mathrm{N\cdot m^2/C^2})\frac{(-1.6\times10^{-19}\ \mathrm{C})(-1.6\times10^{-19}\ \mathrm{C})}{0.010\ \mathrm{m}} = 2.30\times10^{-26}\ \mathrm{J}$$

According to the work-energy principle, the amount of work required is the change in the potential energy:

$$W_{\mathrm{NC}} = \Delta E_P + \cancel{\Delta E_K} = (E_P)_f - (E_P)_i = 2.30\times10^{-26}\ \mathrm{J} - 5.76\times10^{-28}\ \mathrm{J} \cong 2.2\times10^{-26}\ \mathrm{J}$$

Note that there is no change in kinetic energy because we assume the electrons are at rest before and after the motion.

In another example, an electron, initially stationary at an infinite distance away from a carbon nucleus, accelerates toward the nucleus. We are interested in finding the electron's speed when it is 0.0030 m from the nucleus, which we assume is fixed. Again, we use the work-energy principle. Since the electron's motion is caused by the electric force and no other force, $W_{\mathrm{NC}} = 0\ \mathrm{J}$. Thus,

$$\cancel{W_{\mathrm{NC}}} = \Delta E_P + \Delta E_K$$

$$= \left[(9.0\times10^9\ \mathrm{N\cdot m^2/C^2})\frac{\overbrace{6(1.6\times10^{-19}\ \mathrm{C})}^{\text{charge on the carbon nucleus}}(-1.6\times10^{-19}\ \mathrm{C})}{0.0030\ \mathrm{m}} - \underset{\uparrow}{0}\right] + \left[\frac{1}{2}(9.11\times10^{-31}\ \mathrm{kg})v_f^2 - 0\right] = 0$$

Initial PE is zero because r is infinite.

Solve the above equation for the final speed, v_f, and we find that it is $1.0\times10^3\ \mathrm{m/s}$.

8-5. Electric Potential

Every object, including electric charges, prefers moving to a region of lower potential energy to be in a more 'stable' state. For a mass, the altitude can be used to determine the region of lower gravitational potential energy. That is, the lower the altitude of an object, the lower its gravitational potential energy. In order to lose gravitational potential energy, the mass then has to move downward, or fall towards the center of the Earth.

As for electric charges, it is electric potential that can be used to determine the region of lower electric potential energy. **Electric potential** created by charge Q is

$$V = k\frac{Q}{r} \quad [\text{V}] \qquad \text{... (8.7)}$$

← The unit for electric potential is Volt.

Then the electric potential 3.0 m away from a -5.0-μC charge is

$$V = k\frac{Q}{r} = \left(9.0\times10^{9}\ \text{N}\cdot\text{m}^2/\text{C}^2\right)\frac{-5.0\times10^{-6}\ \text{C}}{3.0\,\text{m}} = -1.5\times10^{4}\ \text{V}$$

The potential is negative because the charge is negative.

At 5.0 m away from the same charge, the electric potential increases to

$$V = k\frac{Q}{r} = \left(9.0\times10^{9}\ \text{N}\cdot\text{m}^2/\text{C}^2\right)\frac{-5.0\times10^{-6}\ \text{C}}{5.0\,\text{m}} = -9.0\times10^{3}\ \text{V}$$

As the above calculations show, the electric potential created by a negative charge increases with distance and approaches zero as the distance tends to infinity. On the other hand, the electric potential around a positive charge decreases as we move away from it, as shown in Fig. 8-11.

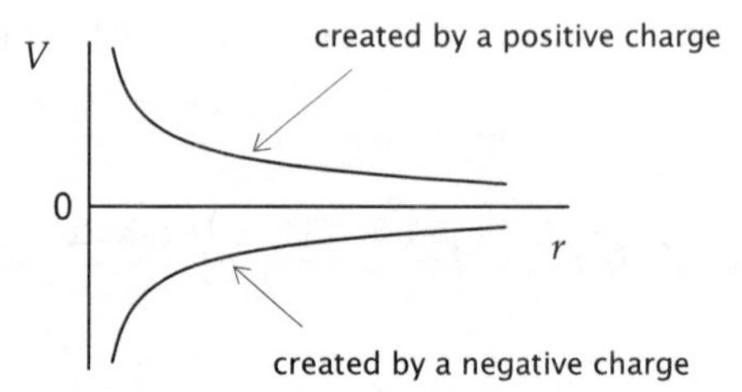

Figure 8-11: Electric potential as a function of distance from the charge

Now that electric potential is defined as in Eq. (8.7), we can express electric potential energy in terms of electric potential as

$$E_P = q\left(k\frac{Q}{r}\right) = qV \quad [\text{J} = \text{C}\cdot\text{V}] \qquad \text{... (8.8)}$$

Hence, if a 3.0-C charge is placed at a point where the electric potential is 2.5 V, the charge has electric potential energy of 7.5 J. The relationship among the units in Eq. (8.8) shows that 1 V is equivalent to 1 J/C. In fact, 1 Volt is defined as the electric potential at a point in an electric field, if 1 J of work is required to bring 1 C of charge from infinity to that point.

Now imagine putting a small test charge around a positive charge, as shown in Fig. 8-12 (a). If the test charge is positive, it will be repelled by and move away from the positive charge, and as it does so, it moves toward the region of lower electric potential. For a negative test charge, the opposite is true: It moves toward the region of higher electric potential, or toward the positive charge.

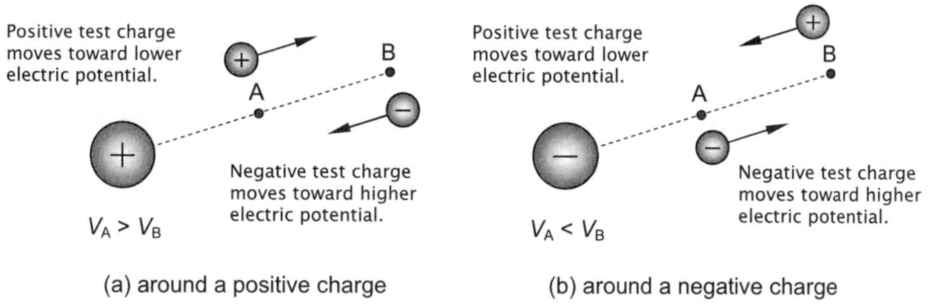

Figure 8-12: Electric potential and the motion of electric charges

Around a negative charge, a positive test charge is attracted by and moves toward the negative charge, but that does not change the fact that it still moves toward the region of lower electric potential. A negative test charge always does the opposite: It moves toward the region of higher electric potential. Therefore, we can make the following analogy: Electric potential is to a positive electric charge as altitude is to a mass. That is, just as a mass always falls from higher altitude to lower altitude, a positive electric charge always moves from higher electric potential to lower electric potential (when no external force acts on the charge.)

One benefit of using electric potential instead of electric field is that the calculation of net electric potential around a group of charges is very simple. This is so because electric potential is a scalar quantity and there is no need to consider the direction. To find the net electric potential, add the individual electric potentials created by the charges. For instance, consider the three charges in Fig. 8-6 (a) (p. 234). The net electric potential at P, the midpoint of the line segment joining the charges B and C, is the algebraic sum of the potentials created by the three charges:

Electric potential created by A: $V_A = \left(9.0\times10^9\,\text{N}\cdot\text{m}^2/\text{C}^2\right)\dfrac{-2.0\times10^{-6}\,\text{C}}{4.0\,\text{m}\cdot\sin 60°} = -5.2\times10^3\,\text{V}$

Electric potential created by B: $V_B = \left(9.0\times10^9\,\text{N}\cdot\text{m}^2/\text{C}^2\right)\dfrac{3.0\times10^{-6}\,\text{C}}{2.0\,\text{m}} = 13.5\times10^3\,\text{V}$

Electric potential created by C: $V_C = \left(9.0\times10^9\,\text{N}\cdot\text{m}^2/\text{C}^2\right)\dfrac{5.0\times10^{-6}\,\text{C}}{2.0\,\text{m}} = 22.5\times10^3\,\text{V}$

Net electric potential at P = $V_A + V_B + V_C = (-5.2+13.5+22.5)\times10^3\,\text{V} = 31\times10^3\,\text{V}$

This method can be used to find electric potential at other points as well. By connecting the points at the same electric potential, we can draw **equipotential lines**. For the charges shown in Fig. 8-8 (p. 236), the equipotential lines are drawn in dashed lines. Note that they are perpendicular to electric field lines and have no starting or ending point. Around a single charge, they form concentric circles, with the charge at the center. The electric potential increases as one moves away from a negative charge or towards a positive charge. Thus, for a pair of charges shown in Fig. 8-8 (a), the electric potential increases, in general, from left to right. The same is true for a pair of parallel plates shown in Fig. 8-8 (d).

8-6. Charges' Motion Between Parallel Plates

Between two oppositely charged parallel plates, electric field lines are equally spaced, parallel to each other, and perpendicular to the plates, as shown in Fig. 8-13. A positive charge, when placed next to the positively charged plate, will experience electric force in the direction of the electric field lines and accelerate toward the negatively charged plate, or to the right, in Fig. 8-13. According to Newton's second law of motion, the magnitude of the charge's acceleration is

The only force the charge experiences is the electric force. → $F_{net} = |q|E = ma \implies a = \dfrac{|q|E}{m}$

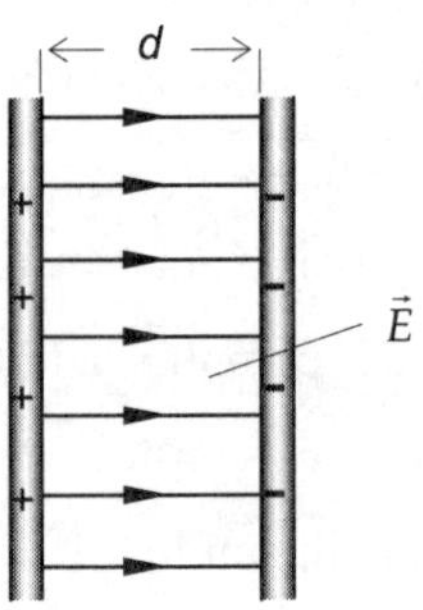

Figure 8-13: Electric field between two oppositely charged plates

Since the electric field strength is constant between the plates, the charge moves with constant acceleration, which allows us to use the kinematic equations to analyze its motion. Suppose the pair of plates shown in Fig. 8-13 is separated by a distance of 12 cm, and the electric field between the plates is 4600 N/C. If an electron is placed next to the negatively charged plate, its acceleration will be

$$a = \frac{|q|E}{m} = \frac{\left|-1.60\times10^{-19}\,\text{C}\right|\times 4600\,\text{N/C}}{9.11\times10^{-31}\,\text{kg}} = 8.1\times10^{14}\ \text{m/s}^2\text{, to the left}$$

To find the time it takes for the electron to reach the positively charged plate, use Eq. (2.7) (on p. 31):

$$\vec{d} = 0.12\,\text{m} = (0\,\text{m/s})\cdot t + \frac{1}{2}\left(8.1\times10^{14}\,\text{m/s}^2\right)\cdot t^2 \implies t = 1.7\times10^{-8}\,\text{s}$$

The electron's speed at the positively charged plate can be found by using Eq. (2.8):

$$v_f^2 = (0\,\text{m/s})^2 + 2\left(8.1\times10^{14}\,\text{m/s}^2\right)(0.12\,\text{m}) \implies v_f = 1.4\times10^{7}\,\text{m/s}$$

An alternative way to analyze a charge's motion between parallel plates is to use the work-energy principle, which for an electric charge, can be expressed as

$$W_{NC} = \Delta E_P + \Delta E_K = \underbrace{\Delta(mgh)}_{\text{change in gravitational potential energy}} + \underbrace{\Delta(qV)}_{\text{change in electric potential energy}} + \Delta\left(\frac{1}{2}mv^2\right) \qquad \ldots (8.9)$$

We usually ignore the change in gravitational potential energy because it is very small compared with the other terms in the equation. Then Eq. (8.9) is simplified to

$$\boxed{W_{NC} = \Delta(qV) + \Delta\left(\frac{1}{2}mv^2\right) = q\,\Delta V + \frac{1}{2}m\,\Delta v^2} \qquad \ldots (8.10)$$

Using Eq. (8.10), we can draw a relationship between the electric field strength and the potential difference between two oppositely charged plates. Once again consider the pair of plates shown in Fig. 8-13, where the distance of separation is d and the electric field strength is E. Then the amount of work required to move a positive charge from the negatively charged plate to the positively charged plate is

$$W_{NC} = F_A\,d\cos\theta = |q|\,E\,d\cos 0^\circ = |q|\,E\,d \qquad \ldots (8.11)$$

Note that the magnitude of the applied force must be that of the electric force on the charge in order to keep the charge moving at constant velocity. Substitute Eq. (8.11) into Eq. (8.10):

$$|q|\,E\,d = q\,\Delta V + \frac{1}{2}m\,\Delta v^2 \qquad \ldots (8.12)$$

Since the charge has been moving at constant velocity, there is no change in its kinetic energy. Thus, the potential difference between the plates is the product of the electric field strength and the distance of separation:

$$|\Delta V| = E\,d \qquad \dots (8.13)$$

Therefore, if two plates are separated by 12 cm and the electric field strength between them is 4600 N/C, the potential difference between the plates is

$$|\Delta V| = (4600\,\text{N/C})\cdot(0.12\,\text{m}) = 552\,\text{V}$$

This means that the electric potential at the positively charged plate is 552 V higher than that at the negatively charged plate. If we let go of an electron at the negatively charged plate, it will accelerate toward the other plate without needing any work done on it. We can find its speed at the positively charge plate by applying the work-energy principle, or Eq. (8.10):

$$\cancel{W_{NC}} = (-1.60\times10^{-19}\,\text{C})(552\,\text{V}) + \frac{1}{2}(9.11\times10^{-31}\,\text{kg})(v_f^2 - 0^2)$$

In the above equation, ΔV is positive because the electron moves toward higher potential. Solving the above equation for the final speed, we find $v_f = 1.4\times10^7\,\text{m/s}$. Note that the answer is the same as the one obtained using the kinematic equations (on the previous page.)

When using Eq. (8.10) to analyze a charge's motion, be careful of the sign of ΔV. When a positive charge moves in space solely under the influence of electric force, the change in potential is always negative since the potential decreases in the direction of its motion. However, for a negative charge, the change is always positive.

8-7. Cathode Ray Tube

A **cathode ray tube** (CRT) is a device in which a beam of electrons (called **cathode rays**) is directed to various parts of a screen to produce an image. It is used in TV's, computer monitors, oscilloscopes, etc. In it, electrons are emitted by the heated cathode (or negative electrode) and are attracted towards the anode (or positive electrode) (see Fig. 8-14). The potential difference between the cathode and the anode is called the **accelerating voltage**, and the greater the accelerating voltage, the greater the speed of the electrons as they pass through a small hole in the anode.

The work-energy principle in Eq. (8.10), when applied to an electron between the cathode and the anode, is written as

$$0 = (-e)\Delta V_A + \frac{1}{2} m_e \left(v_f^2 - 0^2\right)$$

(Annotations: "There is no work done by non-conservative forces." → 0; "the accelerating voltage" → ΔV_A; "electron's speed at the anode" → v_f; "Electrons' speed near the cathode is almost zero." → 0^2)

where m_e is the electron's mass. Then the speed of the electron at the anode can be expressed as a function of the accelerating voltage as

$$v_{\text{at anode}} = \sqrt{\frac{2 e \Delta V_A}{m_e}} \qquad \dots (8.14)$$

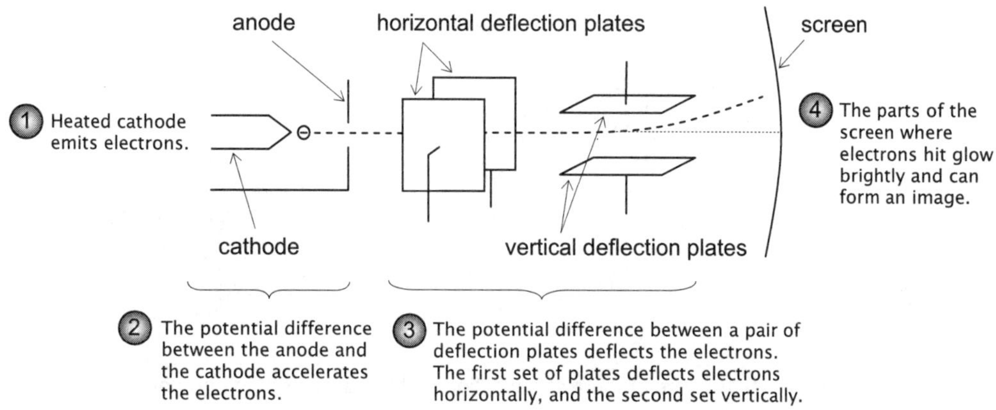

Figure 8-14: Electrons' motion in a cathode ray tube

Thus, if the accelerating voltage is 110 V, the speed of the electron at the anode is

$$v_{\text{at anode}} = \sqrt{\frac{2\left(1.60\times10^{-19}\ \text{C}\right)\left(110\ \text{V}\right)}{9.11\times10^{-31}\ \text{kg}}} = 6.2\times10^{6}\ \text{m/s}$$

Past the anode, the electrons travel at constant velocity until they pass between two pairs of deflection plates, one for horizontal deflection and the other for vertical deflection. While passing between each pair of plates, the electrons accelerate toward the positively charged plate, as shown in Fig. 8-15. To deflect the electrons downward in the figure, we can switch the polarities of the plates.

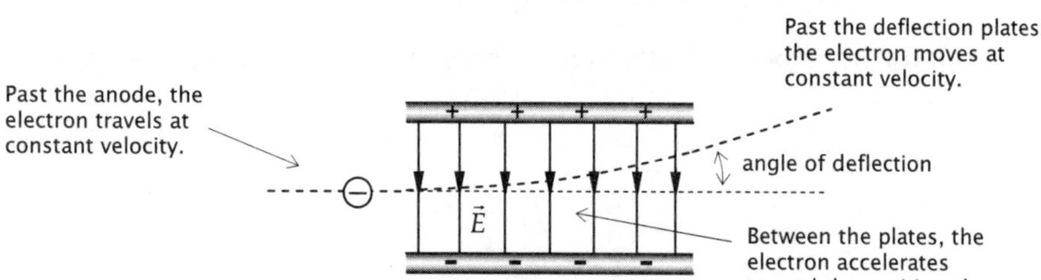

Figure 8-15: Electrons' motion between a pair of deflection plates

The potential difference between the deflection plates is called the **deflecting voltage**, and, obviously, the greater the deflecting voltage, the greater the angle of deflection in Fig. 8-15. However, the greater the accelerating voltage, the faster the electrons' speed before the plates, which reduces the angle of deflection because the electrons spend less time between the deflection plates. In fact, the amount of deflection of the electrons can be expressed as

deflecting voltage between horizontal deflection plates

horizontal deflection

$$D_H \propto \frac{(\Delta V_D)_H}{\Delta V_A} \quad \text{and} \quad D_V \propto \frac{(\Delta V_D)_V}{\Delta V_A} \qquad \ldots (8.15)$$

accelerating voltage

for vertical deflection

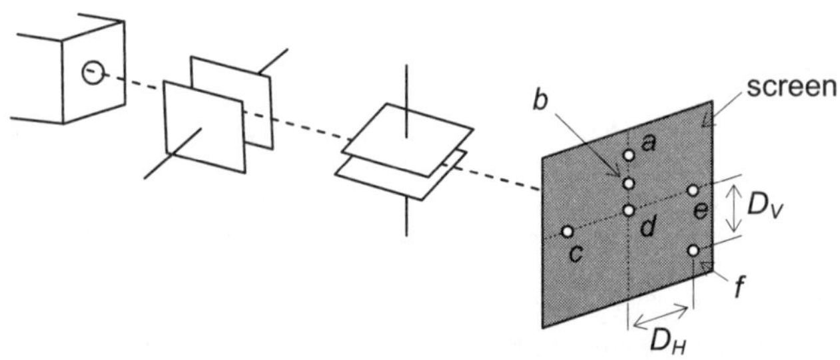

Figure 8-16: The amount of deflection and the accelerating and deflecting voltages

Suppose the deflecting voltage in each pair of deflection plates is zero. Then the electrons will travel straight and hit the screen at the center (point *d* in Fig. 8-16). Now, if we increase the voltage between the vertical deflection plates, the glowing dot will move along the vertical line on the screen, from *d* to *b* to *a* in the figure. Turning the vertical deflecting voltage back to zero returns the dot to the center. Increasing the horizontal deflecting voltage this time, we can move the dot from *d* to *c*. Switching the polarity of the horizontal deflection plates reverses the direction of deflection, and the dot jumps from *c* to *e*. Finally, by increasing the vertical deflecting voltage (but using the opposite polarity for the vertical deflection plates), we can lower the position of the dot vertically from *e* to *f*. (Note that we can also control the amount of deflection by changing the accelerating voltage.)

PHYSICS 12 ANALYZED: ELECTRIC POTENTIAL

8-4. Electric Potential Energy

- Electric potential energy is defined as $E_P = k\frac{qQ}{r}$.
- Electric potential energy between two like charges increases as they approach each other. That between two unlike charges increases as they are separated from each other.

8-5. Electric Potential

Potential is to potential energy as field is to force.

- Electric potential is the electric potential energy per unit charge: $V = \frac{E_P}{q} = k\frac{Q}{r}$
- Electric potential increases as one moves toward a positive charge or away from a negative charge.
- Electric potential is to a positive electric charge as altitude is to a mass. A positive charge always moves from high potential to low potential (in the absence of any external force).
- Equipotential lines are drawn by connecting points at the same potential. They are perpendicular to electric field lines everywhere.

8-6. Charges' Motion Between Parallel Plates

- When analyzing a charge's motion between two parallel plates...
  1. you may use the kinematic equations with the charge's acceleration equal to

$$a = \frac{|q|E}{m}$$

  2. you may also use the work-energy principle, which can be written as

$$W_{NC} = q\,\Delta V + \frac{1}{2}m\,\Delta v^2$$

- The potential difference (ΔV) between two parallel plates is the product of the electric field strength (E) and the distance of separation (d): $|\Delta V| = E\,d$

8-7. Cathode Ray Tube

- Cathode ray tube is a device used to create images by directing a beam of electrons to various parts of a screen.
- The deflection of the beam is controlled by the accelerating voltage between the cathode and the anode and the deflecting voltage between each pair of deflection plates.
- The amount of deflection is directly proportional to the deflecting voltage and inversely proportional to the accelerating voltage.

EXERCISES ON CHAPTER 8: ELECTRIC POTENTIAL

Exercise 8-16

A 15μC charge is located 5.8 m from a -48 μC charge, as shown. Find the electric potential at point P.

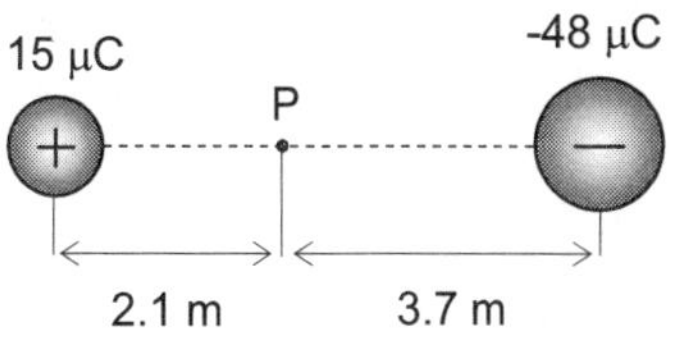

Exercise 8-17

A 17 μC charge is initially 12 m away from a stationary 88 μC charge. Find the minimum amount of work required to bring the 17 μC charge to a point 1.2 m from the stationary charge.

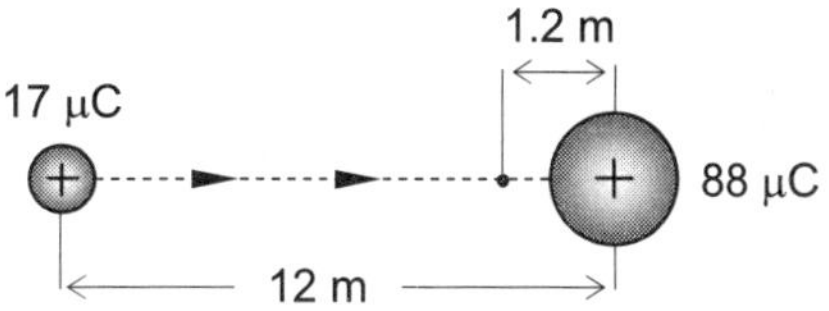

Exercise 8-18

A charged object of mass 6.4×10^{-24} kg is accelerated through a potential difference of 350 V. If its speed increases from zero to 8.5×10^{6} m/s as a result, what is the magnitude of the charge on the object?

Exercise 8-19

The electric potential at point P is zero. Find the magnitude and polarity of the unknown charge Q.

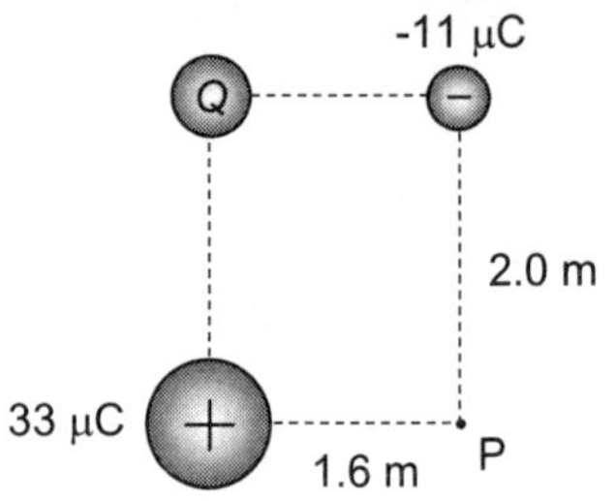

Exercise 8-20

Two oppositely charged parallel plates are separated by 5.3 cm, and the electric field strength between them is 2400 N/C. If point P is located midway between the plates, what is the potential difference between point P and one of the plates?

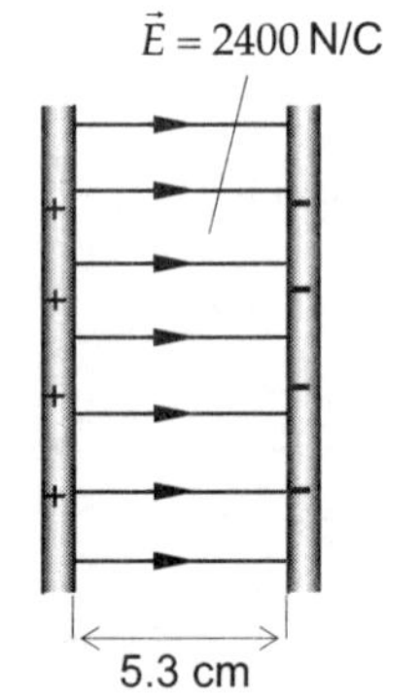

Exercise 8-21

An electron is initially located 2.5 m from a fixed -1.7 μC charge, as shown. Find (a) the change in electric potential energy as the electron moves to a point 8.5 m from the fixed charge and (b) the electron's speed when it is infinitely far from the fixed charge. (The electron is initially at rest.)

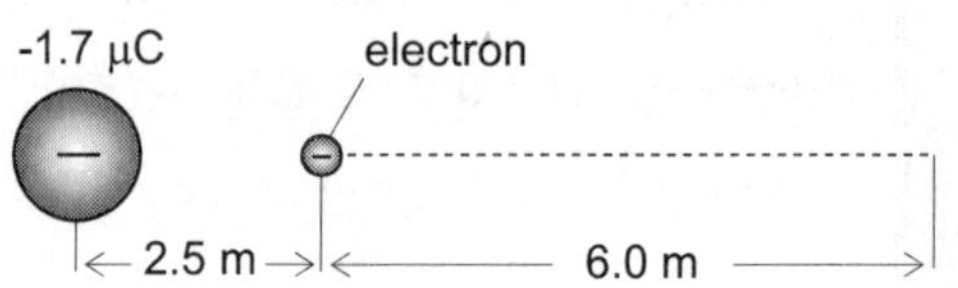

Exercise 8-22

An alpha particle initially at rest is accelerated between two oppositely charged parallel plates as shown. If the potential difference between the plates is 630 V, find the maximum speed of the particle.

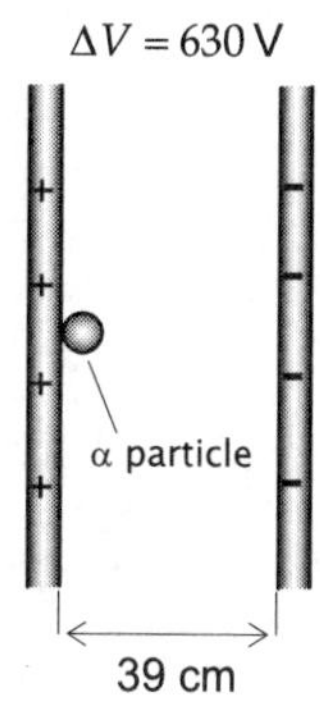

Exercise 8-23

An unknown charge q is initially 2.8 m away from a stationary 67 μC charge. If 0.054 J of work is required to move the charge to point P, find the magnitude and polarity of the charge.

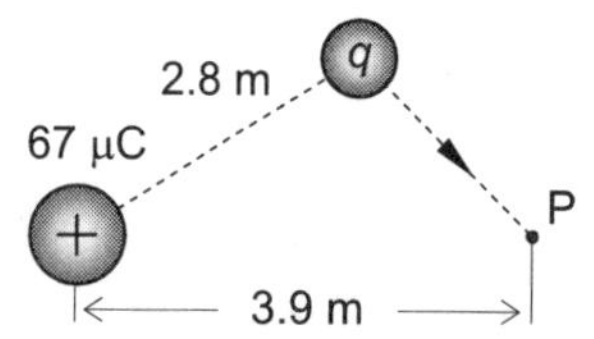

Exercise 8-24

A proton moving at 5.4×10^5 m/s passes through a small hole in the left plate and enters an electric field of strength 3800 N/C as shown. The two plates are separated by a distance of 27 cm.

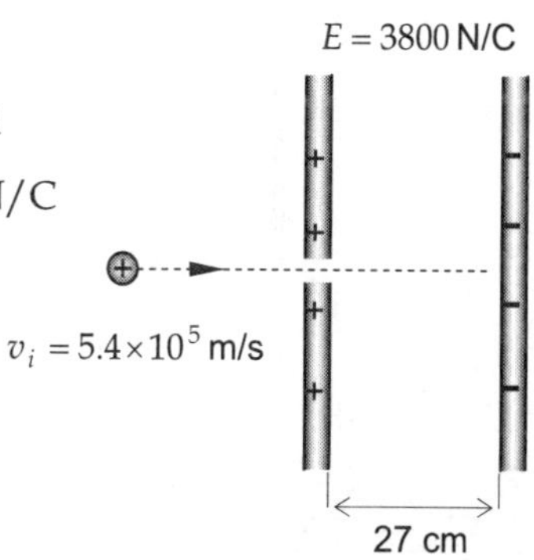

a) Find the impact speed of the proton on the right plate.

b) If the polarities of the plates are reversed, will the proton be able to reach the right plate? If so, what is its impact speed?

Exercise 8-25

A 2.3 μC charge and a 3.2 μC charge are positioned as shown. Find the amount of work required to move a proton from position P to position R.

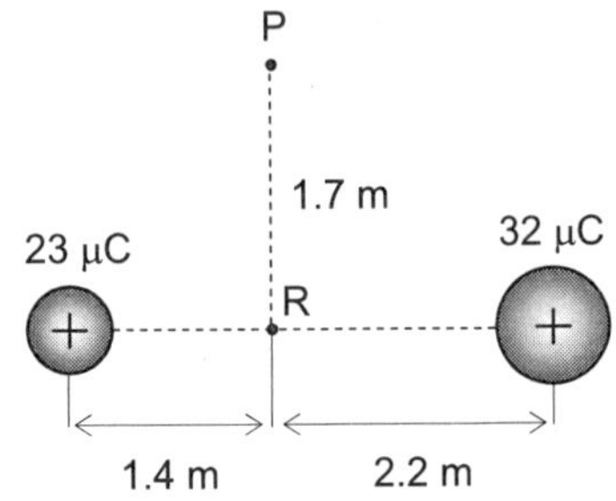

Exercise 8-26

An electron passing between a pair of vertical deflection plates experiences an upward electric force of 1.7×10^{-15} N. If the plates are separated by 8.5 cm, what is the potential difference between the plates?

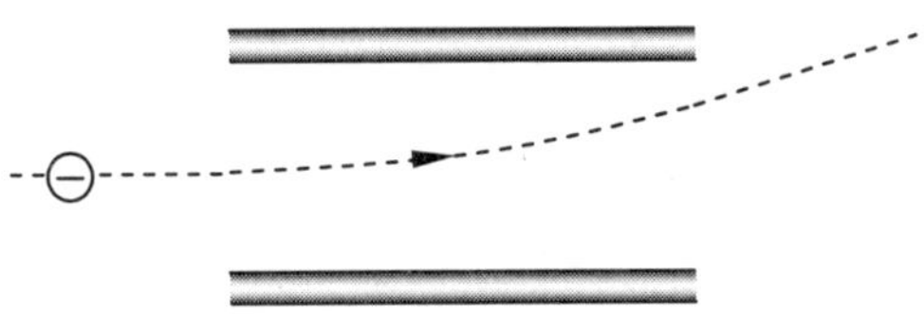

Exercise 8-27

An alpha particle initially moving with 8.5×10^{-16} J of kinetic energy is accelerated by two oppositely charged parallel plates as shown. If the electric field strength between the plates is 750 N/C, find the kinetic energy of the particle after it emerges from the parallel plates.

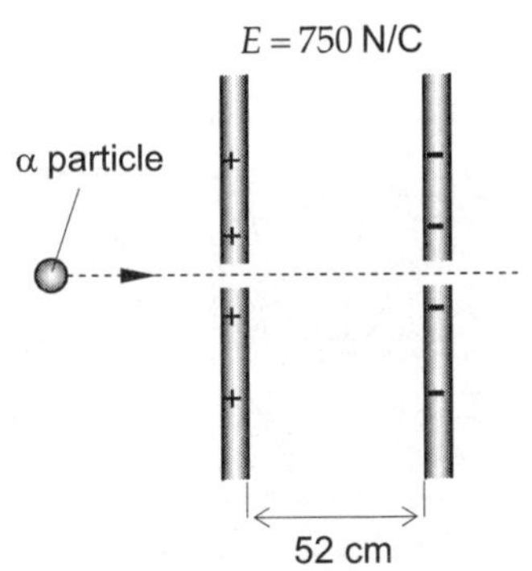

Exercise 8-28

An electron accelerated through a potential difference of 120 V enters an electric field between a pair of deflection plates as shown. Find (a) the electron's speed before entering the electric field and (b) the magnitude and direction of its acceleration within the electric field between the deflection plates.

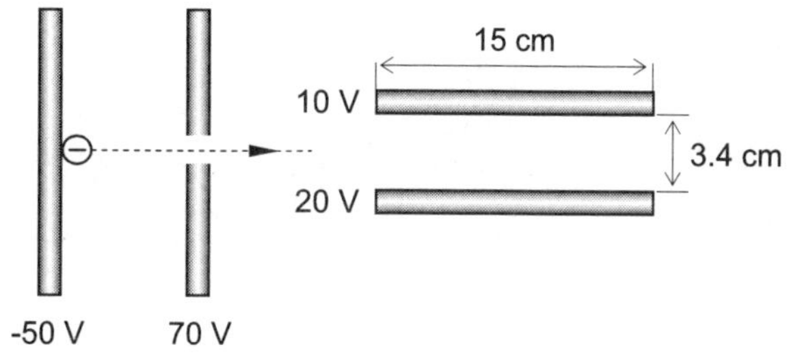

Exercise 8-29

A proton is accelerated through a series of charged parallel plates as shown. Calculate the distance between the proton and the fixed 2.5 μC charge when the proton momentarily comes to a stop.

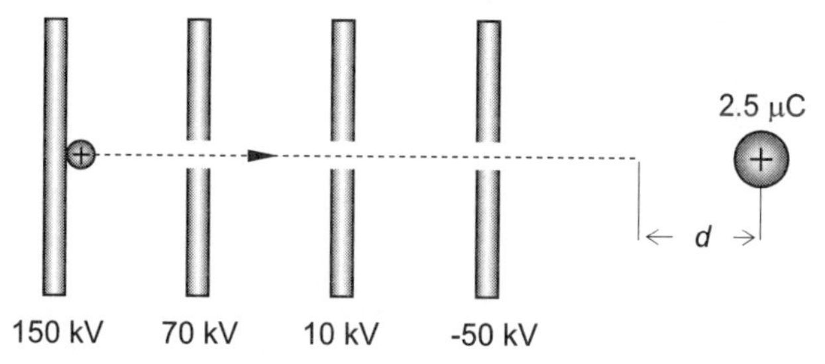

Exercise 8-30

In a CRT, the beam of electrons hits the screen at point *a* for the voltages shown below.

accelerating voltage: $\Delta V_A = 90\text{ V}$

horizontal deflecting voltage: $(\Delta V_D)_H = 75\text{ V}$

vertical deflecting voltage: $(\Delta V_D)_V = 60\text{ V}$

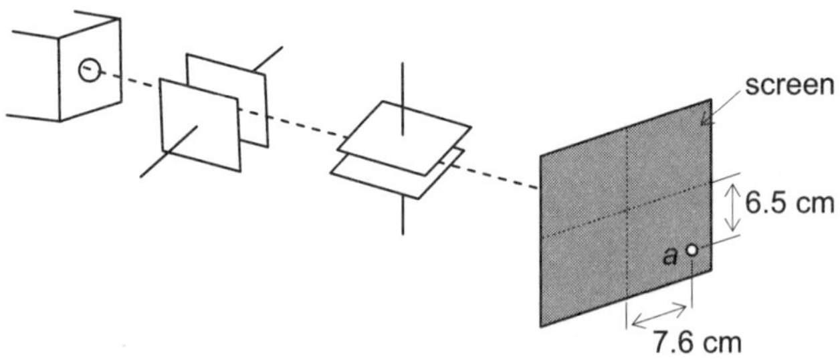

Find the horizontal and vertical deflections of the beam if

a) the vertical deflecting voltage is reduced to 30 V.

b) the accelerating voltage is doubled and the horizontal deflecting voltage is increased to 100 V.

PROBLEMS ON CHAPTER 8: ELECTROSTATICS

1 - 7. Fill in the blanks

1. The charge carried by a single proton is known as the __________ __________.

2. __________ __________ is to electric force as Newton's law of gravitation is to gravitational force.

3. Electric force per unit charge is ________ __________.

4. Drawing smooth lines so that the electric field vectors become tangent to them, we can construct a set of ________ __________ lines to visualize the electric field around an electric charge or a group of charges.

5. Electric potential energy per unit charge is ________ __________.

6. We can draw __________ lines by connecting points at the same electric potential.

7. A ________ __________ __________ is a device in which a beam of electron is directed to various parts of a screen to produce an image.

8 - 20. True/False questions

8. _____ If the net charge on an object is $+2e$, it indicates that the object has two protons and no electrons.

9. _____ A negative charge placed in an electric field heading north experiences electric force directed south.

10. _____ The electric field created by a positive charge points away from the charge.

11. _____ Electric field lines have no starting or ending points.

12. _____ One Volt is equivalent to one Newton per Coulomb.

13. _____ Electric potential energy between two unlike charges is always negative and increases as the distance between the charges increases.

14. _____ With no external force applied, protons always move toward a higher potential.

15. _____ Electric field strength is zero where the electric potential is zero.

16. _____ Equipotential lines are perpendicular to electric field lines everywhere.

17. _____ No work is required to move an electric charge along an equipotential line.

18. _____ An electric charge's acceleration is constant within the electric field between two charged parallel plates.

19. _____ If the distance between two parallel plates is increased while maintaining the electric field strength between them, the potential difference between the plates increases.

20. _____ The greater the accelerating voltage in a CRT, the greater the deflection of the cathode rays.

21 – 34. Multiple choice questions

21. When solving problems involving charges, we often do not consider the force of gravity because

a) charges do not experience the force of gravity.
b) the magnitude of the gravitational force is very small compared with that of the electric force.
c) the charges' motion becomes too complex if we consider the force of gravity.
d) the gravitational force disappears in the presence of electric force.

22. A proton, when placed at a certain point in space, experiences an electric force of 4.8×10^{-16} N and accelerates east. The proton is then replaced by an alpha particle. Which of the following statements is true?

a) The electric force on the alpha particle is 4.8×10^{-16} N.
b) The electric force on the alpha particle is 1.9×10^{-15} N.
c) The alpha particle has the same initial acceleration as the proton.
d) The alpha particle's initial acceleration is one half the proton's.

23. Two unlike charges are located as shown. At which point could the electric field be zero?

a) A b) B c) C d) D

24. An electron is initially near a fixed negative charge as shown. When released, the electron is pushed away from the fixed charge. Which of the following descriptions on the electron's motion is true?

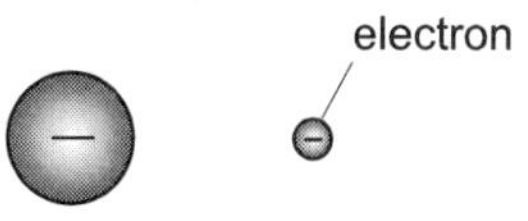

a) Its speed increases, and its acceleration is constant.
b) Its speed increases, and the magnitude of its acceleration increases.
c) Its speed increases, but the magnitude of its acceleration decreases.
d) Its speed increases at first but decreases as the magnitude of its acceleration decreases.

25. The diagram below shows the electric field lines around two unknown charges, q_1 and q_2. Which of the following could be the charges?

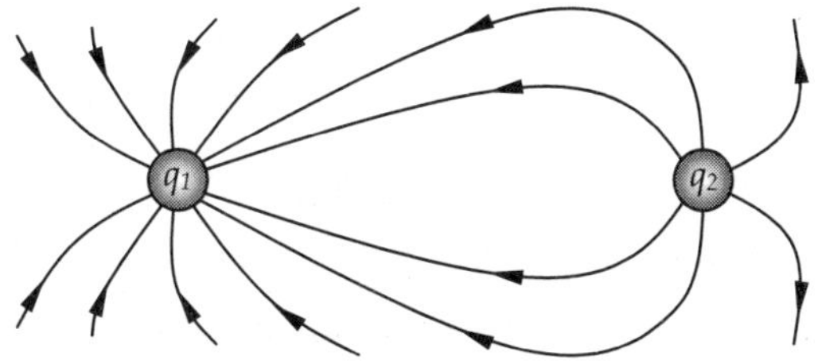

a) q_1 = 10 μC, q_2 = -20 μC b) q_1 = 20 μC, q_2 = -10 μC
c) q_1 = -10 μC, q_2 = 20 μC d) q_1 = -20 μC, q_2 = 10 μC

26. Electric field lines (solid lines) and equipotential lines (dashed lines) near a group of electric charges are shown. Find the direction of the electric field at point P.

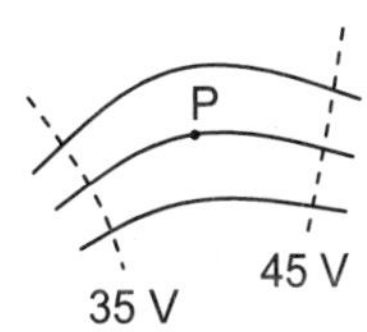

a)  b)  c)  d) 

27. An electric charge is to be moved from point A and point B, where the electric potentials are 15 V and -27 V, respectively. Which of the following paths requires the least amount of work on the electric charge?

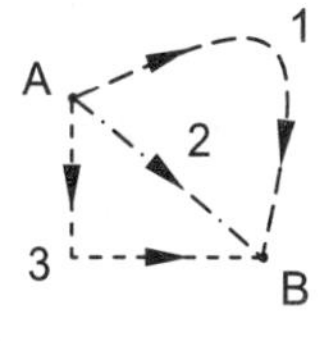

a) 1 b) 2
c) 3 d) They all require the same amount of work

28. An imaginary line is straight from *a* to *b* near a fixed negative charge, as shown. Which of the following graphs best represents the electric potential from a to b?

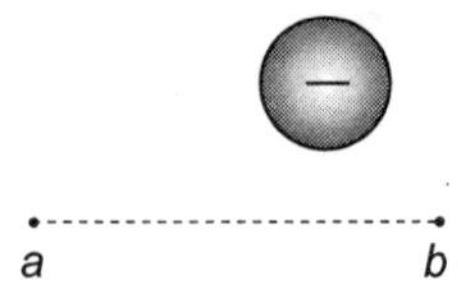

a) 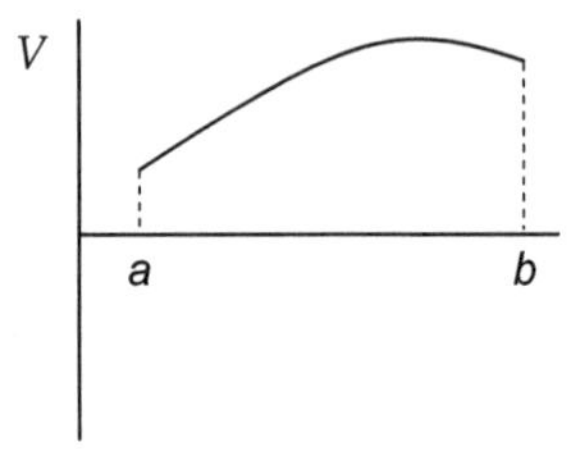

b) 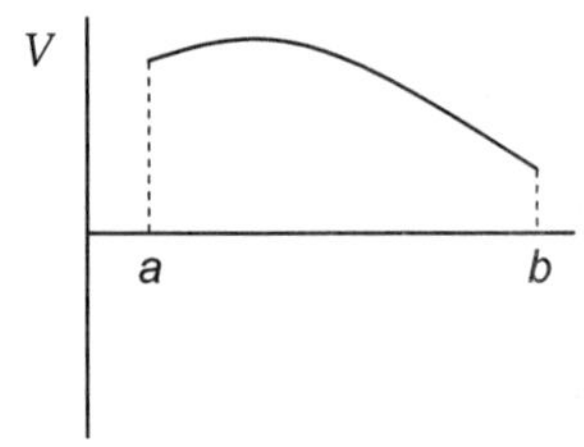

c) 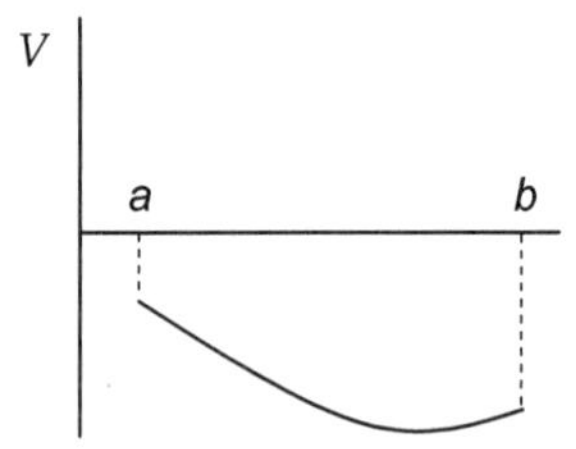

d) 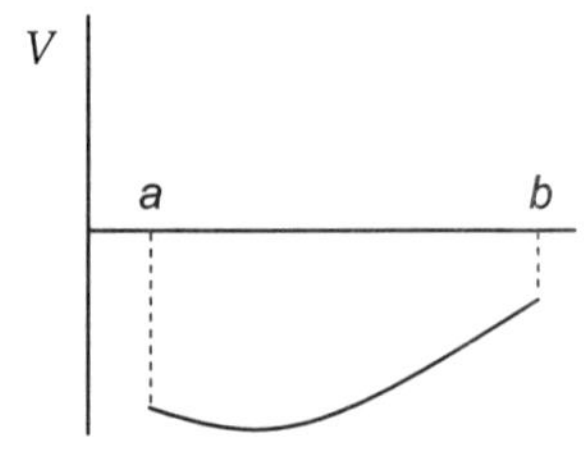

29. Consider Problem 28. Now an electron is made to follow the straight path from *a* to *b*. Which of the following graphs best represents the electric potential energy between the charges as the electron travels from *a* to *b*?

a) 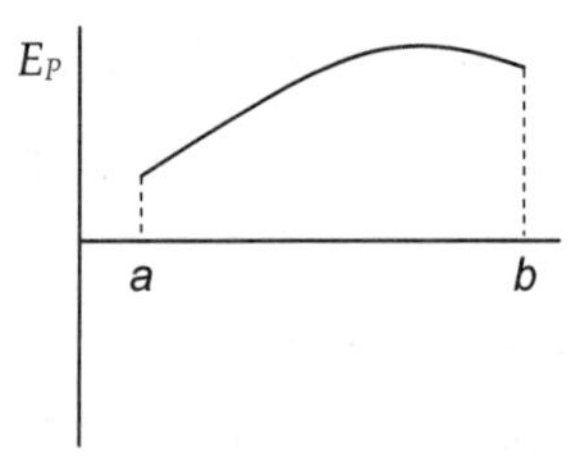

b) 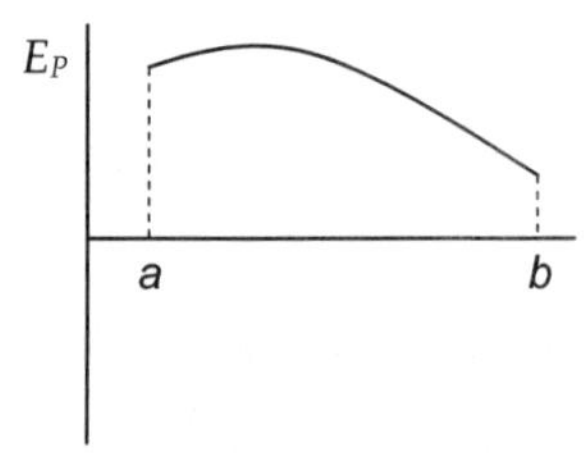

c) 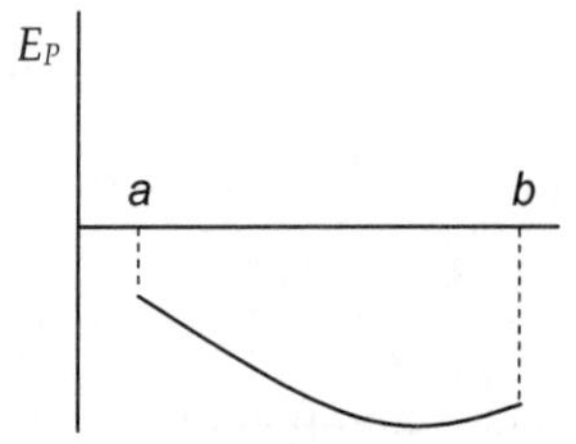

d) 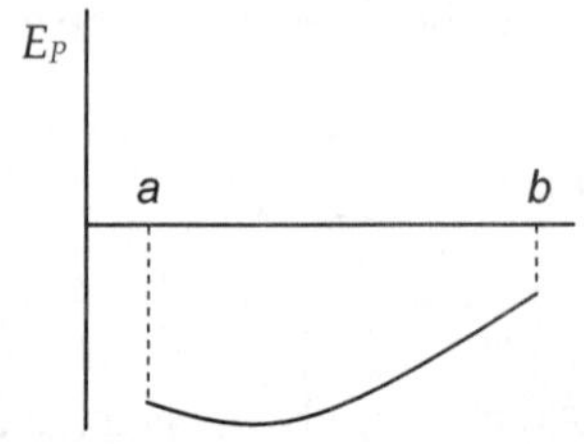

30. An electron is to be moved near a fixed, negatively charged object. Which displacement requires the most net work done on the electron?

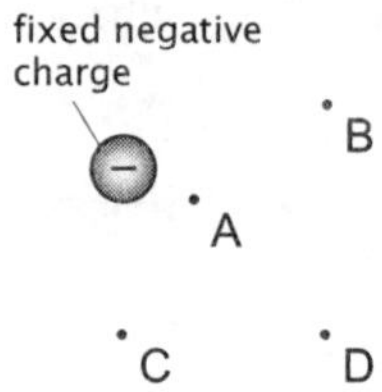

a) from D to A
b) from C to B
c) from A to C to D
d) from B to D to C to A to B

31. Which of the following is <u>false</u> for two oppositely charged parallel plates?

a) The electric field lines are perpendicular to the plates.

b) The electric potential increases from the positively charged plate to the negatively charged plate.

c) The equipotential lines are parallel to the plates.

d) The electric field strength is constant between the plates.

32. A proton and an electron are released from each of two oppositely charged parallel plates as shown. Which of the following is true?

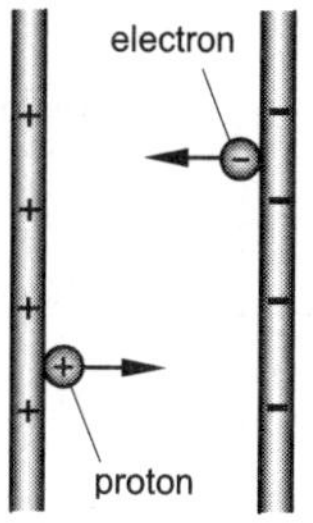

a) They both arrive at the opposite plate at the same time.

b) They both have the same speed when they arrive at the opposite plate.

c) They both have the same momentum when they arrive at the opposite plate.

d) They both have the same kinetic energy when they arrive at the opposite plate.

33. A CRT deflects a beam of electrons to the location shown in Figure A for a certain combination of accelerating and deflecting voltages. If the accelerating voltage is reduced and the polarity of the vertical deflection plates is reversed, which point in Figure B is the most likely position for the electrons?

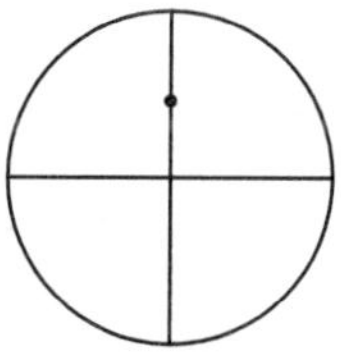

Figure A

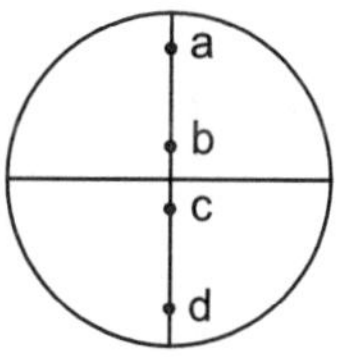

Figure B

a) a b) b c) c d) d

34. Which combination of the accelerating and deflecting voltages will cause the largest deflection of an electron beam in a CRT?

a) $\Delta V_A = 120$ V and $\Delta V_D = 20$ V b) $\Delta V_A = 180$ V and $\Delta V_D = 20$ V

c) $\Delta V_A = 120$ V and $\Delta V_D = 40$ V d) $\Delta V_A = 180$ V and $\Delta V_D = 40$ V

35 – 50. Written-response questions

35. A proton and an electron are 1.0 m away from each other. (a) Find the magnitude of the gravitational force between them. (b) Find the magnitude of the electric force between them. (c) Compare the magnitudes of the two forces.

36. Two spherical objects, initially having charges of 45 μC and -75 μC, are brought into contact for a brief moment and hung by two 63-cm cables from the same point on the ceiling. The objects repel each other and soon establish equilibrium positions as shown. If the objects have the same mass, what is each object's mass? (Hint: Each object has the same net charge after the contact.)

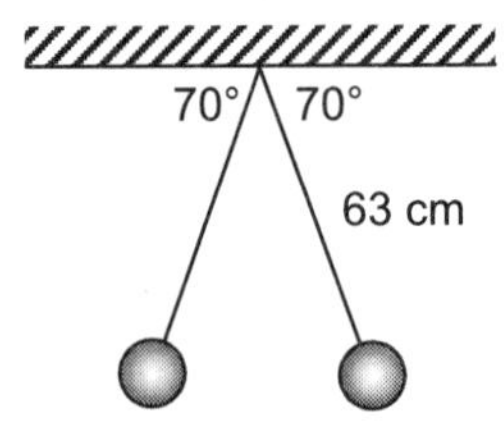

37. The direction of the net electric field at point P is 65° north of west, as shown. Find the magnitude and polarity of the charge q.

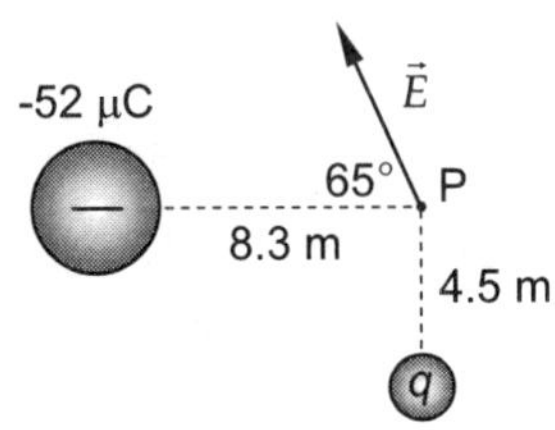

38. In the diagram shown right, the net electric field at point P is 3.6×10^{4} N/C, left. Find the magnitude and polarity of the unknown charge q.

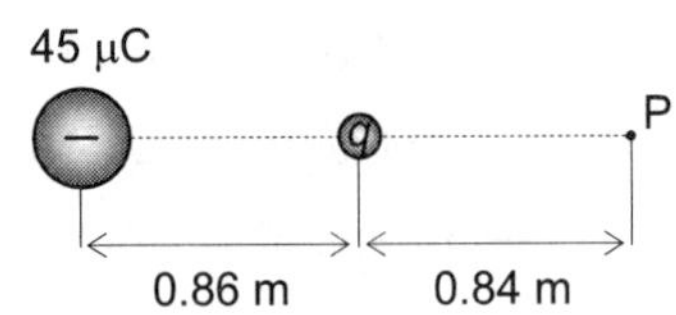

39. Find the electric field and the electric potential at point P.

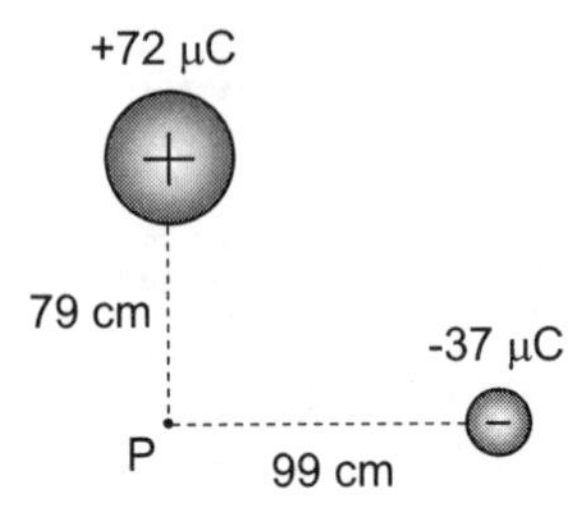

40. An electron moving at high speed passes through a small hole in the left plate and enters an electric field of strength 2600 N/C as shown. The two plates are separated by a distance of 9.5 cm. If the electron comes to a stop 4.1 cm before the negative plate, what was its initial speed?

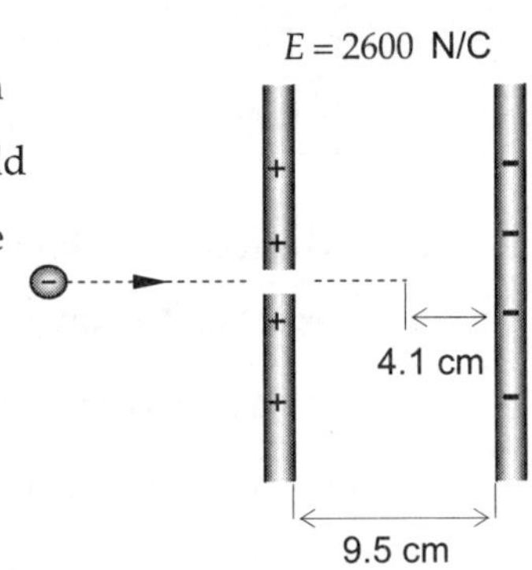

41. A charged object of mass 0.016 kg is suspended between two oppositely charged plates as shown. If the potential difference between the plates is 850 V and the distance of separation between them is 3.8 cm, what is the magnitude and polarity of the charge on the object?

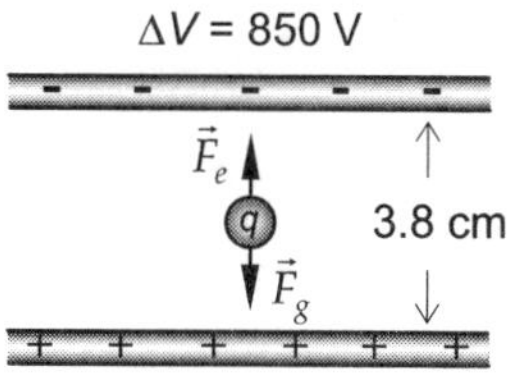

42. Two 360-g objects are initially at rest, separated by a distance of 4.9 m. The charges on the object are, respectively, 320 μC and -770 μC.

a) If one of the objects is released while the other object is held fixed, what is the speed of the moving object when it is 75 cm away from the fixed object?

b) If both objects are released simultaneously, what are their speeds when the distance between the objects is 75 cm?

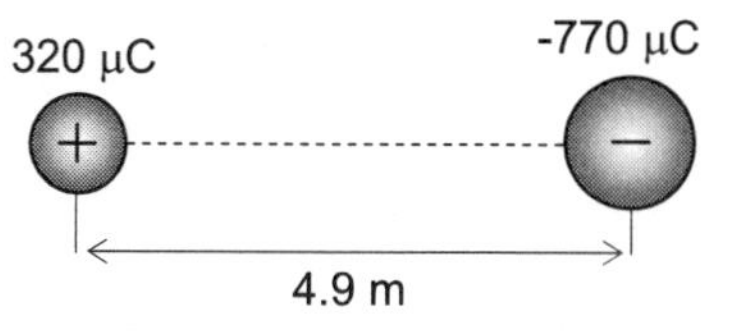

43. Determine the amount of work required to bring four charges to the vertices of a square as shown. Assume each charge was initially an infinite distance away from the others. (q = 3.5 μC, and d = 35 cm)

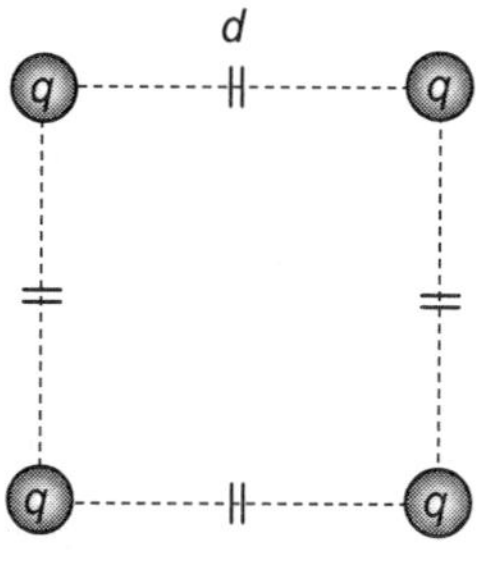

44. An electron accelerated through a potential difference of 140 V enters an electric field between a pair of deflection plates as shown. Find (a) the electron's speed before entering the electric field and (b) the magnitude and direction of its acceleration within the electric field between the deflection plates.

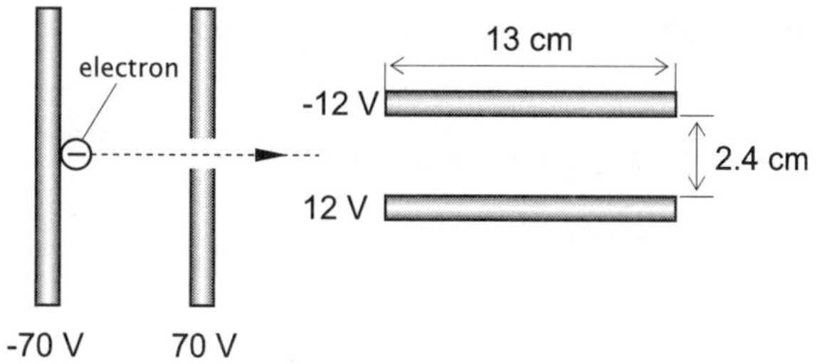

45. Consider the electron in Problem 44. (a) For how long is the electron within the electric field between the deflection plates? (b) What is the magnitude and direction of the electron's velocity as it leaves the electric field between the deflection plates? (Hint: Use the kinematic equations.)

46. An object with the net charge of -160 μC is accelerated from rest through a potential difference of 660 V as shown. After passing through the small hole, it collides and sticks to a neutral object with twice its mass, and the combined objects initially travel at 380 m/s. Find the mass of the charged object.

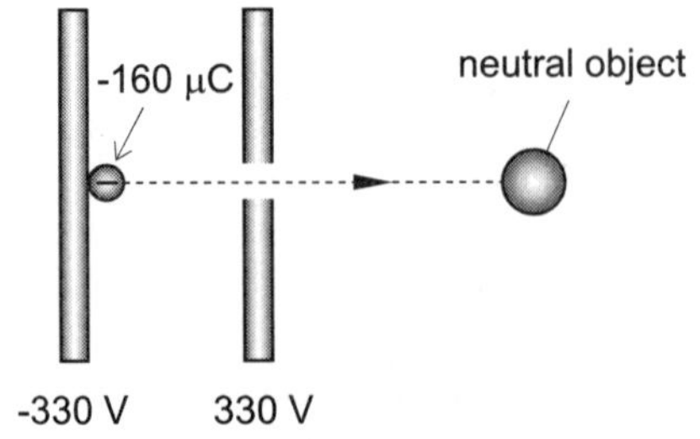

47. A 3.2-kg object with the net charge of 45 μC is dropped from 1.8 m above the ground. It is drawn towards the ground not only by the force of gravity but also by an attractive electric force exerted by the negatively charged object on the ground. Find its speed just as it lands on top of the negatively charged object. (The radius of each object is 12 cm. Assume the negatively charged object remains motionless.)

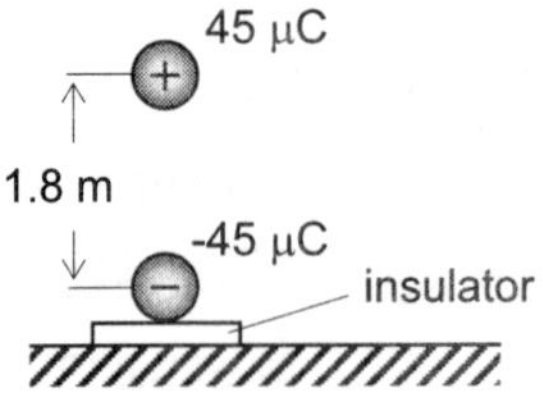

48. Two oppositely charged objects are initially at rest, 2.8 m away from each other. The mass of the negatively charged object is 1.7 kg, and that of the other object is 6.8 kg. Their radii are 16 cm and 26 cm, respectively. Released simultaneously, they are attracted toward each other and eventually collide.

(a) Find the speed of each object just before the collision.

(b) If the collision is perfectly inelastic, find their velocity after the collision.

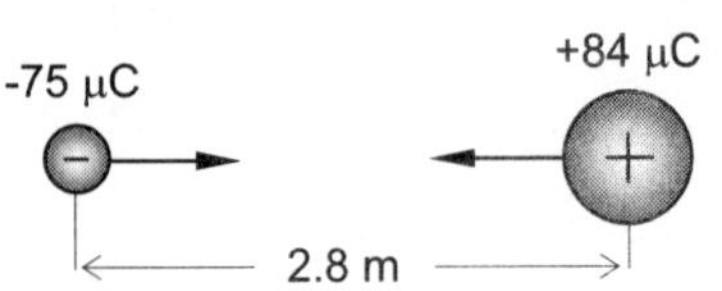

49.* An object with the net charge of 760 μC is released from 9.6 m above an object on the ground with the same net charge. The mass of each object is 25 kg.

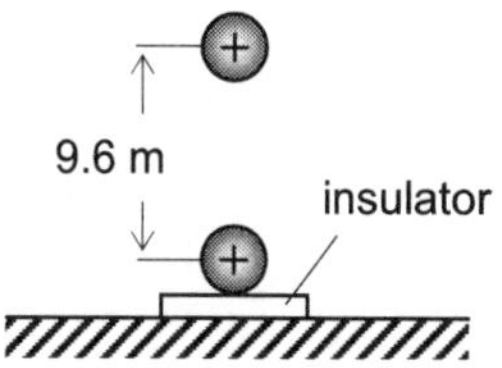

a) Does the released object rise or fall? What is its initial acceleration?

b) What is the speed of the released object when it has traveled 4.8 m?

c) Do the objects collide? If so, what is the impact speed? If not, how close do they get before the dropped object momentarily comes to a stop?

50. A 2.4-kg beam is kept in static equilibrium by an electric force between two charged objects, one of which has the net charge of 89 μC and is pulling up the right end of the beam as shown. The mass of the positively charged object is 1.7 kg, and the other object has the net charge of -130 μC. Find the distance between the two charged objects.

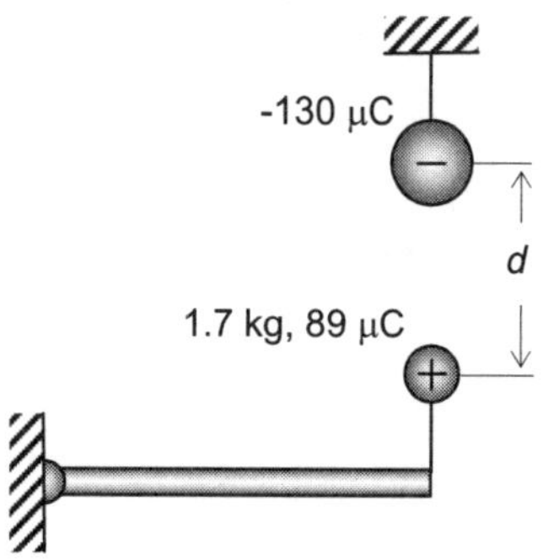

9

CHAPTER 9

DC CIRCUITS

Recall from the previous chapter that positive electric charges 'fall' toward lower electric potential (unless positive work is done on them). As they do so, their potential energy is converted to kinetic energy. This is what happens to every object falling to lower altitude because of gravity. In a hydro plant, useful work is extracted out of water's kinetic energy by resisting its motion with turbines that turn the work into electricity. Similarly, once we make electric charges flow by creating a potential difference, we can extract useful work out of them by resisting their motion.

In electric circuits within cell phones, IPods, and calculators, electric charges perform work by carrying out various functions associated with each device. Those circuits are very complex, and analyzing them is not an easy task unless you first attain a solid understanding of how an electric circuit works and how to build and analyze one. This chapter serves as an introduction to the electric circuit, and we will study the basic elements of a circuit and consider some 'relatively' simple circuits.

In this chapter, you will:

- ✓ learn about the basic elements of a circuit - wire, current, battery, and resistor
- ✓ learn about Ohm's law and Kirchhoff's rules
- ✓ learn how to calculate the equivalent resistance of a circuit
- ✓ learn about the terminal voltage of a battery and electric power
- ✓ learn about measuring devices - ammeters and voltmeters
- ✓ solve exercise problems on DC circuits
- ✓ test your knowledge of DC circuits

9-1. Building a DC Circuit

Consider a piece of wire sitting on a table. Being a conductor, the wire has many free electrons on its surface. However, the charges would not move in any one direction unless there is a potential difference between the ends of the wire. With no potential difference across the wire, the electrons are like round balls sitting on level ground; unless they are on a sloping surface, they will not move, as shown in Fig. 9-1 (b).

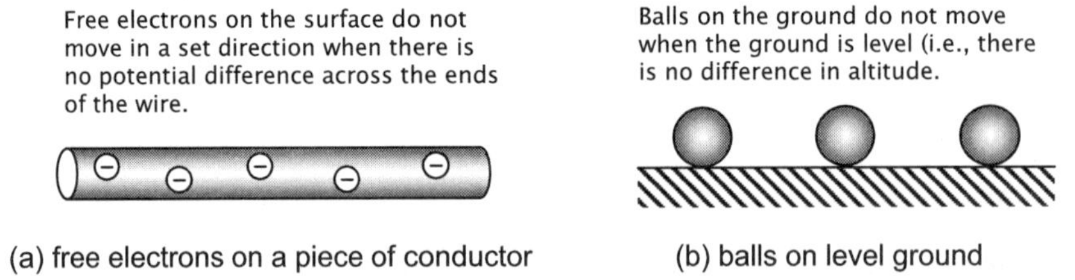

Figure 9-1: Movement of free electrons and balls

Now suppose that one end of the wire is connected to a positive plate and the other end to a negative one, as shown in Fig. 9-2 (a). This creates a potential difference across the wire, and the electrons are drawn towards the positive plate. Such a flow of electric charges is called **current**, defined as the amount of charge transferred per unit time:

The SI unit for electric current is the Ampere.

$$I = \frac{Q}{t} \quad [C/s = A] \qquad \ldots (9.1)$$

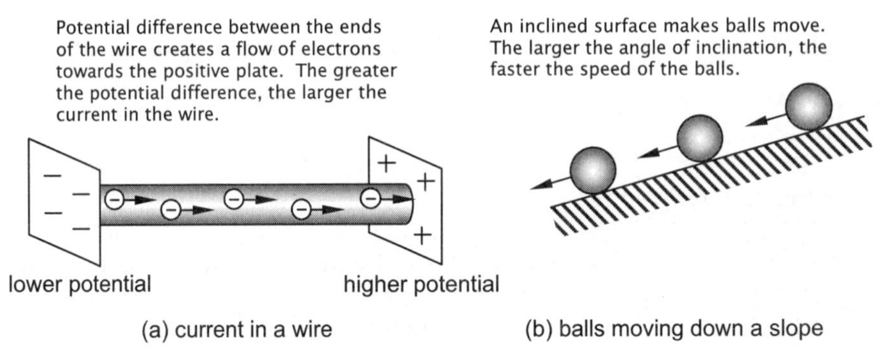

Figure 9-2: Movement of free electrons and balls

As more and more electrons arrive at the positive plate in Fig. 9-2 (a), the potential difference across the wire is reduced, and the current decreases over time, eventually becoming zero when there is no potential difference between the plates. So, in order to sustain a constant flow of electric charges in the wire, we must use a source of electrical energy (that is, a **battery**), which keeps the potential difference between the ends of the wire constant by transforming chemical energy into electric energy, as shown in Fig. 9-3.

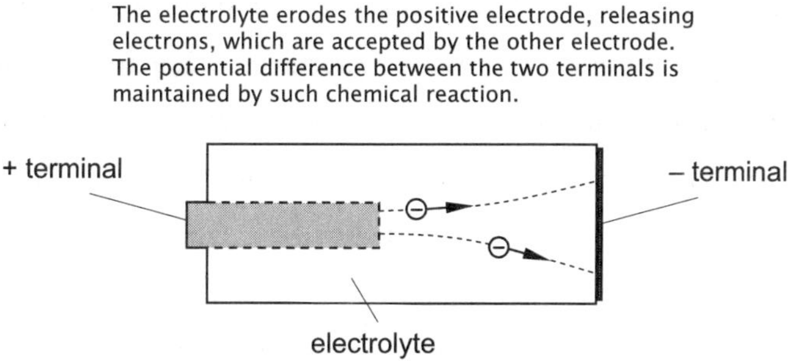

Figure 9-3: Diagram of a battery

A battery is represented by the symbol —|⊢ in a circuit, where the longer line indicates the positive terminal and the shorter line the negative terminal. By bending a piece of wire and connecting one end to each terminal of a battery, we can construct the simplest circuit, whose schematic diagram is shown in Fig. 9-4.

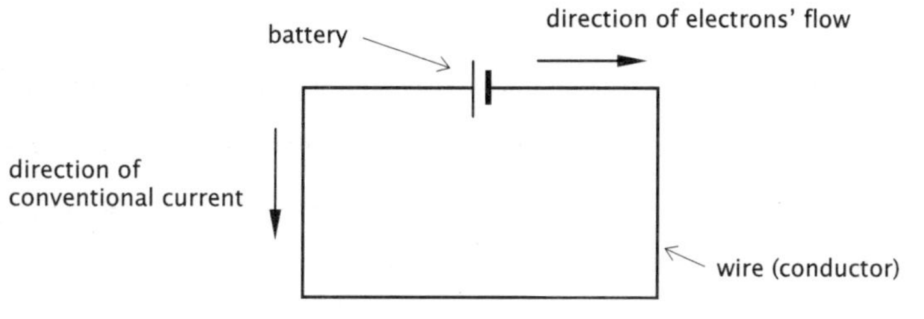

Figure 9-4: Diagram of a circuit consisting of a battery and a piece of wire

Although electric current is a scalar quantity, we often need to find the direction in which it flows. While it is electrons that actually move in a conducting wire, from the lower to the higher potential, the direction of the **conventional current** is defined to be the opposite to the direction in which electrons move. For example, in the circuit shown in Fig. 9-4, the electrons flow in a clockwise direction as they move along the wire and head toward the positive terminal. According to the definition, we then say the current in that circuit flows counterclockwise. In fact, in a circuit with a single battery, the direction of the conventional current is always from the battery's positive terminal toward the negative terminal. Such electric circuits, in which the current flows in only one direction, are called DC (**direct current**) circuits. (The other type of circuit is called the AC (**alternating current**) circuit.)

While the use of a battery allows us to generate a steady current in a circuit, each battery eventually reaches the end of its lifecycle; the chemical reaction described in Fig. 9-3 cannot go on forever because the positive electrode has only a finite number of electrons to give up. If the current in a circuit is too high, the battery will be drained quickly, and its life will be shortened. Therefore, we need something to control the amount of current drawn from a battery, and **resistors** (or loads) are used for that purpose. A resistor is represented by the symbol —\/\/— in a circuit. In Fig. 9-5, a resistor has been added to the circuit shown in Fig. 9-4.

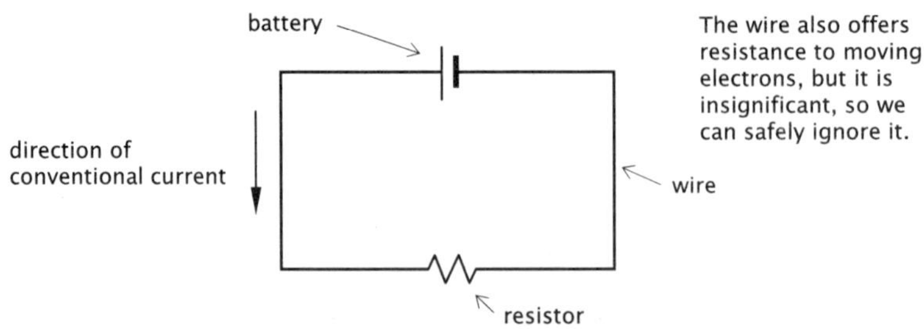

Figure 9-5: Diagram of a circuit consisting of a battery, wire, and a resistor

Resistors, as the name suggests, resist the motion of electrons. So, when electrons pass through a resistor, they slow down, and since the circuit forms a loop, all the electrons in the circuit move at a reduced speed. This is how a resistor reduces the amount of current in a circuit. As you can imagine, the greater the resistance within a circuit, the smaller the current.

9-2. Ohm's Law

Having learned about the basic components of a circuit, we will soon be considering various DC circuits and analyzing them. Before that, we first need to learn Ohm's law. It states that the potential difference across a device is equal to the product of the current flowing through it and its resistance. Applied to a resistor, **Ohm's law** can be written as

Resistance is measured in Ohms [Ω].

$$\left|\Delta V_{\text{across resistor}}\right| = I_{\text{through resistor}} \times R_{\text{of resistor}} \quad [\text{V} = \text{A}\cdot\Omega] \qquad \dots (9.2)$$

The resistance of a resistor is measured in Ohms [Ω], which is equivalent to Volts per Ampere.

Suppose a 5.0-A current is flowing through a 3.0-Ω resistor as shown in Fig. 9-6. The potential difference across that resistor is

15 J is used in the resistor when 1 C of charge flows through.

$$\left|\Delta V_{\text{across } 3.0-\Omega \text{ resistor}}\right| = (5.0\,\text{A})(3.0\,\Omega) = 15\,\text{V} = 15\,\text{J/C}$$

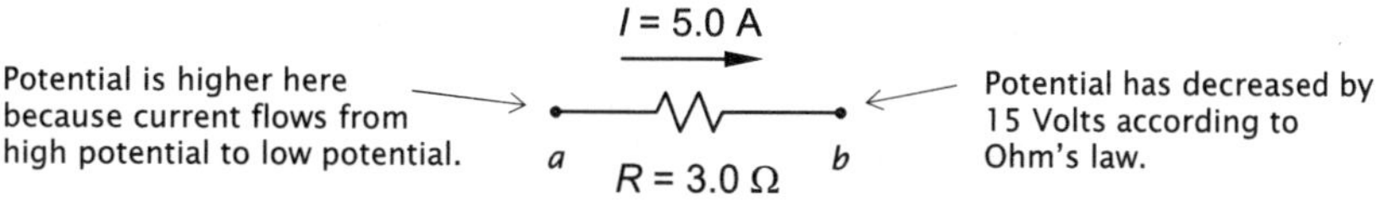

Figure 9-6: Potential drop across a resistor

Electric potential always decreases in the direction of the conventional current. So the potential at point *b* is 15 V lower than that at point *a* in Fig. 9-6. Had the current been flowing in the opposite direction, the potential at point *b* would have been higher than the potential at point *a* by 15 V.

Now suppose a 4.0-A current flows through two resistors connected one after the other, or in series, as shown in Fig. 9-7. Then the potential drops by 6.0 Volts across the 1.5-Ω resistor and by 8.0 Volts across the 2.0-Ω resistor. In total, the potential decreases by 14 Volts across the pair of resistors.

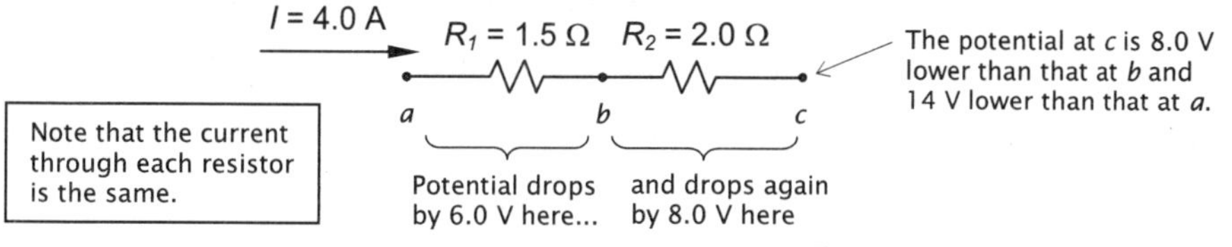

Figure 9-7: Potential drop across a pair of resistors in series

9-3. Kirchhoff's Rules

Kirchhoff's rules are the tools we can use to analyze a circuit systematically. The first rule, called the **junction rule**, states that the sum of the currents leaving a junction equals the sum of the currents entering it. That is,

This rule is based on the law of conservation of charge.

$$\sum I_{\text{into junction}} = \sum I_{\text{out of junction}} \quad \text{... (9.3)}$$

A **junction**, also called a node, is a point where three or more wires meet. The circuit shown in Fig. 9-8 has two junctions, *a* and *b*. Then at junction *a*, the current into it, I_0, must equal the sum of the currents leaving it, I_1 and I_2:

$$I_0 = I_1 + I_2$$

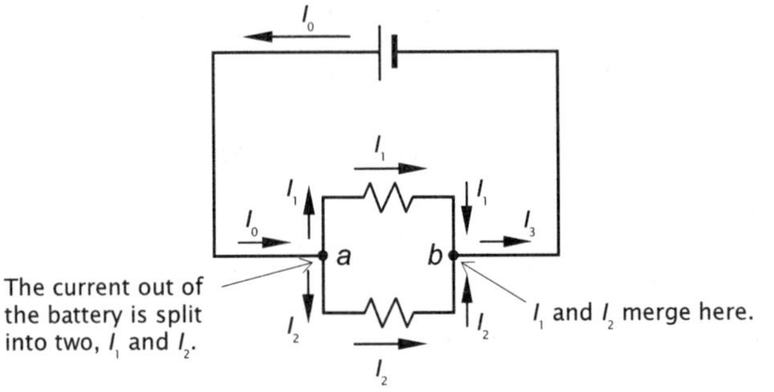

Figure 9-8: Kirchhoff's junction rule

At junction *b*, we have

$$I_1 + I_2 = I_3$$

Combining these two equations, we can see that the current into junction *a* is equal to the current out of junction *b*. (That is, $I_0 = I_3$.) In fact, the current at any point past junction *b* will be I_0 because the current does not change unless there is a junction. In other words, without a junction, the current in the wire stays constant.

To understand this rule better, think of electric current as a river. When a river branches off into several smaller streams, the amount of water entering that diverging point must equal the sum of the amount leaving it. Otherwise the water level at the diverging point will rise or fall over time, which makes no sense. Later, when those streams converge again, the amount of water leaving the point of convergence will be the same as that entering it.

Kirchhoff's second rule is called the **loop rule**. It states that the sum of the changes in electric potential around a loop is zero. That is,

This rule is based on the law of conservation of energy. →

$$\sum \Delta V_{\text{around a loop}} = 0 \qquad \ldots (9.4)$$

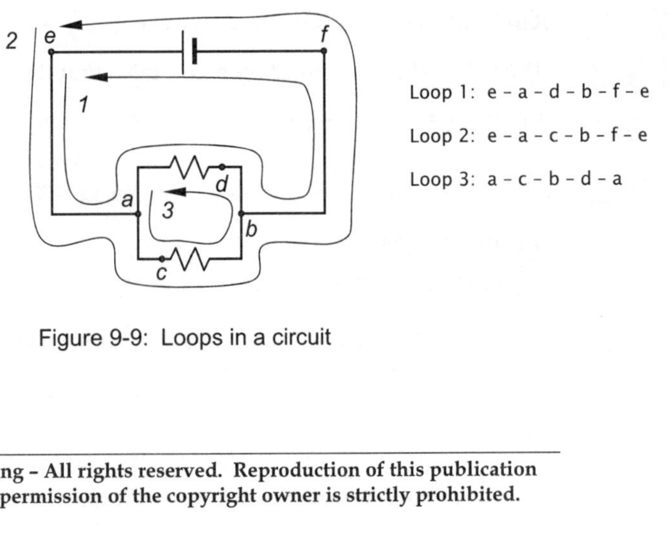

Figure 9-9: Loops in a circuit

A **loop** is any closed path in a circuit. The circuit shown in Fig. 9-9 has three loops, labeled 1, 2, and 3. You may start a loop at any point in a circuit and complete a loop in a clockwise or a counterclockwise direction, but you may not visit the same point more than once in a loop.

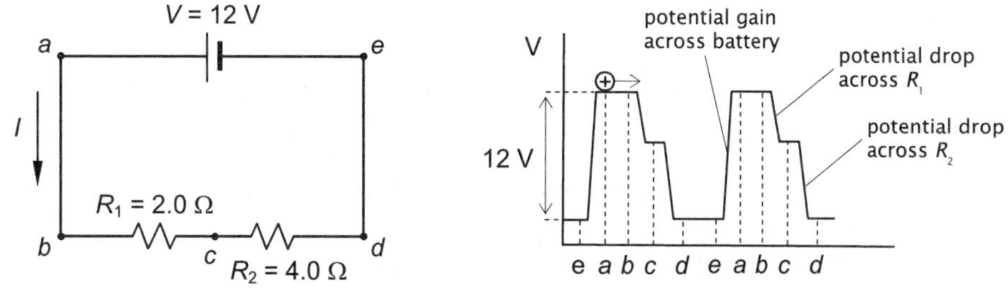

(a) a circuit with two resistors in series (b) potential along the loop

Figure 9-10: Applying the loop rule to a simple circuit

In the circuit shown in Fig. 9-10 (a), a 12-V battery is connected to two resistors in series. Usually the voltage of a battery and the resistance of resistors are known when they are used in a circuit. The amount of current in the circuit, however, depends on how we arrange those elements, and thus we are usually interested in finding how much current flows in a circuit. There is only one loop in the circuit. If we start from point *a* and follow the current in a counterclockwise direction, Eq. (9.4) becomes

$$\underbrace{(-IR_1)}_{\text{potential drop across } R_1}+\underbrace{(-IR_2)}_{\text{potential drop across } R_2}+\overbrace{V}^{\text{potential gain across the battery}}=0 \qquad \text{... (9.5)}$$

Note that, as we cross the battery from its negative terminal to its positive one, the electric potential increases by the battery's voltage, V. Substituting the known values in the above equation, we can find the current:

$$-I\cdot(2.0\Omega)-I\cdot(4.0\Omega)+12\,\text{V}=0 \Longrightarrow I=2.0\,\text{A}$$

Thus, the potential drops by 4.0 V across the first resistor and drops again by 8.0 V across the next. The total drop in potential, 12 V, is recovered as we cross the battery. This change in electric potential around the loop is plotted in Fig. 9-10 (b). Recall that a positive charge is like mass and electric potential is like height. Hence once the charge is set in motion by rolling it to the right, it falls twice as it crosses the resistors, and gets picked up by the battery, which provides the work required to return it to the initial height. The charge goes through the same motion as it moves around the circuit.

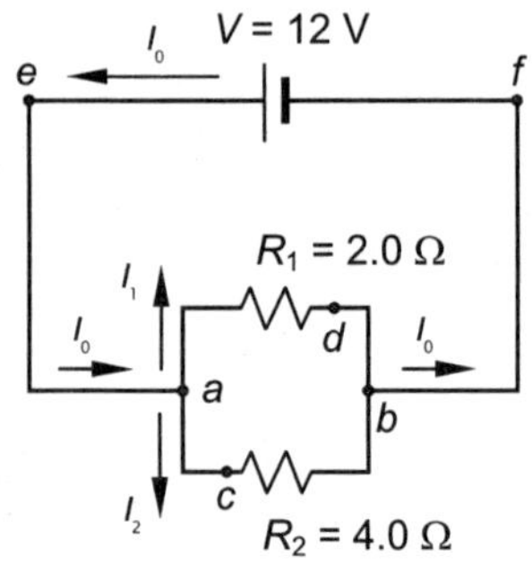

Figure 9-11: Analyzing a circuit with two resistors in parallel

The next circuit, shown in Fig. 9-11, consists of the same set of elements as shown in Fig. 9-10 (a), but arranged differently. Namely, the two resistors are now in parallel. At junction *a*, the current out of the battery, I_0, is split into two currents, I_1 and I_2:

$$I_0 = I_1 + I_2 \qquad \text{... (9.6)}$$

To find the current I_1, apply Kirchhoff's loop rule to the loop *e* – *a* – *d* – *b* – *f* - *e*, which yields the following equation:

$$\underbrace{-I_1R_1}_{\text{potential drop across } R_1} + \overbrace{V}^{\text{potential gain across the battery}} = 0 \qquad \text{... (9.7)}$$

Substitute 2.0 Ω and 12 V for R_1 and *V*, respectively, and we find that I_1 = 6.0 A. Similarly, applying the loop rule to the loop *e* – *a* – *c* – *b* – *f* - *e* produces

$$-I_2R_2 + V = 0 \qquad \text{... (9.8)}$$

Substitute 4.0 Ω and 12 V for R_2 and *V*, respectively, and we find that I_2 = 3.0 A. Then according to the junction rule, the total current out of the battery, I_0, is 9.0 A.

9-4. Calculation of Equivalent Resistance

When a circuit consists of a battery and a resistor, as shown in Fig. 9-12, the potential difference across the battery is equal to that across the resistor (this can be verified by applying Kirchhoff's loop rule). That is,

$$V = I \times R \qquad \text{... (9.9)}$$

voltage of battery → V; current out of battery → I; resistance of resistor → R

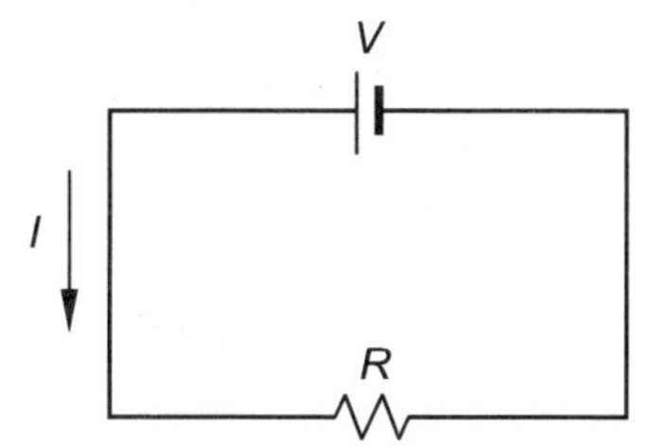

Figure 9-12: A circuit consisting of a battery and a resistor

Normally, Eq. (9.9) cannot be applied to more complex circuits containing more than one resistor, such as those shown in Figs. 9-10 (a) and 9-11. However, if we can represent all the resistors in a circuit with a single resistor with **equivalent resistance**, we will be able to utilize Eq. (9.9) for faster analysis of a circuit. For example, consider the circuit shown in Fig. 9-13 (a). Applying Kirchhoff's loop rule, we obtain Eq. (9.5), which, when rearranged, can be written as

$$V = IR_1 + IR_2 = I(R_1 + R_2) = IR_{eq}$$

Therefore, we can replace the two resistors in the circuit with a single resistor whose resistance is R_{eq}, the sum of the two resistances. The simplified circuit is shown in Fig. 9-13 (b).

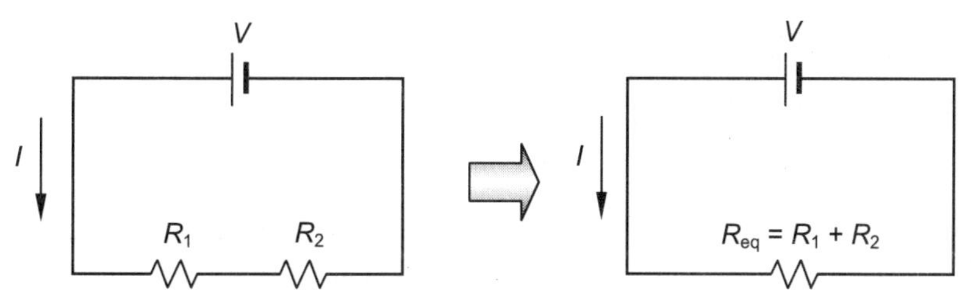

(a) a circuit with two resistors in series (b) a circuit with equivalent resistance

Figure 9-13: Finding the equivalent resistance of resistors connected in series

Using the same approach, we can find that the equivalent resistance of resistors connected in series is the sum of their resistances. That is,

$$R_{eq} = R_1 + R_2 + R_3 + \cdots \quad \text{(resistors in series)} \qquad \ldots (9.10)$$

Thus, adding more resistors in series increases the overall resistance of the circuit, and the current decreases. We can now easily find the current in the circuit shown in Fig. 9-14 as follows. First, we find that the equivalent resistance is 6.0 Ω. Then using Eq. (9.9), the current in the circuit is

$$I = \frac{V}{R_{eq}} = \frac{12\text{ V}}{6.0\,\Omega} = 2.0\text{ A}$$

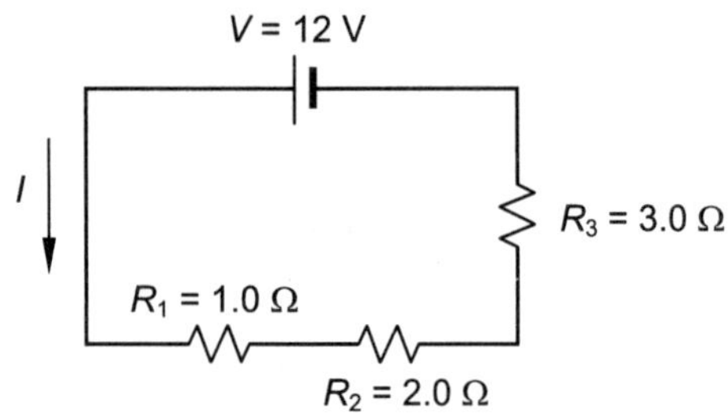

Figure 9-14: Analyzing a circuit with three resistors in series

When resistors are connected in parallel, the current must split into separate branches to yield the same voltage drop across each resistor. For example, while analyzing the circuit shown in Fig. 9-11, we had obtained Eqs. (9.6), (9.7), and (9.8). Solving the last two equations for I_1 and I_2, respectively, we get

$$I_1 = \frac{V}{R_1} \quad \text{and} \quad I_2 = \frac{V}{R_2}$$

Substitute the above equations into Eq. (9.6):

$$I_0 = \frac{V}{R_1} + \frac{V}{R_2} = V\left(\frac{1}{R_1} + \frac{1}{R_2}\right) = \frac{V}{R_{eq}}$$

Therefore, we can replace the two resistors in the circuit with a single resistor where the reciprocal of its resistance is the sum of the reciprocal of each resistance. In general, when there are many resistors connected in parallel, the equivalent resistance can be found from the following equation:

$$\frac{1}{R_{eq}} = \frac{1}{R_1} + \frac{1}{R_2} + \frac{1}{R_3} + \cdots \text{ (resistors in parallel)} \quad \ldots (9.11)$$

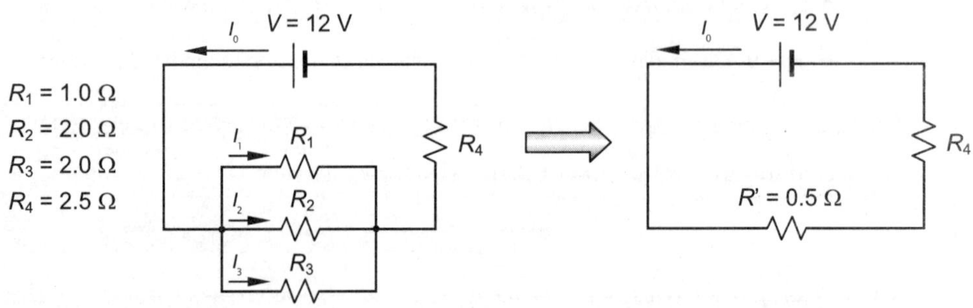

(a) before reduction (b) after reduction

Figure 9-15: Analyzing a circuit with many resistors

Consider the circuit in Fig. 9-15 (a). It has four resistors, three of which are connected in parallel. Let's find the total current in the circuit and the current through each resistor. First, we compute the equivalent resistance of the three resistors in parallel:

$$\frac{1}{R'}=\frac{1}{1.0\,\Omega}+\frac{1}{2.0\,\Omega}+\frac{1}{2.0\,\Omega}=2.0\,\Omega^{-1} \Longrightarrow R'=0.50\,\Omega$$

The equivalent resistance of resistors in parallel is always less than the smallest of the resistances.

The circuit can now be modeled as shown in Fig. 9-15 (b). The overall resistance of this circuit is

$$R_{eq}=R_4+R'=2.5\,\Omega+0.50\,\Omega=3.0\,\Omega$$

Using Eq. (9.9), we can calculate the total current out of the battery:

$$I_0=\frac{V}{R_{eq}}=\frac{12\text{ V}}{3.0\,\Omega}=4.0\text{ A}$$

According to Ohm's law in Eq. (9.2) (on p. 272), the potential drop across R_4 is

$$I_0R_4=4.0\text{ A}\cdot(2.5\,\Omega)=10\text{ V}$$

Since Kirchhoff's loop rule requires that the sum of the changes in potential around any loop be zero, the potential must drop across each resistor in parallel by an additional 2.0 V. Thus,

$$I_1=\frac{2.0\text{V}}{R_1}=\frac{2.0\text{V}}{1.0\Omega}=2.0\text{ A};\quad I_2=\frac{2.0\text{V}}{R_2}=\frac{2.0\text{V}}{2.0\Omega}=1.0\text{ A};\quad I_3=\frac{2.0\text{V}}{R_3}=\frac{2.0\text{V}}{2.0\Omega}=1.0\text{ A}$$

You will find that, for circuits containing many resistors, some in series and some in parallel, using the equivalent resistance of the circuit makes the analysis of the circuit easier than dealing with many equations derived from Kirchhoff's rules. For instance, applying Kirchhoff's rules to the circuit in Fig. 9-15 yields the following four equations with four unknowns (I_0, I_1, I_2, and I_3):

from applying the junction rule → $I_0=I_1+I_2+I_3$

from applying the loop rule:

$$\begin{cases} -I_1R_1-I_0R_4+V=0 \Longrightarrow -I_1(1.0\,\Omega)-I_0(2.5\,\Omega)+12\text{ V}=0 \\ -I_2R_2-I_0R_4+V=0 \Longrightarrow -I_2(2.0\,\Omega)-I_0(2.5\,\Omega)+12\text{ V}=0 \\ -I_3R_3-I_0R_4+V=0 \Longrightarrow -I_3(2.0\,\Omega)-I_0(2.5\,\Omega)+12\text{ V}=0 \end{cases}$$

Although this one is not too bad, solving a system of four equations in general is no easy task. It usually takes a long time because you have to make a lot of substitutions and do the arithmetic. Solve the above system for yourself and confirm that you do obtain the same set of solutions as the one we found above.

So, when simplifying a circuit to find equivalent resistance, be careful of the order in which you combine the resistors. For example, the circuit shown in Fig. 9-16 (a) has to be simplified by combining R_2 and R_3 first (Fig. 9-16 (a)), then the three resistors in parallel (Fig. 9-16 (b)), and finally all of the resistors together (Fig. 9-16 (c)). If you use any other order, the value you obtain will be wrong.

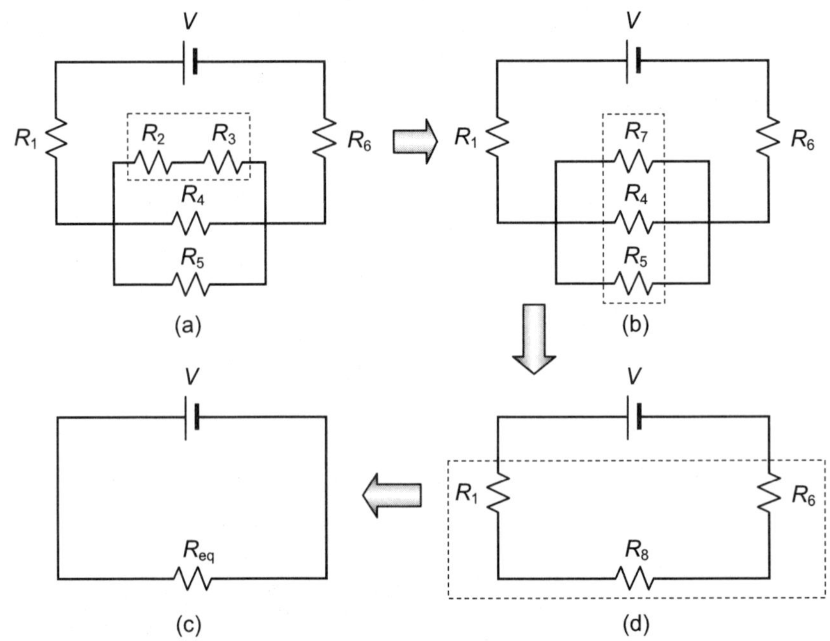

Figure 9-16: Simplifying a circuit containing many resistors

9-5. Terminal Voltage

In real life, no machine and device operates ideally, and batteries are no exception. Friction and other resistive mechanisms inside them reduce the speed at which electrons are delivered to the negative terminal. In effect, these mechanisms have the same effect on the amount of current in the circuit as a resistor. Thus, in a more realistic representation of a battery, they are represented with the same symbol as a resistor inside a battery, as shown in Fig. 9-17. The symbol lumps all the resistive mechanisms inside a battery and is called **internal resistance**.

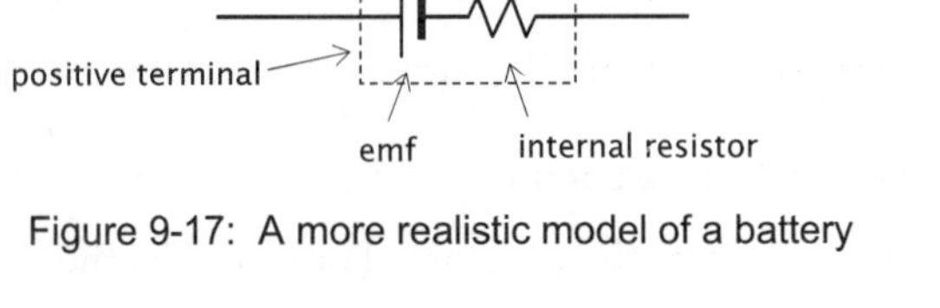

Figure 9-17: A more realistic model of a battery

The symbol we have been using for a battery now represents what is called **electromotive force**, or emf for short. Recall from Section 9-1 (on p. 271) that chemical reactions inside a battery create a potential difference between its two terminals. **Emf** is this influence to drive electrons from a lower to a higher potential and is the maximum potential difference across the terminals of a cell (that is, the potential difference across the terminals when no current is being drawn from the cell and the cell is not doing any work separating the charge). Note that emf is not a force. The standard unit for emf is the Volt [V = J/C].

Due to the presence of internal resistance, the potential difference between the terminals of a battery, called the **terminal voltage**, is not the same as its emf. To find the relationship between them, consider the battery shown in Fig. 9-18 (a), with the current flowing out of its positive terminal. Counting the changes in potential from point *b* to point *a*, we find that there is a potential drop across the internal resistance ($-Ir$) and a potential gain across the emf ($+\mathcal{E}$):

$$\Delta V = \underbrace{V_a - V_b}_{\text{potential difference between two terminals}} = -Ir + \mathcal{E}$$

Thus, the potential at the positive terminal is higher than that at the negative terminal by $-Ir + \mathcal{E}$, and this is the terminal voltage (V_T) of the battery. That is,

$$V_T = \mathcal{E} - Ir \qquad \text{... (9.12)}$$

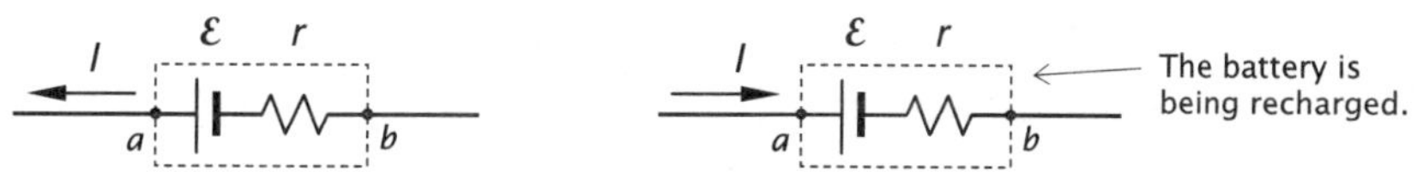

(a) current flowing out of battery (b) current forced into battery

Figure 9-18: Terminal voltage

Sometimes electric current can be forced into a battery, as shown in Fig. 9-18 (b). This occurs, for example, when two batteries are facing each other and the one with greater emf forces current into the other. In this case, the terminal voltage becomes

$$V_T = \mathcal{E} + Ir \qquad \text{... (9.13)}$$

Note that, as we move against the current from point *b* to point *a*, the potential increases across the internal resistance.

When finding the terminal voltage of a battery, we often have to find the current flowing through it first. Then treat internal resistance as if it were an external resistor and calculate the current using the method described in the previous section. For example, consider the battery in the circuit shown in Fig. 9-19 (a). To find its terminal voltage, we first simplify the circuit as in Fig. 9-19 (b). The equivalent resistance of the resistors in parallel is 1.5 Ω. We then combine the three resistors in series, R_1, R', and r, to find the overall equivalent resistance of 6.0 Ω. Then the total current in the circuit is 2.0 A, and using Eq. (9.12), we can obtain the battery's terminal voltage:

$$V_T = 12\,\text{V} - (2.0\,\text{A})(0.5\,\Omega) = 11\,\text{V}$$

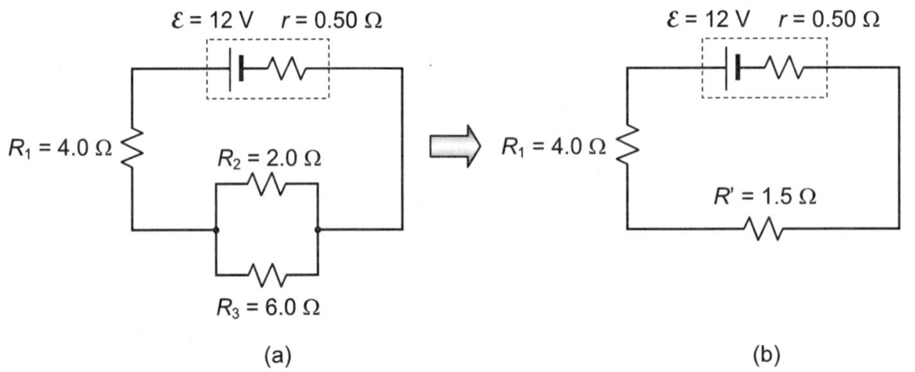

Figure 9-19: Finding terminal voltage

Sometimes we represent a battery without internal resistance when it is insignificant. So when you see only the symbol for the emf, —|⊢, in a battery, assume the internal resistance is zero. Then the terminal voltage of the battery is equal to the emf.

9-6. Electric Power

Electrons moving in a circuit towards the positive terminal of a battery constantly give up their electric potential energy and perform work equivalent to the decrease in their potential energy. That is,

$$W_{\text{by an electron}} = e\Delta V \quad [\text{J} = \text{C}\cdot\text{V}] \qquad \text{... (9.14)}$$

For instance, suppose 3.2-A current is flowing through a 2.5-Ω resistor as shown in Fig. 9-20. The potential drop across the resistor is 8.0 V. Therefore, the work done by each electron as it crosses the resistor is

$$W = (1.6\times10^{-19}\,\text{C})(8.0\,\text{V}) = 1.28\times10^{-18}\,\text{J} \quad (\text{or } 8.0\,\text{eV})$$

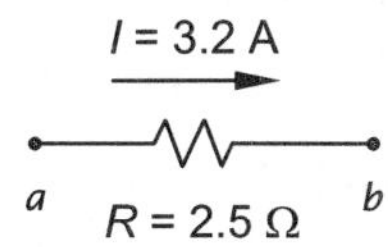

Figure 9-20: Work done by current through a resistor

According to Eq. (9.1) (on p. 270), the amount of charge crossing the resistor in one second is 3.2 C, which means that the number of electrons passing the resistor in one second is 2.0×10^{19}. The work done by this many electrons in one second is then

$$W = \underbrace{\left(2.0 \times 10^{19}\right)\left(1.6 \times 10^{-19}\,\mathrm{C}\right)}_{Q}\underbrace{\left(8.0\,\mathrm{A}\right)}_{\Delta V} \cong 26\,\mathrm{J}$$

From the above equation, we can show that the power dissipated in a resistor is

$$P_{\text{used in a resistor}} = \frac{Q\Delta V}{t} = \frac{Q}{t}\Delta V = I_{\text{through resistor}} \times \Delta V_{\text{across resistor}} \qquad \text{... (9.15)}$$

With Ohm's law, we can find other expressions for the power dissipated:

$$P_{\text{used in a resistor}} = I^2 R = \frac{(\Delta V)^2}{R} \qquad \text{... (9.16)}$$

Let's consider the circuit shown in Fig. 9-21, and find the power dissipated in each light bulb. A light bulb is essentially a resistor that converts electrons' work into light. Therefore, this circuit is basically the same as the one we considered earlier, shown in Fig. 9-19 (a). While calculating the battery's terminal voltage, we had calculated the total current in the circuit to be 2.0 A. Then according to Ohm's law, the potential drop across R_1 is 8.0 V and that across the internal resistance is 1.0 V.

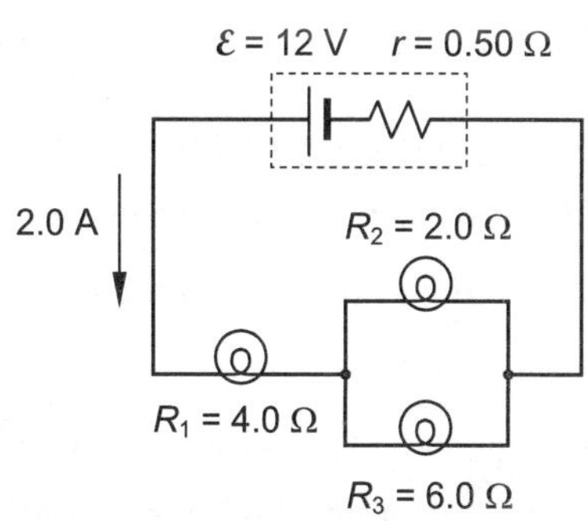

Figure 9-21: Finding power dissipated in each light bulb

Now, apply Kirchhoff's loop rule, and we find that the potential must drop by 3.0 V across each light bulb connected in parallel. Therefore, the current through the

2.0-Ω light bulb is 1.5 A, and that through the 6.0- Ω bulb is 0.50 A. Having found the current through each and every bulb, we can use Eq. (9.16) to calculate the power dissipated in each bulb:

Power dissipated in 4.0-W light bulb: $P_1 = (2.0\,\text{A})^2(4.0\,\Omega) = 16\text{ W}$

Power dissipated in 2.0-W light bulb: $P_2 = (1.5\,\text{A})^2(2.0\,\Omega) = 4.5\text{ W}$

Power dissipated in 6.0-W light bulb: $P_3 = (0.50\,\text{A})^2(6.0\,\Omega) = 1.6\text{ W}$

The total power dissipated by the three bulbs is 22 W, which is provided by the battery. In general, the power supplied by a battery is given as

$$P_{\text{supplied by a battery}} = I_{\text{out of battery}} \cdot V_T \qquad \ldots (9.17)$$

In this circuit, the battery's terminal voltage is 11 V. Thus,

$$P_{\text{supplied by a battery}} = (2.0\,\text{A})(11\,\text{V}) = 22\text{ W}$$

This is the same as the sum of the power dissipated in the three bulbs.

In any circuit, the sum of the electric power consumed by the resistors is equal to the power supplied by the battery, or else the conservation of energy is violated. Thus,

$$P_{\text{supplied by battery}} = \sum P_{\text{dissipated in resistors}} \qquad \ldots (9.18)$$

Note that the thermal energy, or heat, generated in the wires and resistors increases at a rate proportional to the current squared (I^2). This implies that, if too much current is drawn from the power source, electric devices may eventually become hot enough to start a fire. In order to prevent such overloading, fuses or circuit breakers are installed to cut off current in the circuit.

9-7. Measuring Devices: Ammeters and Voltmeters

Designing an electric circuit often begins with drawing a diagram such as the one shown in Fig. 9-21. We can then analyze it theoretically by using the analysis techniques introduced in this chapter. Electric circuits used in today's electronic devices are much more complex than the ones we studied in this chapter, and the equations used in an analysis can become very complicated. Therefore, after building an actual circuit, we must check the current and the potential difference across each resistor using ammeters and voltmeters.

An **ammeter**, represented by the symbol 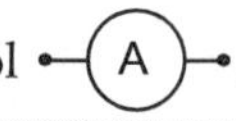, is an instrument that measures the current in a wire. It must be installed in series with a battery or a resistor, as shown in Fig. 9-22. For accurate measurements, ammeters have very low resistance. (Otherwise, the total current in the circuit will decrease substantially after an ammeter is installed.)

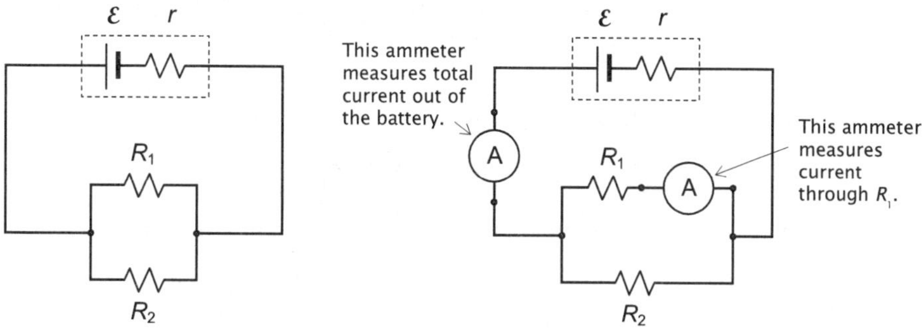

(a) original circuit (b) after installing ammeters

Figure 9-22: Using ammeters to measure current

A **voltmeter**, represented by the symbol 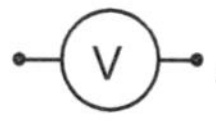, measures the potential difference between two points in a circuit. (No machine can measure electric potential. We can only measure potential <u>difference</u> using a voltmeter.) Unlike an ammeter, it has very high resistance and must be installed in parallel to the other elements, as shown in Fig. 9-23.

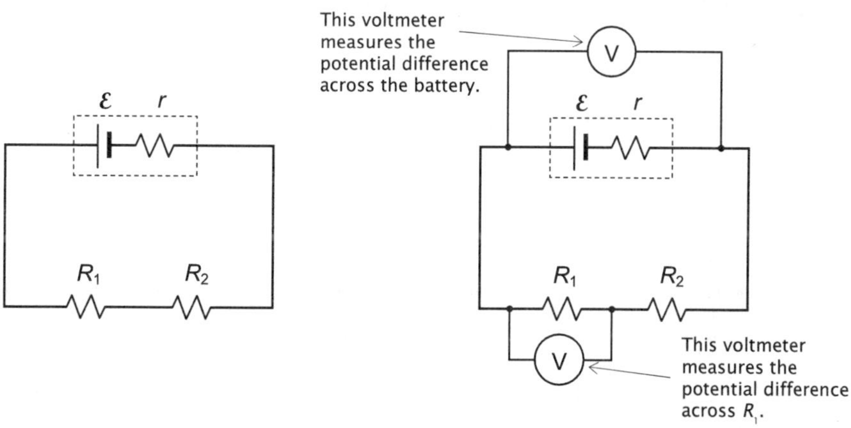

(a) original circuit (b) after installing voltmeters

Figure 9-23: Using voltmeters to measure potential difference

PHYSICS 12 ANALYZED: DC CIRCUITS

9-1. Building a DC Circuit

- The battery, wires, and resistors (or loads) are the basic building blocks of a DC circuit.
- The wire is a pathway for electrons. When there is a potential difference between the ends of a wire, the electrons move. ⇨ electric current
- In a circuit with a single battery, the conventional current always flows from the battery's positive terminal to its negative one.
- The resistor is an element used to control the amount of current in a circuit by resisting the motion of electrons. The amount of current is inversely proportional to the amount of resistance within a circuit.

9-2. Ohm's Law

- Ohm's law states that a voltage drop across a device is equal to the product of the current through it and the resistance of the device. That is,

$$\left|\Delta V_{\text{across resistor}}\right| = I_{\text{through resistor}} \times R_{\text{of resistor}}$$

- Ohm's law can be applied to the whole of the external circuit or to any part of it.

9-3. Kirchhoff's Rules

- Kirchhoff's junction rule: The sum of all currents entering a junction must equal he sum of all currents leaving it.
- Kirchhoff's loop rule: The sum of changes in electric potential around any loop is zero.
- The junction rule is based on the conservation of charge; the loop rule is based on the conservation of energy.

9-4. Calculation of Equivalent Resistance

- Equivalent resistance is obtained from…

1. $R_{eq} = R_1 + R_2 + R_3 + \cdots$ for resistors in series
2. $\frac{1}{R_{eq}} = \frac{1}{R_1} + \frac{1}{R_2} + \frac{1}{R_3} + \cdots$ for resistors in parallel

Then we can use $V = I_{total} \times R_{eq}$

- Adding a resistor in series increases the overall resistance; adding one in parallel decreases it.

9-5. Terminal Voltage

- Electromotive force ($\mathcal{E}$) causes a potential difference between the terminals of a battery and drives current through the whole circuit.
- Internal resistance (r) refers to friction and other resistive mechanisms inside a battery.
- Terminal voltage (V_T) is the potential difference between two terminals of a battery. It drives current through the external circuit.
- When current flows out of the positive terminal, $V_T = \mathcal{E} - Ir$.
- When current is forced into the positive terminal, $V_T = \mathcal{E} + Ir$.

9-6. Electric Power

- Electric power consumed in a resistor: $P_{\text{dissipated in a resistor}} = I\Delta V = I^2 R = \frac{(\Delta V)^2}{R}$
- Electric power supplied by a battery: $P_{\text{supplied by a battery}} = I V_T$
- Power supplied by a battery must equal total power dissipated by resistors.

9-7. Measuring Devices: Ammeters and Voltmeters

- An ammeter measures current in a wire. It has very low resistance and must be connected in series with a battery or a resistor.
- A voltmeter measures the potential difference between two points in a circuit. It has very high resistance and must be connected in parallel to the other elements.

EXERCISES ON CHAPTER 9: DC CIRCUITS

Exercise 9-1

Part of an electric circuit is shown right. Find the magnitude and direction of the current passing through R_1.

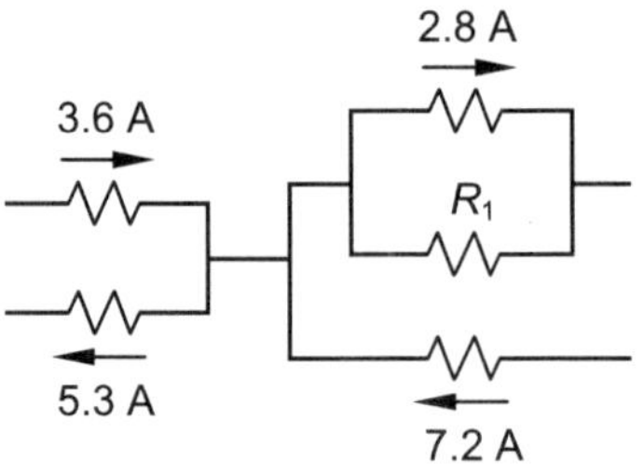

Exercise 9-2

If the equivalent resistance of the circuit needs to be 8.1 Ω, what should be the value of R_5?

$R_1 = 1.5\ \Omega$	$R_2 = 2.5\ \Omega$	$R_3 = 3.5\ \Omega$
$R_4 = 4.5\ \Omega$	$R_5 = ?$	$R_6 = 5.7\ \Omega$

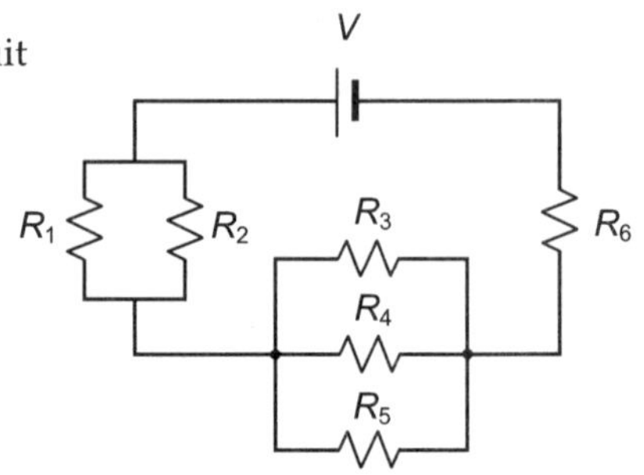

Exercise 9-3

A 2.1-A current is supplied to a battery whose emf is 3.0 V. If the terminal voltage of the battery is 3.6 V, what is the internal resistance?

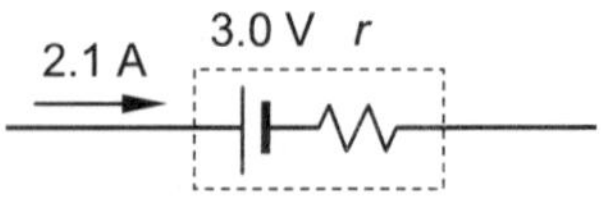

Exercise 9-4

In the circuit shown right, find:

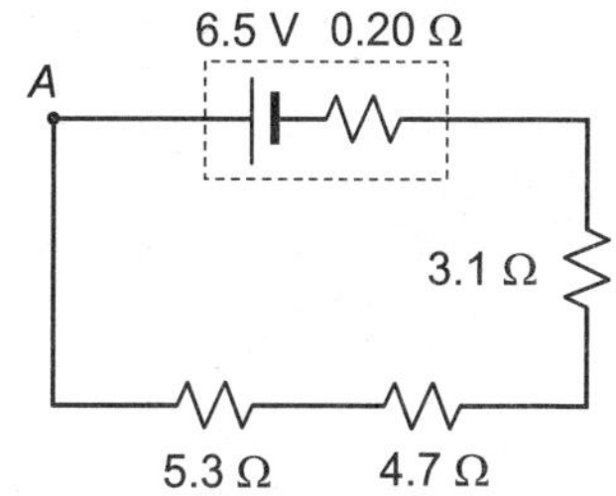

(a) the equivalent resistance of the whole circuit

(b) the current in the circuit

(c) the potential drop across the 5.3-Ω resistor

(d) the terminal voltage of the battery

(e) the amount of charge that flows through point *A* in 15 seconds

Exercise 9-5

Part of an electric circuit is shown right. Find

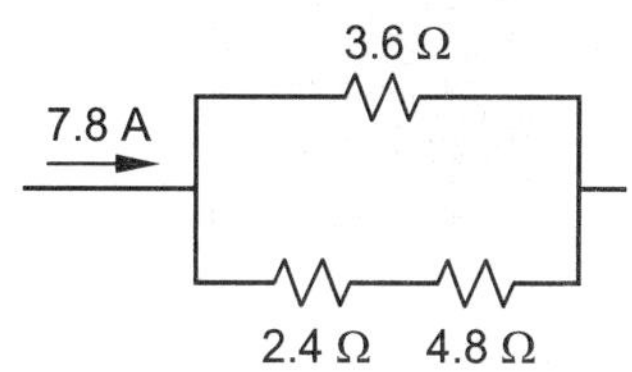

(a) the current through the 3.6-Ω resistor

(b) the voltage drop across the 2.4-Ω resistor

Exercise 9-6

Four identical resistors are connected as shown right. The total current in the circuit is 0.96 A. If the potential at point *a* is 7.0 V, find the potential at points *b*, *c*, and *d*.

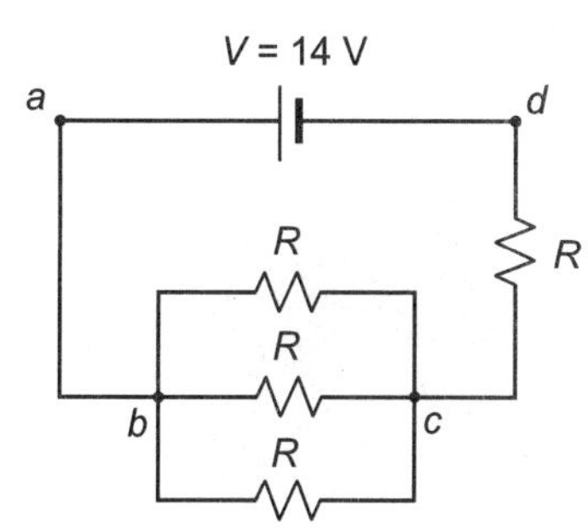

Exercise 9-7

For the circuit shown right, find the current through each resistor by

(a) applying Kirchhoff's rules

(b) finding the equivalent resistance of the circuit and then applying Kirchhoff's loop rule

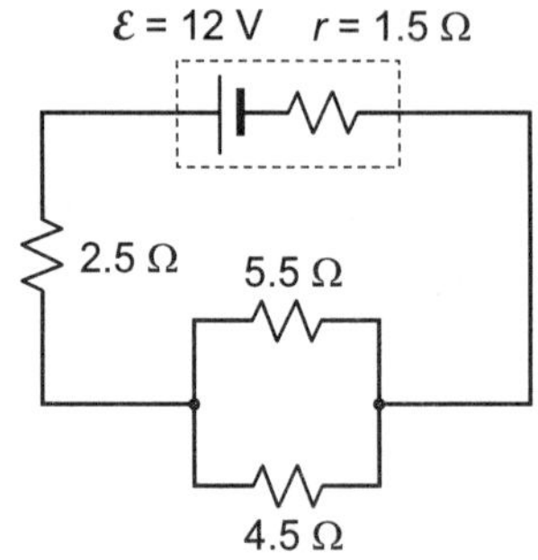

Exercise 9-8

A 1.5 V battery is used to light a 35-W light bulb. If the battery is completely drained after 24 hours of use, find

(a) the total energy supplied by the battery during the 24 hours

(b) the total charge that passed through the bulb during the 24 hours

Exercise 9-9

A battery supplies a 3.7-A current to the light bulbs as shown. Find

(a) the power supplied by the battery

(b) P_2, the power dissipated in the lower bulb

(c) the current through each bulb

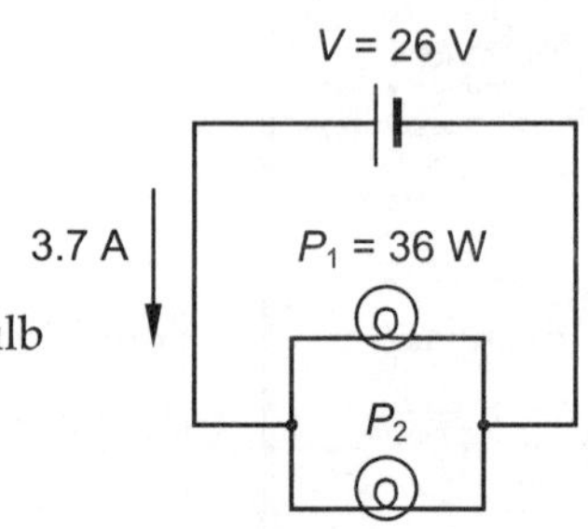

Exercise 9-10

In the circuit shown right, the total current is 2.1 A. Find the potential difference between

(a) a and b (b) c and d

(c) c and f (d) d and f

(e) g and h (f) a and g

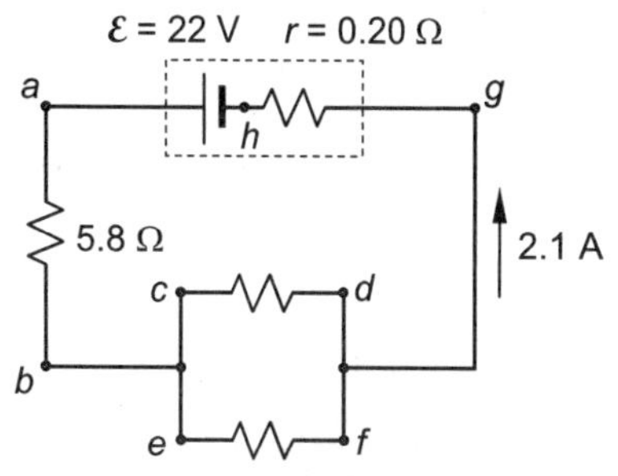

Exercise 9-11

In the circuit shown right, the current measured by the ammeter is 0.78 A.

(a) Find the terminal voltage of the battery.

(b) Find the reading of the voltmeter.

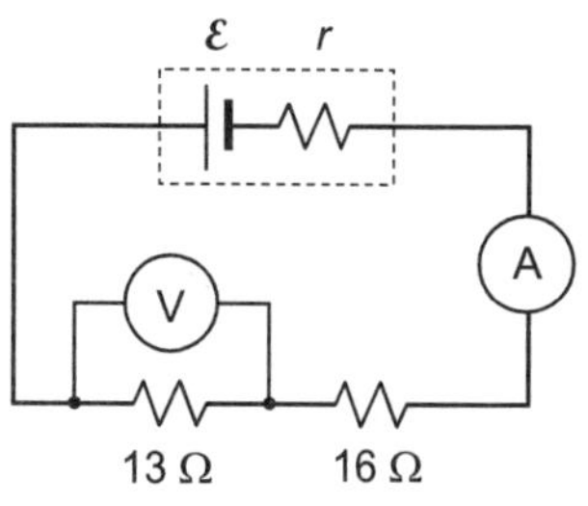

Exercise 9-12

The circuit shown in Exercise 9-11 is modified by adding a resistor as shown. The ammeter now reads 0.99 A. Find:

(a) the current through each resistor

(b) the power dissipated in each resistor

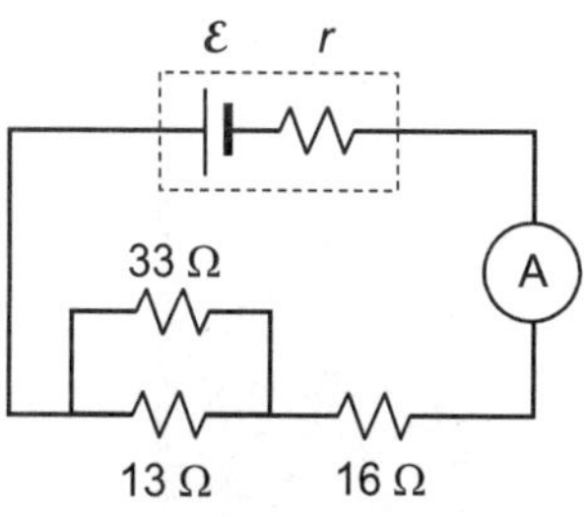

Exercise 9-13

Part of an electric circuit is shown right. The terminal voltage of the battery is 2.7 V. Find

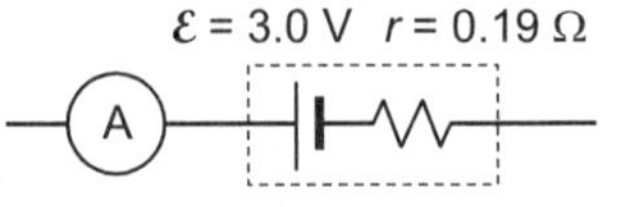

(a) the magnitude and direction of the current

(b) the amount of charge passing through the ammeter in 2.2 seconds

(c) the equivalent resistance of the whole circuit (There is only one battery in the circuit.)

(d) the power supplied by the battery

Exercise 9-14

In the circuit shown right, the power dissipated in the 26-Ω resistor is 230 W. Find the power dissipated in the other resistors.

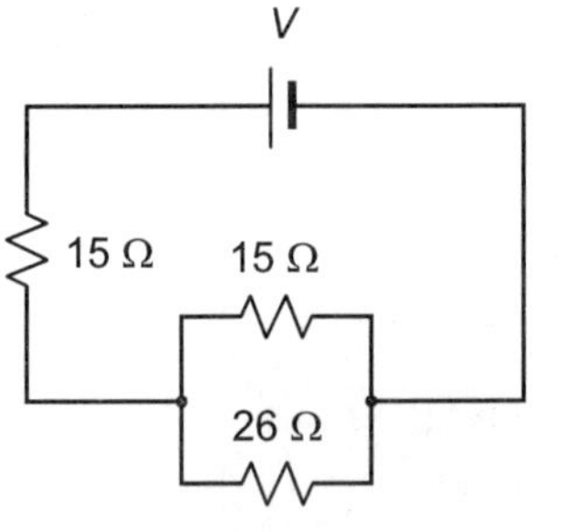

Exercise 9-15

In the circuit shown right, the reading of the ammeter is 1.8 A. Find

(a) the voltage of the battery

(b) the equivalent resistance of the circuit

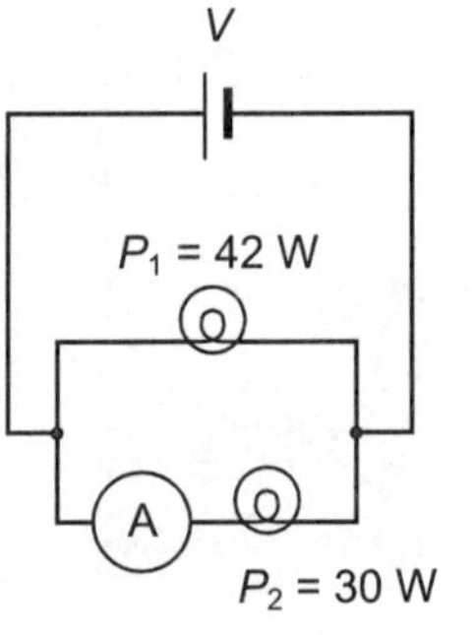

Exercise 9-16

In the circuit shown right, find

(a) the current through the 6.3-Ω resistor

(b) the terminal voltage of the battery

(c) the voltage drop across R_1

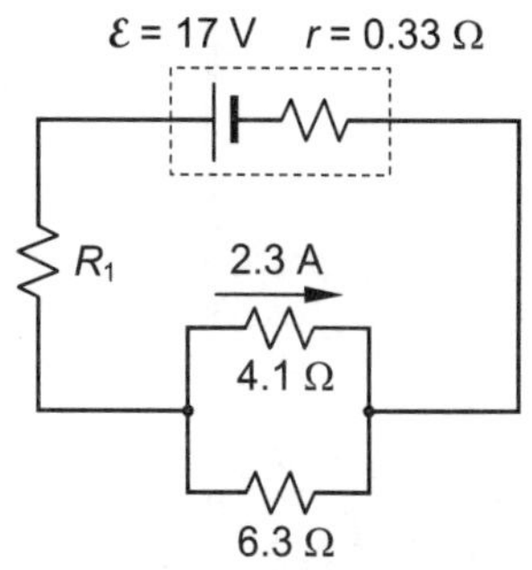

Exercise 9-17

In the circuit shown right, find

(a) the equivalent resistance of the circuit

(b) the current through R_1, R_3, and R_6

(c) the power dissipated in R_2, R_4, and R_5

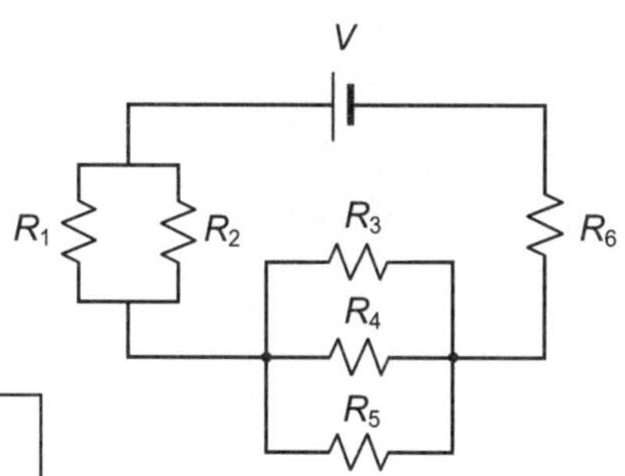

$V = 15$ V		
$R_1 = 3.6\ \Omega$	$R_2 = 2.5\ \Omega$	$R_3 = 6.4\ \Omega$
$R_4 = 1.6\ \Omega$	$R_5 = 4.9\ \Omega$	$R_6 = 8.1\ \Omega$

Exercise 9-18

In the circuit shown right, switch S is initially left open.

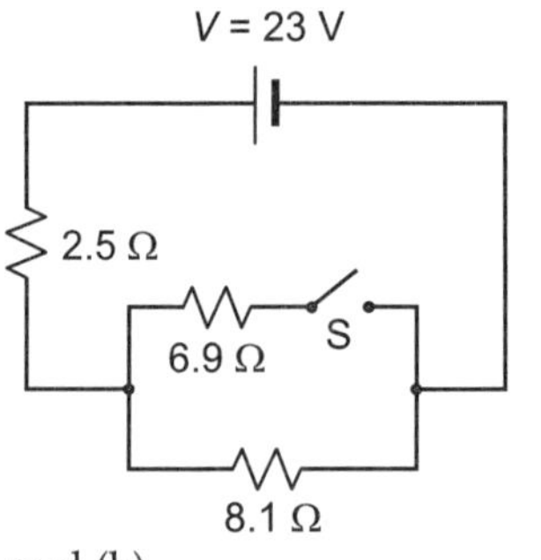

(a) Find the current through the 8.1-Ω resistor.

(b) Find the potential drop across the 8.1-Ω resistor.

(c) Now the switch is closed. Repeat parts (a) and (b).

Exercise 9-19

In the circuit shown right, the power dissipated in the 7.3-Ω resistor is 5.8 W. Find R_2.

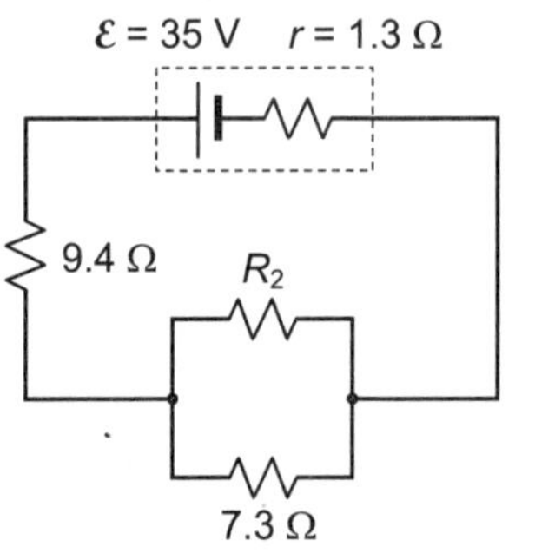

Exercise 9-20

In the circuit shown right, two batteries are connected in series, facing each other. Find

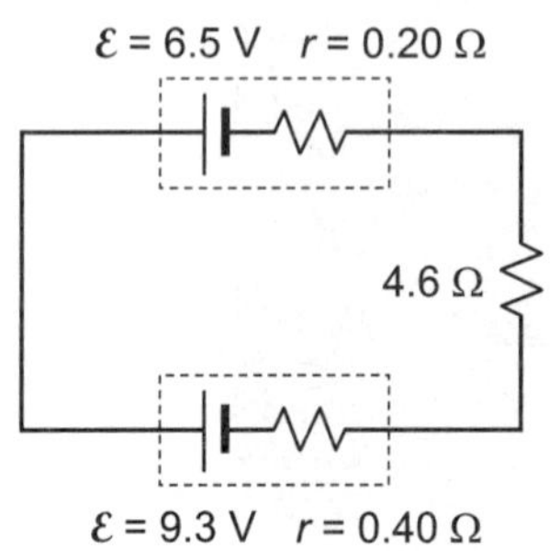

(a) the current through the 4.6-Ω resistor (Hint: Use Kirchhoff's loop rule.)

(b) the terminal voltage of each battery

Exercise 9-21

In the circuit shown right, the current through R_6 is 0.75 A. Find

(a) R_{eq} of the whole circuit

(b) the current through each resistor

(c) the terminal voltage of the battery

(d) the power dissipated in the battery's internal resistance

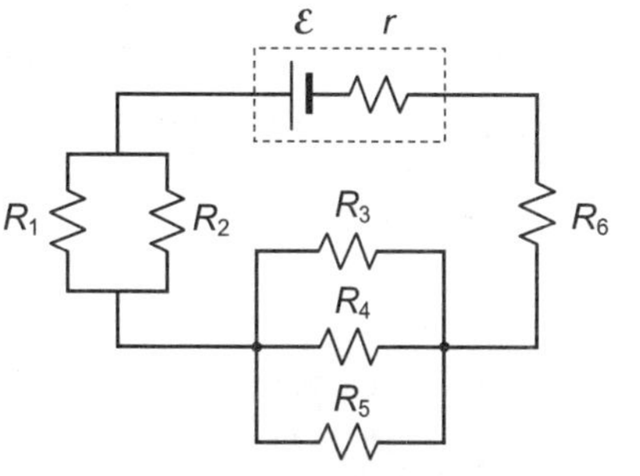

$\mathcal{E}$ = 4.9 V
R_1 = 6.3 Ω R_2 = 5.2 Ω R_3 = 4.6 Ω
R_4 = 3.7 Ω R_5 = 9.4 Ω R_6 = 1.8 Ω

Exercise 9-22

In the circuit shown right, find

(a) R_{eq} of the whole circuit

(b) the current through each resistor

(c) the terminal voltage of the battery

(d) the reading of the voltmeter

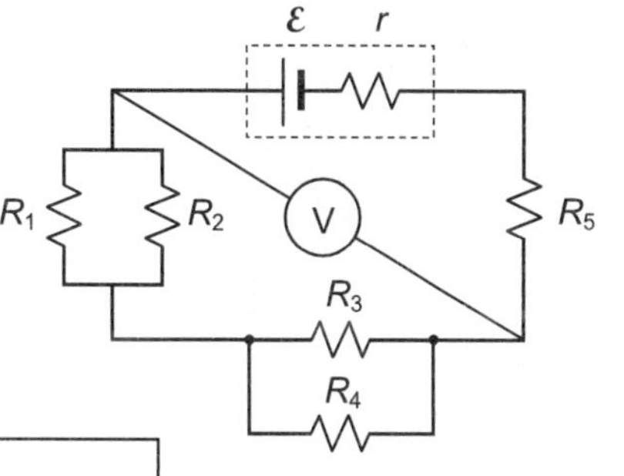

$\mathcal{E}$ = 32 V r = 1.5 Ω
R_1 = 12 Ω R_2 = 18 Ω R_3 = 24 Ω
R_4 = 15 Ω R_5 = 11 Ω

Exercise 9-23

In the circuit shown right, find

(a) the value of resistance R_2

(b) the voltage of the battery

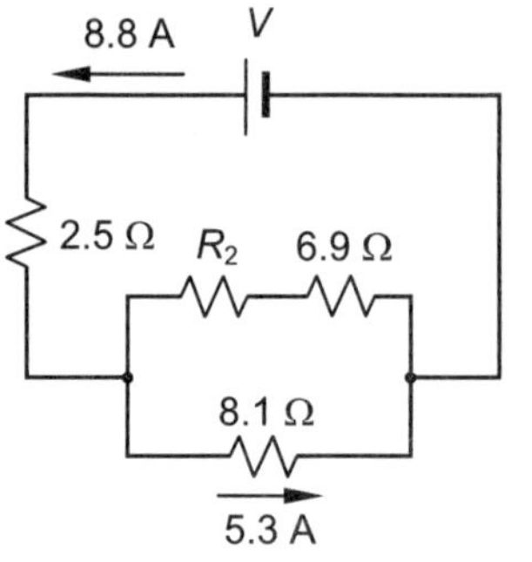

Exercise 9-24

In the circuit shown right, the voltmeter reading is 10 V when $R_2 = R_1$. When one of the resistors is replaced with a resistor with different resistance, the voltmeter reading drops to 8.4 V. Find the internal resistance within the battery and the total current in the circuit after the replacement.

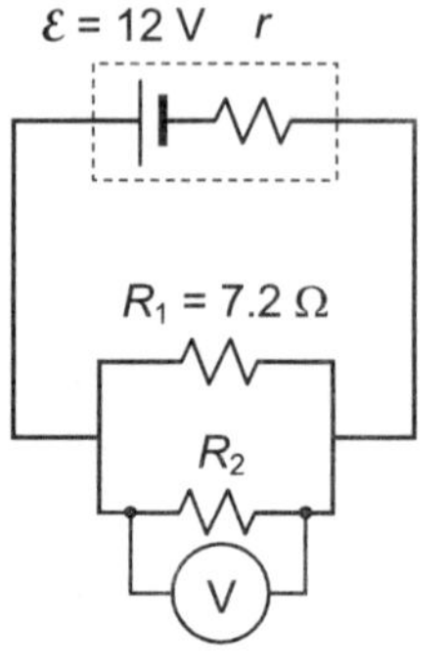

Exercise 9-25

The light bulb in the circuit consumes 11 W when connected to the battery shown. Find two possible values of the bulb's resistance.

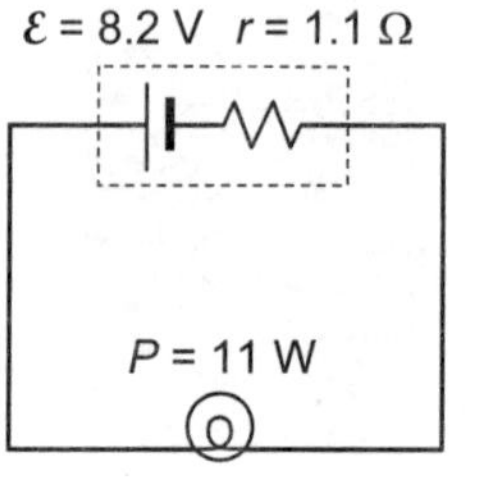

PROBLEMS ON CHAPTER 9: DC CIRCUITS

1 – 9. Fill in the blanks

1. ________ ________ is the amount of charge transferred per unit time.

2. The ________ is an element used to control the amount of current by resisting the motion of electrons.

3. A ______ is a point where three or more conductors meet in a circuit.

4. ______ ________ rule states that the sum of the changes in electric potential around a loop is zero.

5. ______ law states that the potential difference across an element is equal to the product of the current flowing through it and its resistance.

6. ______ ________, or emf, of a battery causes a potential difference between its two terminals and drives current through a circuit.

7. ______ ________ represents all resistive mechanisms inside a battery.

8. The ______ ________ of a battery is the potential difference between its two terminals.

9. An ______ is used to measure current in a wire, and a ______ is used to measure the potential difference between two points in a circuit.

10 – 20. True/False questions

10. ____ Current is a vector because it has a direction.

11. ____ The direction of conventional current is the opposite to electrons' flow.

12. ____ Using too many resistors in a circuit shortens the battery's life.

13. ____ Kirchhoff's junction rule is based on the law of conservation of energy.

14. ____ The electric potential always drops across a resistor when following the direction of conventional current.

15. ____ Adding additional resistors in series increases the equivalent resistance of the circuit.

16. ____ Adding additional resistors in parallel increases the overall current in the circuit.

17. _____ The terminal voltage of a battery can never be greater than its emf.

18. _____ The emf of a battery is measured in Newtons [N].

19. _____ The energy dissipated in a resistor or a light bulb is eventually turned into thermal energy.

20. _____ A voltmeter must be installed in parallel to other elements.

21 – 35. Multiple choice questions

21. Which of the following is incorrect?

a) Ampere = Coulomb/second
b) Ohm = Volt × Ampere
c) Volt = Joule/Coulomb
d) Watt = $(\text{Volt})^2/\text{Ohm}$

22. For the following arrangements of four identical resistors shown below, which of the following correctly lists the arrangements in order of greatest equivalent resistance?

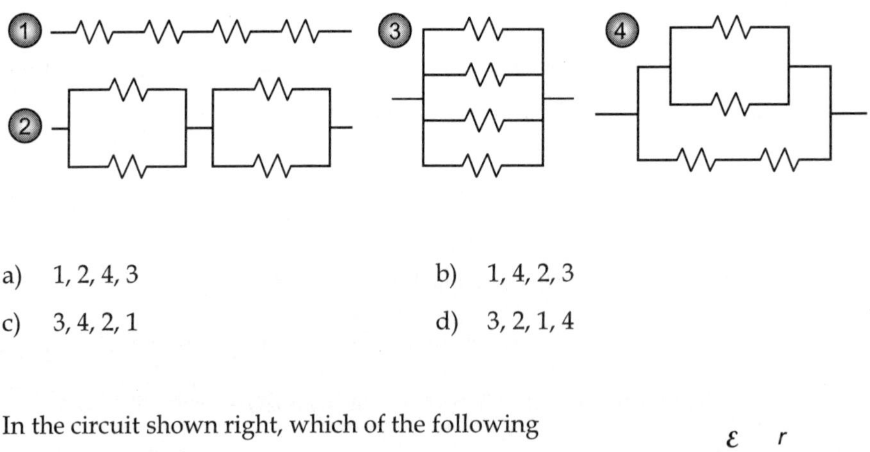

a) 1, 2, 4, 3
b) 1, 4, 2, 3
c) 3, 4, 2, 1
d) 3, 2, 1, 4

23. In the circuit shown right, which of the following equations is valid?

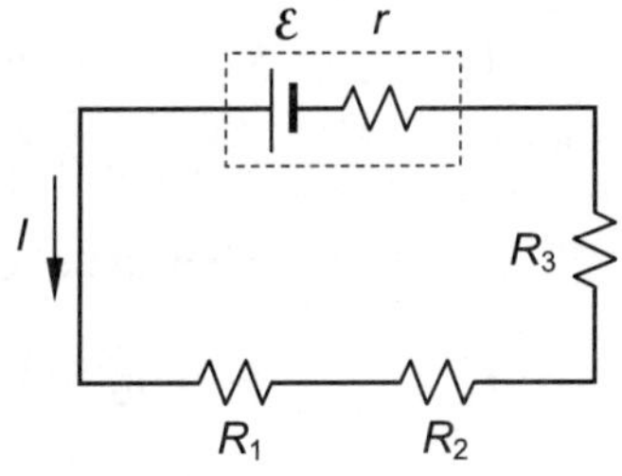

a) $IR_1 = IR_2 = IR_3 = \mathcal{E}$

b) $IR_1 + IR_2 + IR_3 = \mathcal{E} + Ir$

c) $IR_1 + IR_2 + IR_3 = \mathcal{E} - Ir$

d) $\mathcal{E}\left(\frac{1}{R_1} + \frac{1}{R_2} + \frac{1}{R_3} + \frac{1}{r}\right) = I$

For Problems 24 to 26, refer to the electric circuit shown below.

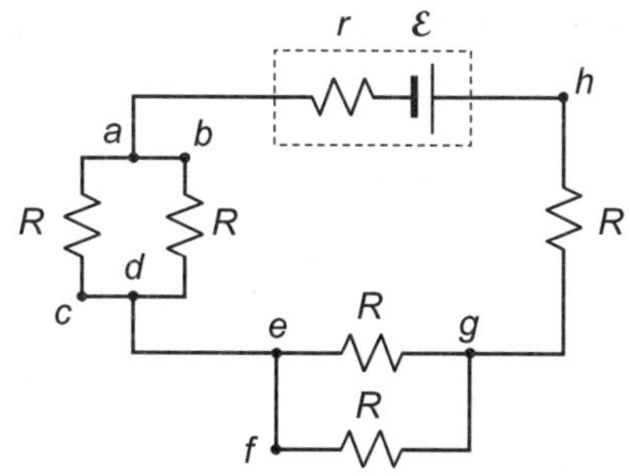

24. What are the directions of conventional current and electrons' flow?

   a) Conventional current flows clockwise and electrons flow counterclockwise.
   b) Conventional current flows counterclockwise and electrons flow clockwise.
   c) Both flow clockwise.
   d) Both flow counterclockwise.

25. Which of the following is <u>not</u> a loop?

   a) $a-b-d-c-a$
   b) $e-f-g-e$
   c) $g-f-e-d-c-a-b-d-e-g$
   d) $h-a-c-d-e-f-g-h$

26. For the loop $h-g-e-d-b-a-h$, which of the following is the correct application of Kirchhoff's loop rule? (I is the total current in the circuit, and all resistors have the same resistance of R.)

   a) $-IR-IR-IR-Ir+\mathcal{E}=0$
   b) $IR+IR+IR+Ir-\mathcal{E}=0$
   c) $-IR-\frac{I}{2}R-\frac{I}{2}R-Ir+\mathcal{E}=0$
   d) $-IR-I(2R)-I(2R)-Ir+\mathcal{E}=0$

27. In the circuit shown right, switch S is initially left open. If the switch is closed, which of the following happens?

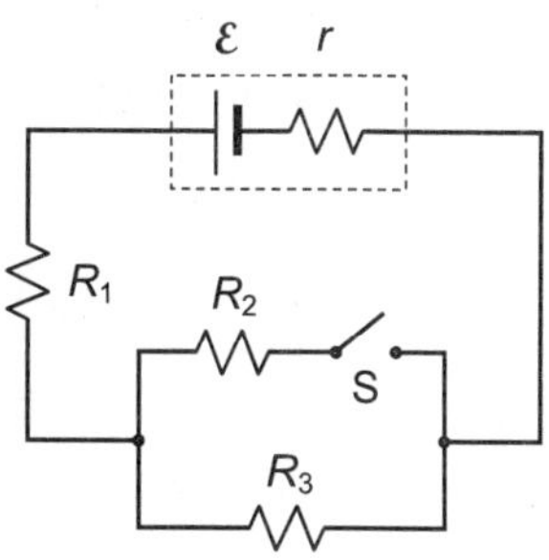

   a) The current through R_1 increases.
   b) The power dissipated in R_3 increases.
   c) The equivalent resistance of the circuit increases.
   d) The internal resistance in the battery increases.

28. Three identical resistors are connected as shown in a DC circuit. Which of the following I vs. R graphs is correct?

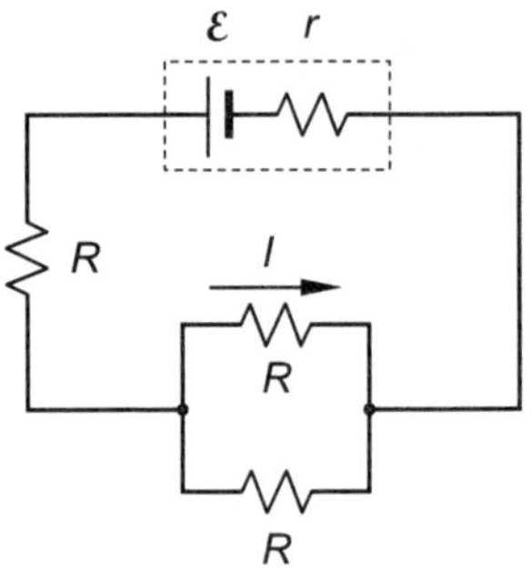

a) 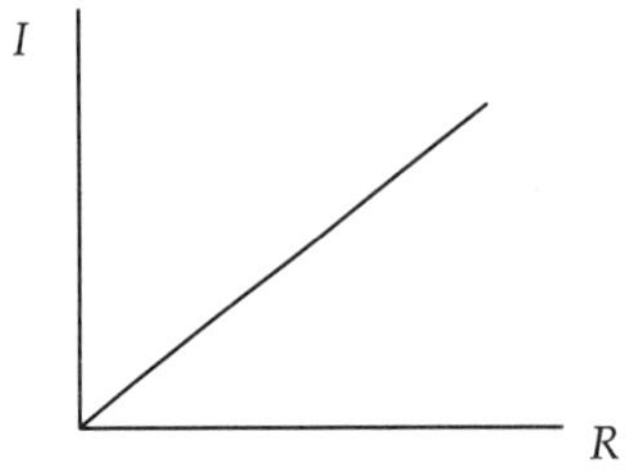

b) 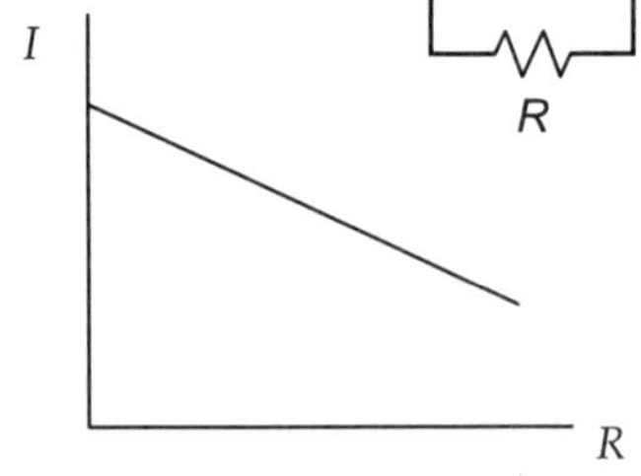

c) 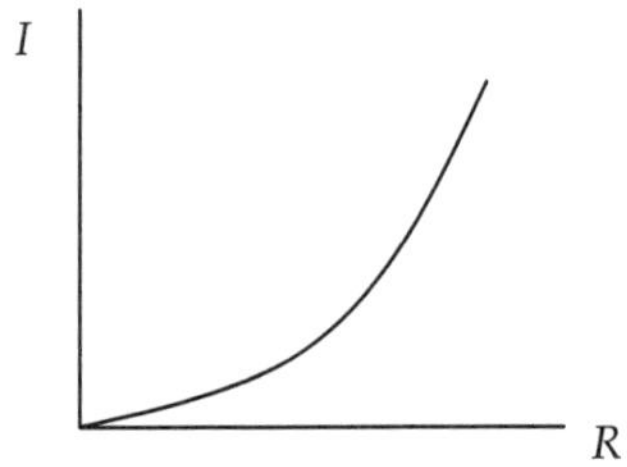

d) 

29. Four identical light bulbs are connected as shown. Which bulb would be the brightest?

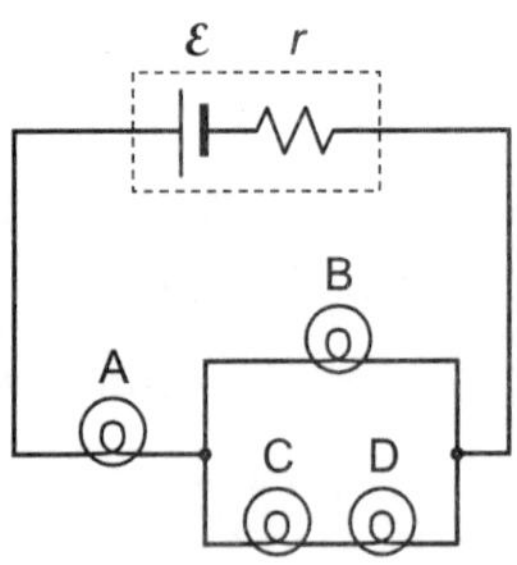

a) A
b) B
c) C
d) D

30. When electricity generated at a power generating station is distributed through transmission lines, it is sent at very high voltages, (typically 250 kV to 500 kV), and at relatively low currents of a few thousand amperes. What is the purpose of sending electric power this way instead of sending it at low voltages and high currents?

a) To send more electric energy
b) To send electric energy faster
c) To minimize energy loss
d) To keep animals away from transmission lines

31. In the circuit shown right, switch S is initially left open. If the switch is closed, which of the following happens?

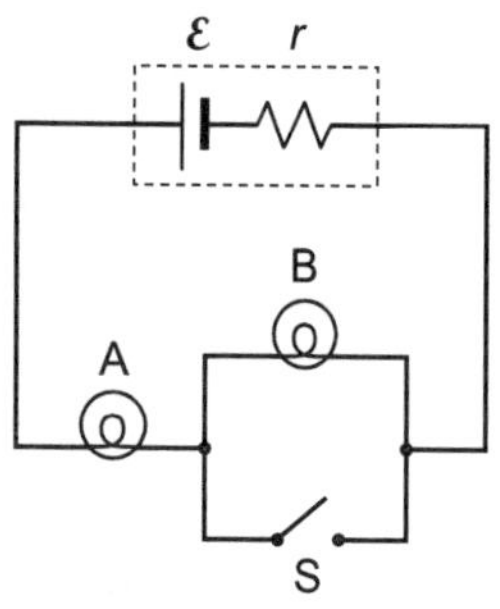

a) Light bulb A becomes brighter.
b) Light bulb B becomes brighter.
c) Both light bulbs become brighter.
d) We don't know what happens unless the resistance of each bulb is given.

32. Which of the following statements on ammeters and voltmeters is correct?

a) For accurate measurements, ammeters need to have little resistance.
b) For accurate measurements, voltmeters need to have little resistance.
c) For accurate measurements, only one ammeter must be used in a circuit at a time.
d) For accurate measurements, only one voltmeter must be used in a circuit at a time.

33. In the circuit shown right, what happens to the readings of the ammeter and the voltmeter when switch S is closed?

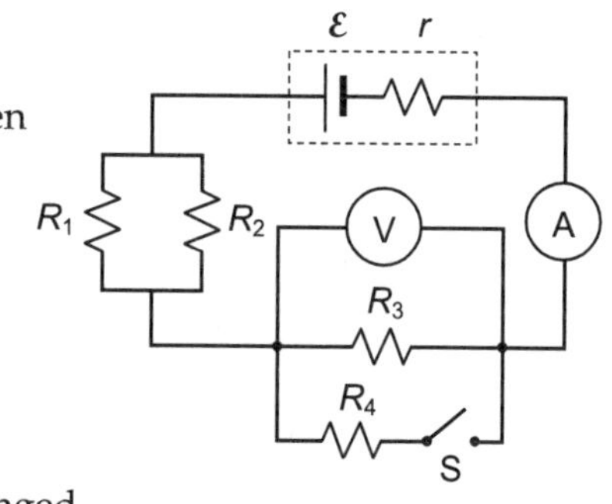

a) The ammeter reading decreases.
b) The voltmeter reading decreases.
c) The readings of both meters decrease.
d) The readings of both meters remain unchanged.

34. Which of the following electrical appliances consumes the most energy when turned on for equal amount of time?

a) cell phone b) computer c) oven d) TV

35. Too much current flowing in a circuit is dangerous because

a) the battery may explode.
b) electrons may escape the wire and cause an electric shock.
c) the wire may get hot enough to cause a fire.
d) any light bulb in the circuit may be shattered.

36 – 50. Written-response questions

36. A 130-W vacuum machine is designed to operate from a 110-V source. If the machine is operated using a 220-V source, what will be the power consumed by the machine?

37. Part of an electric circuit is shown right. Find the magnitude and direction of the current passing through R_1 and R_2.

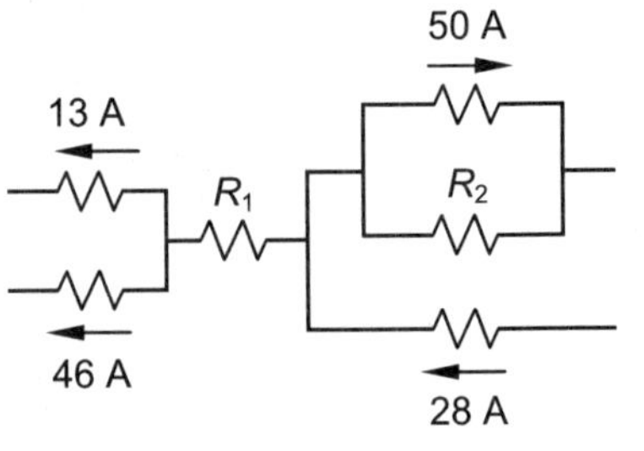

38. A battery with the internal resistance of 0.77 Ω supplies a 1.8-A current to the circuit. If the terminal voltage of the battery is 6.1 V, what is the battery's emf?

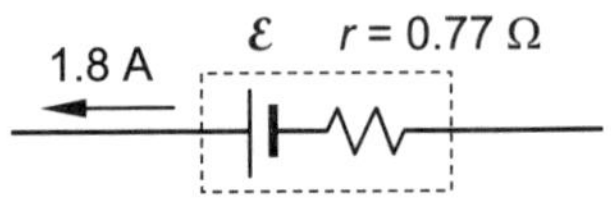

39. In the circuit shown right, what is the internal resistance of the battery?

$\mathcal{E}$ = 8.2 V I = 1.2 A
R_1 = 3.2 Ω R_2 = 4.2 Ω R_3 = 4.6 Ω
R_4 = 5.6 Ω R_5 = 6.6 Ω R_6 = 2.9 Ω

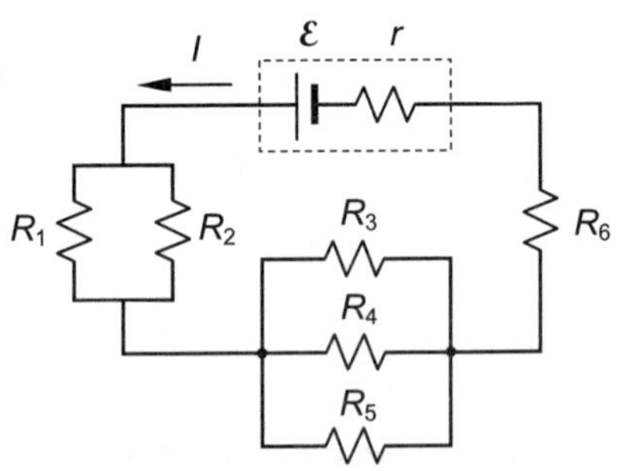

40. Part of an electric circuit is shown right. Find
a) the current through the 25-Ω resistor
b) the voltage drop across the 19-Ω resistor.

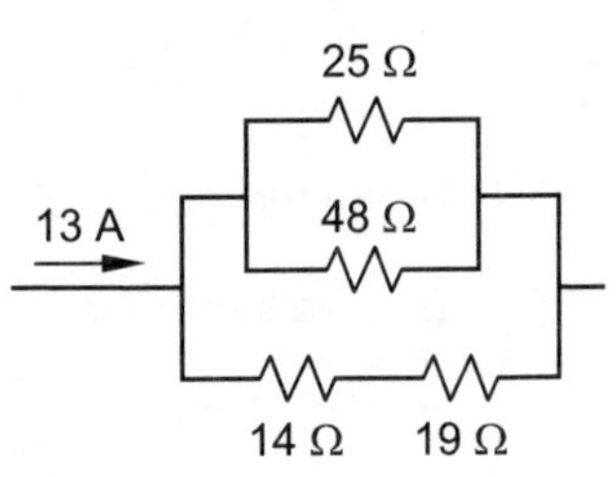

41. Four identical resistors are connected as shown. Find

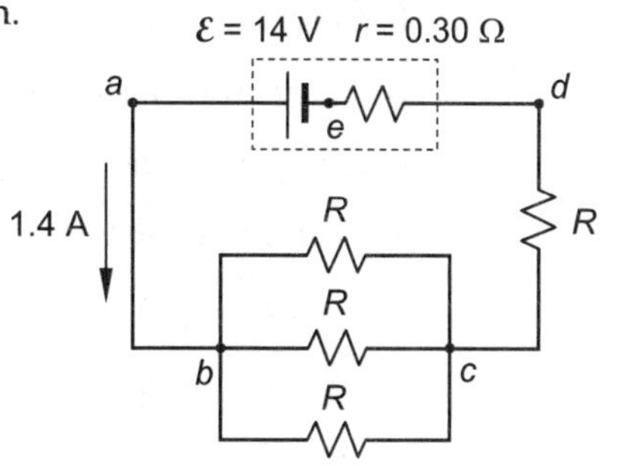

a) the terminal voltage of the battery
b) R_{eq} of the whole circuit
c) the resistance of each resistor, R
d) the power dissipated by each resistor
e) the potential difference between b and c
f) the potential difference between c and e
g) the number of electrons that pass through point d in 3.7 seconds.

42. Two light bulbs are connected in parallel as shown. If one bulb dissipates 5.3 W of power and the other twice that, find

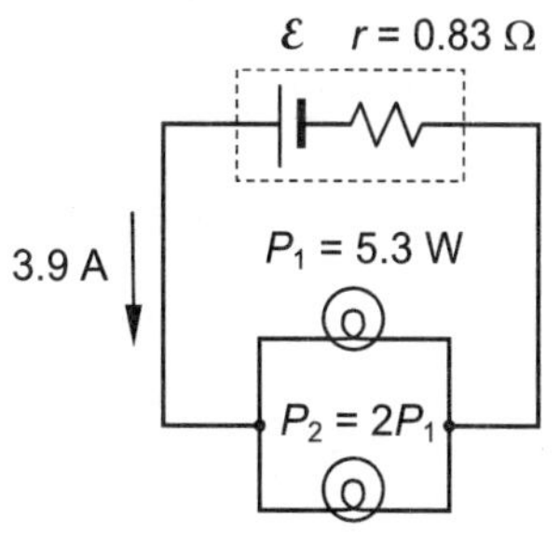

a) the resistance of each bulb
b) the terminal voltage of the battery
c) the emf of the battery

43. In the circuit shown right, the voltmeter reading is 2.9 V.

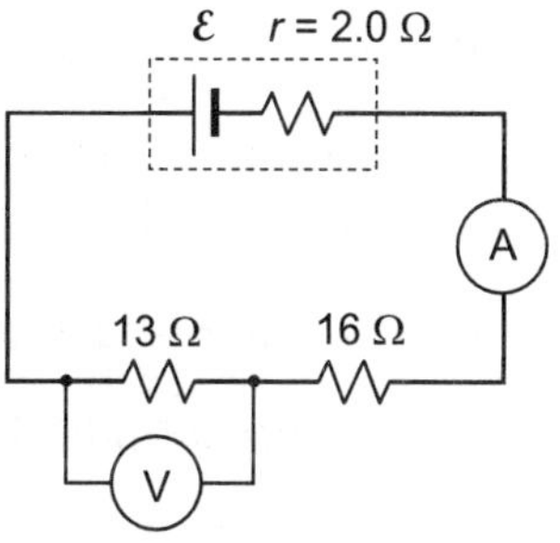

a) Find the reading of the ammeter.
b) Find the emf of the battery.

44. In the circuit shown right, the power dissipated in the 7.3-Ω resistor is 170 W. Find the power supplied by the battery.

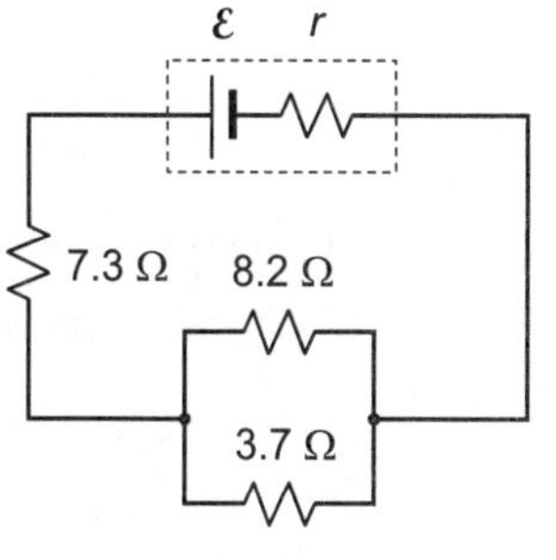

45. In the circuit shown right, two identical light bulbs are connected in parallel, and each bulb consumes 23 J of energy per second.

 a) Let the current out of the battery be I, and express the power supplied by the battery in terms of I.

 b) Setting the power supplied by the battery equal to the total power dissipated by both bulbs, find the possible values of I.

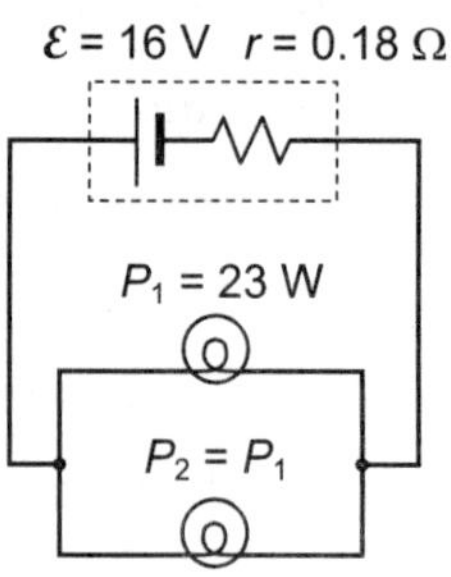

46. In the circuit shown right, find

 a) the equivalent resistance of the circuit

 b) the current through each resistor

 c) the terminal voltage of the battery

 d) the potential difference between points a and b

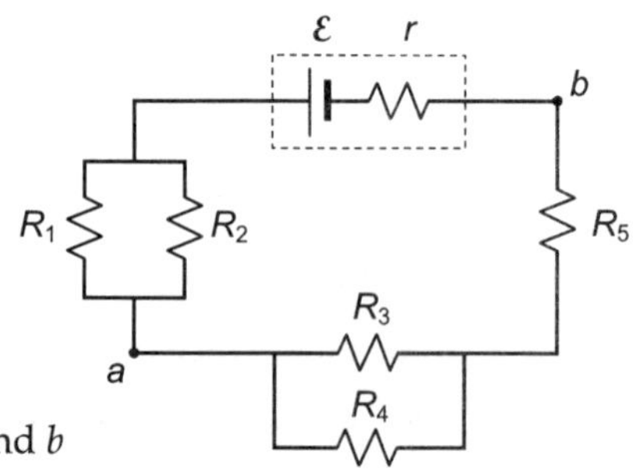

$\mathcal{E}$ = 15 V r = 1.1 Ω
R_1 = 8.5 Ω R_2 = 2.3 Ω R_3 = 5.6 Ω
R_4 = 9.7 Ω R_5 = 4.9 Ω

47. In the circuit shown in Problem 46, points a and b are connected by a copper wire as shown. Find

 a) R_{eq} of the whole circuit

 b) the total current out of the battery

 c) the potential difference between a and b

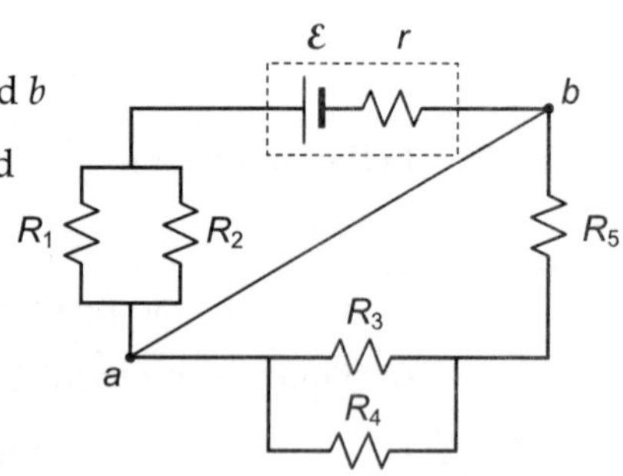

48. In the circuit shown right, switch S is initially left open, and the ammeter reading is 0.84 A.

 a) Find the internal resistance within the battery.

 b) Now the switch is closed. Find the terminal voltage of the battery.

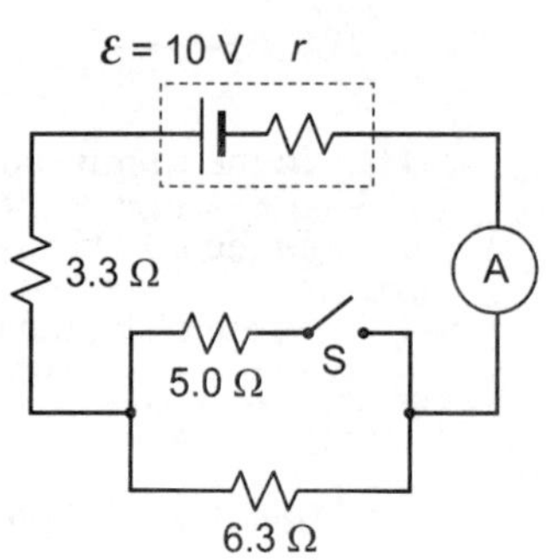

49. In the circuit shown right, the voltmeter reading is 22 V when $R_2 = 2R_1$. When the lower resistor (R_2) is replaced with a different resistor, the voltmeter reading drops to 12 V. Find the internal resistance within the battery and the total current in the circuit after the replacement.

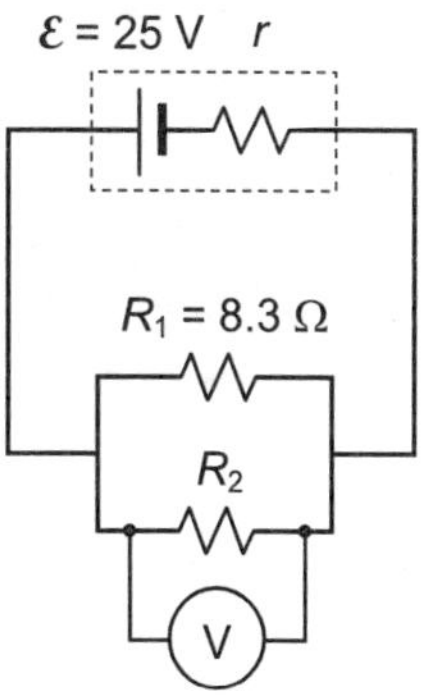

50.* The light bulb in the circuit consumes 19 W when connected to the battery shown. If the battery is replaced with another battery with twice the emf but the same internal resistance, find the power consumed by the bulb.

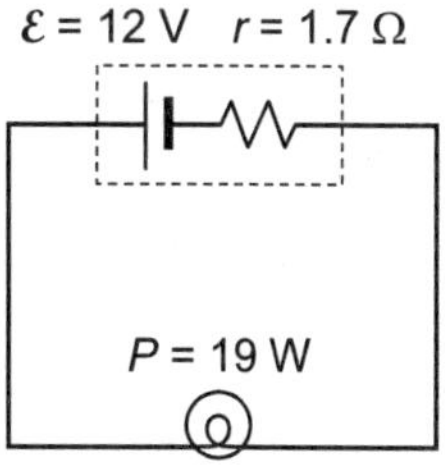

10

CHAPTER 10

ELECTROMAGNETISM

In 1820 the Danish physicist Oersted demonstrated the relationship between electricity and magnetism, and this opened the door to huge changes in the scientific world and in everyday life. Because of our understanding of electromagnetism, we now enjoy instant communication using radio, television, cell phones, etc. Using coils of wire rotating in magnetic fields, we have developed generators, which provide electrical energy to people throughout the industrialized world. In this chapter we will look at the principles underlying electromagnetic effects, and discuss motors, generators, transformers, and other devices that we take for granted today – but our great grandparents would have marveled at!

In this chapter, you will:

- ✓ learn about magnetic fields produced by moving charges, electric currents, and solenoids
- ✓ learn about magnetic force acting on a moving charge or a current-carrying wire
- ✓ solve exercise problems on magnetic field and magnetic force
- ✓ learn Lenz's law and Faraday's law
- ✓ learn about devices that utilize electromagnetic induction: generator, transformer, and motor
- ✓ solve exercise problems on electromagnetic induction
- ✓ test your knowledge of electromagnetism

10-1. Magnet and Magnetic Field

Magnetism has been known for thousands of years and used for navigation since the tenth century. The Earth's magnetic field is like that of a huge bar magnet, with a magnetic north pole near the geographic south pole, as shown in Fig. 10-1. Because like poles repel and opposite poles attract each other, a bar magnet will settle in the north-south direction if allowed to swing freely. Its north-seeking pole will point toward the Earth's geographic north pole (which is near the Earth's magnetic south pole).

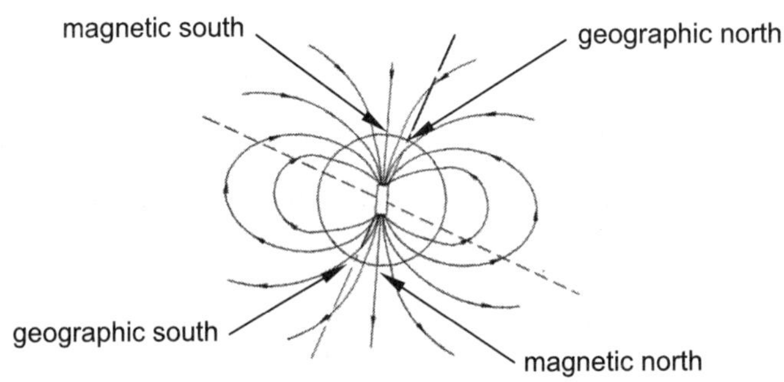

Figure 10-1: Earth's magnetic field

To determine the direction of a magnetic field, we can use a compass or a bar magnet. Namely, it is the direction in which the north pole of a compass or bar magnet points. Figure 10-2 (a) shows the magnetic field vectors determined this way at various points around a bar magnet.

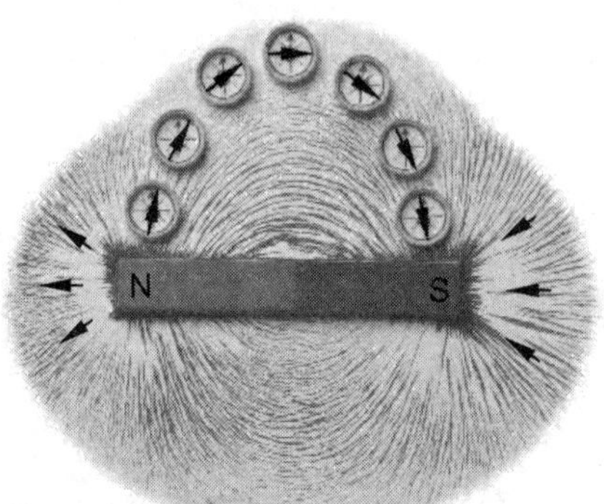

(a) magnetic field vectors

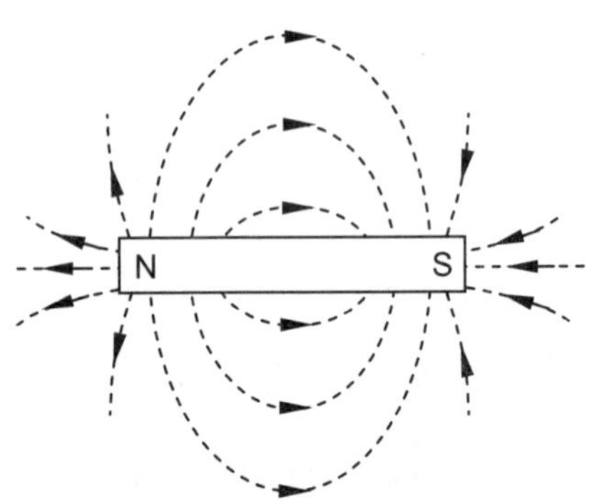

(b) magnetic field lines

Figure 10-2: Magnetic field around a bar magnet

Using the field vectors, we can draw **magnetic field lines** so that the vectors become tangent to the lines. (Recall that we drew electric field lines using the same approach in Chapter 8.) The magnetic field lines for a bar magnet are shown in Fig. 10-2 (b). Notice that they look similar to the electric field lines drawn for a pair of unlike charges (see Fig. 8-8 (a), p. 236). The two sets of lines indeed share many characteristics. For example, the closer the lines are together, the greater the field strength. The magnetic field is thus strongest near the poles. However, there is one important difference between the two sets of lines. Magnetic field lines are continuous and do not have a starting or ending point. (Note that electric field lines start from positive charges and terminate at negative ones.) Although the magnetic field lines appear to start from the north pole and end at the south pole, they actually continue through the magnet. In Fig. 10-2 (b), the magnetic field within the magnet is in general towards the left, from the south pole to the north.

10-2. The Magnetic Field around Moving Charges or Current-Carrying Wires

When Oersted realized that the current flowing in a wire produced its own magnetic field around the wire, the connection between moving electric charges and magnetism was established. We now believe that all magnetism is a result of moving charges – electrons moving around the nucleus of atoms and spinning on their axes. When a positive charge moves in a given direction, or when a conventional current flows through a straight conductor, the magnetic field that it produces consists of field lines that form concentric circles with the charge or the wire at the center. The diagrams in Fig. 10-3 show magnetic field lines around a moving positive charge and a wire carrying electric current.

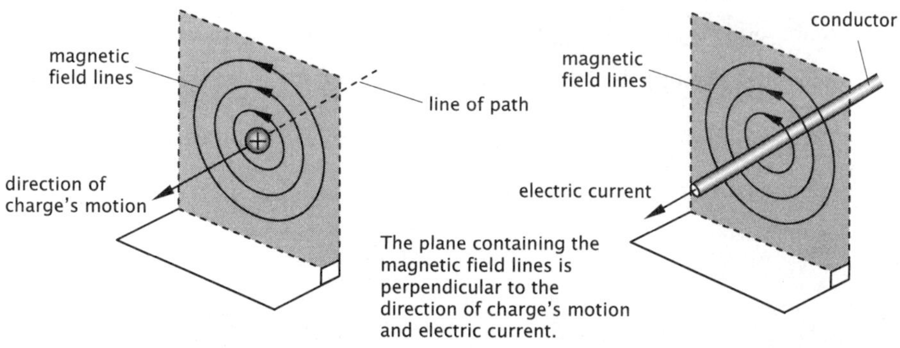

(a) around a moving positive charge (b) around a current-carrying wire

Figure 10-3: Magnetic field lines

The magnetic field, denoted by $\vec{B}$, is a vector as are all other field quantities. The direction of magnetic field lines around a moving positive charge is found by applying the following 'right-hand rule' (RHR #1):

> You grasp the line of path (shown in Fig. 10-3) with your right hand so that your thumb points in the direction of charge's motion. Your fingers will now encircle the line of path in the direction of the magnetic field.

There are three right-hand rules introduced in this chapter. This is the first of them.

Using this right-hand rule, we can find the magnetic field around a positive charge moving out of this page or into the page. To denote the vectors out of and into the page, we use the symbols '⊙' and '⊗', which represent the tip and the tail end of an arrow, respectively. Then for a positive charge moving out of the page, you must point your right thumb toward yourself and find the magnetic field to be in a counterclockwise direction. As for a positive charge going into the page, your thumb now points away from yourself, and the magnetic field is in a clockwise direction, as shown in Fig. 10-4 (b).

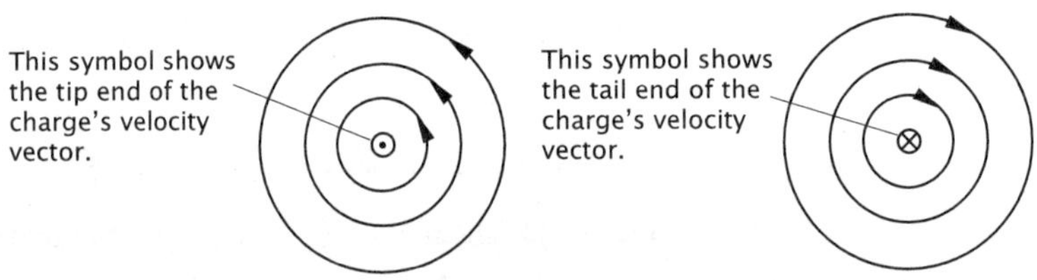

(a) moving out of page (b) moving into page

Figure 10-4: Magnetic field lines around a moving positive charge

The direction of the field lines around a moving negative charge is the opposite to that around a positive charge. Thus, a negative charge coming out of this page generates a magnetic field in a clockwise direction.

The magnitude of the magnetic field is measured in units of Tesla (T). Around a current-carrying conductor, it is given as

$$B = \frac{\mu_0}{2\pi}\frac{I}{r} \quad \ldots (10.1)$$

where I is the current, and r is the distance from the conductor. μ_0 is a proportionality constant called the **permeability of free space**, whose value is

$$\mu_0 = 4\pi \times 10^{-7}\ \mathrm{T{\cdot}m/A} \quad \ldots (10.2)$$

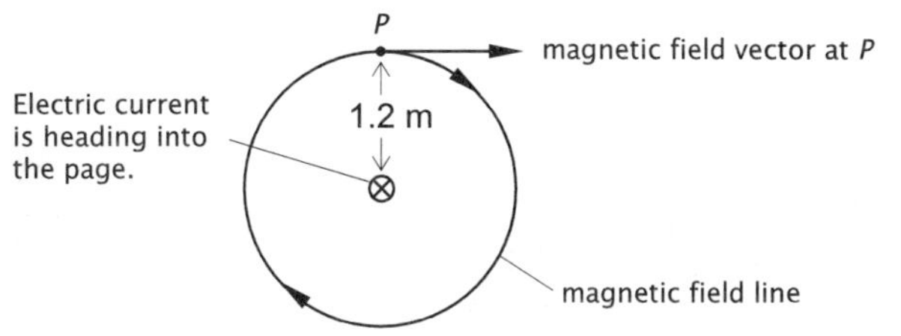

Figure 10-5: Finding magnetic field around a current-carrying wire

Suppose a wire carries a current of 0.33 A into the page, as shown in Fig. 10-5. At point P, 1.2 m directly above the wire, the magnetic field strength is

$$B = \frac{4\pi \times 10^{-7}\ \mathrm{T \cdot m/A}}{2\pi}\frac{0.33\,\mathrm{A}}{1.2\,\mathrm{m}} = 5.5 \times 10^{-8}\,\mathrm{T}$$

The direction of the magnetic field vector at P is to the right since the field vectors are always tangent to the field lines.

10-3. Magnetic Field around a Current-Carrying Loop of Wire and a Solenoid

Along the length of a current-carrying conductor, the magnetic field is a series of concentric cylinders. Because they are three-dimensional, we can only represent them on this two-dimensional page as shown in Fig. 10-6.

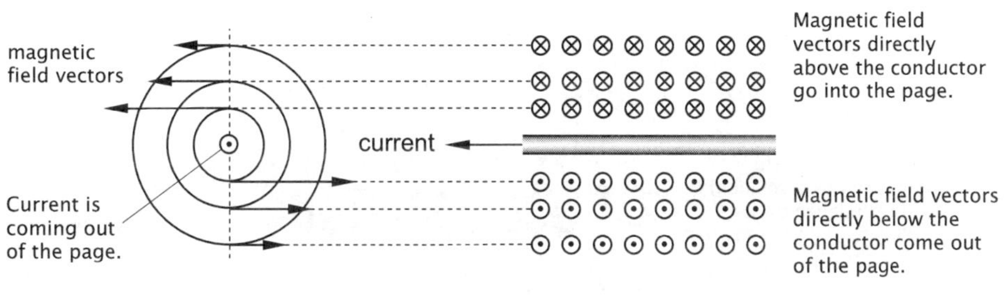

Figure 10-6: Two-dimensional representation of the magnetic field around a conductor

Now, if we form a loop of wire by bending it as shown in Fig. 10-7 (a), the magnetic field generated by the current inside the loop will be heading into the page. Figure 10-7 (b) shows the magnetic field lines going through the loop. The lines are curved and form circular loops themselves, but their direction at the plane of the wire loop is to the right.

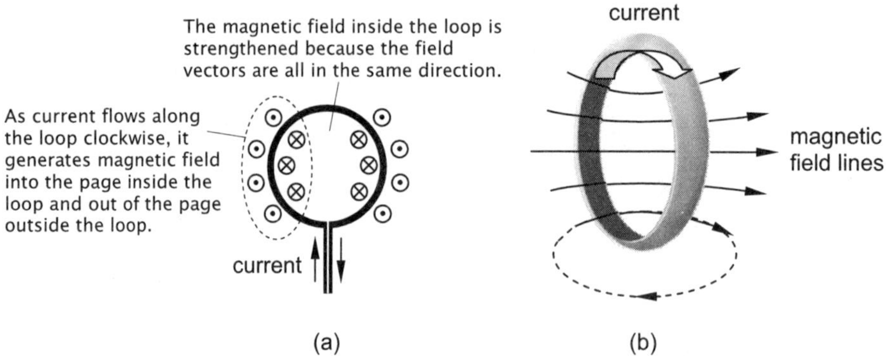

Figure 10-7: Magnetic field inside a loop of current-carrying wire

We can then take this idea one step further. Imagine a wire forming many loops, called a **solenoid**, as shown in Fig. 10-8. When a current flows through this solenoid, each loop produces a magnetic field as shown in Fig. 10-7 (b). Inside those loops, the magnetic field will be directed to the right. In fact, when the loops are spaced close together, the field inside them becomes parallel to the axis, as shown in Fig. 10-8. The field strength inside the solenoid is uniform and is given by

$$B = \mu_0 \left(\frac{N}{L}\right) I \qquad \text{... (10.3)}$$

where N is the number of loops and L is the distance over which the loops are wound. Hence, the more tightly wound the coils, the greater the magnitude of the magnetic field inside the solenoid.

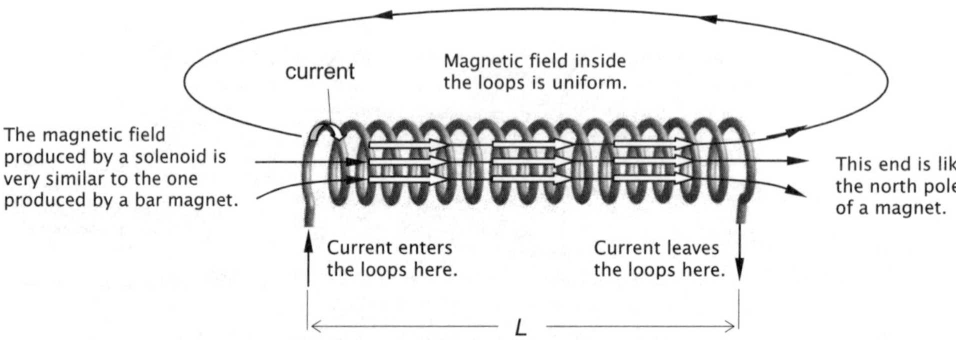

Figure 10-8: Magnetic field inside a solenoid

For instance, the solenoid in Fig. 10-8 has 16 loops. If it carries a 0.44-A current and the length between its ends is 0.87 cm, the magnetic field strength inside it is

$$B = \left(4\pi \times 10^{-7}\ \text{T}\cdot\text{m/A}\right)\left(\frac{16}{0.87 \times 10^{-2}\ \text{m}}\right)(0.44\ \text{A}) = 1.0 \times 10^{-3}\ \text{T}$$

With the current entering the solenoid as shown in Fig. 10-8, the direction of the magnetic field inside the solenoid is to the right. If the direction of the current is reversed, the direction of the magnetic field will be reversed as well. To obtain the direction of the magnetic field inside a solenoid easily, use the following right-hand rule (RHR #2):

> You follow the current along the loops with your fingers. If the current is coming toward you over the top of the loops, your thumb will point to the right (see Fig. 10-9 (a)). On the other hand, if the current is going away from you over the top of the loops, your thumb will point to the left (see Fig. 10-9 (b)). Your thumb indicates the direction of the magnetic field inside the solenoid.

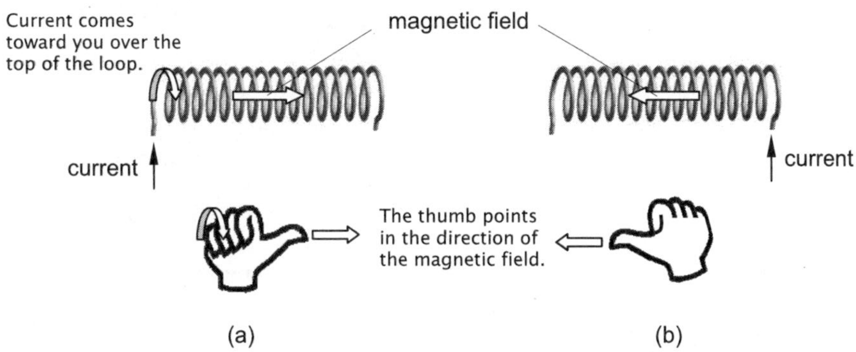

Figure 10-9: Right-hand rule for solenoids

When a solenoid carries a current, it produces a magnetic field similar to the one produced by a bar magnet, and, for this reason, a solenoid is also called an **electromagnet**. Often 'soft' iron is placed inside a solenoid to increase the strength of the magnetic field. Electromagnets are used in bells, switches, motors, and generators.

10-4. Magnetic Force on a Moving Charge

In Chapter 8 we learned that every charge creates an electric field around itself, and when a charge is within the electric field generated by another charge, it experiences an electric force, which arises from the interaction between two electric fields. Similarly, two magnetic fields interact to produce a magnetic force. This occurs, for example, when a charge moves in an external magnetic field (generated by a magnet, a current, or another moving charge). In this case, the moving charge generates its own magnetic field but is at the same time within the field of another. Suppose a positive charge is moving into the page in a magnetic field directed to the right, as shown in Fig. 10-10. Directly above the charge, the magnetic field generated by the charge is in the same direction as the external field, which produces a strong downward push on the charge. Directly below the charge, however, the magnetic field generated by the charge is in the opposite direction to the external field, and such interaction produces a weak upward push against the charge. The net effect on the charge is a downward force.

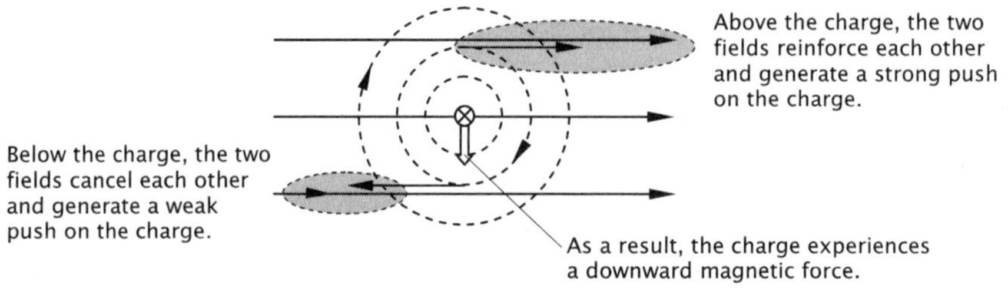

Figure 10-10: Magnetic force on a moving charge

We can find the direction of the magnetic force on a moving charge systematically by applying a third right-hand rule (RHR #3), shown in Fig. 10-11. The direction of the magnetic force is always perpendicular to both the magnetic field and the charge's velocity vector. Take caution in that this right-hand rule applies only to positive charges. Negative charges always experience a magnetic force in the opposite direction. Therefore, for a negative charge, you may assume it to be a positive charge, find the direction of the magnetic force using RHR #3, and then reverse the direction of the force vector at the end.

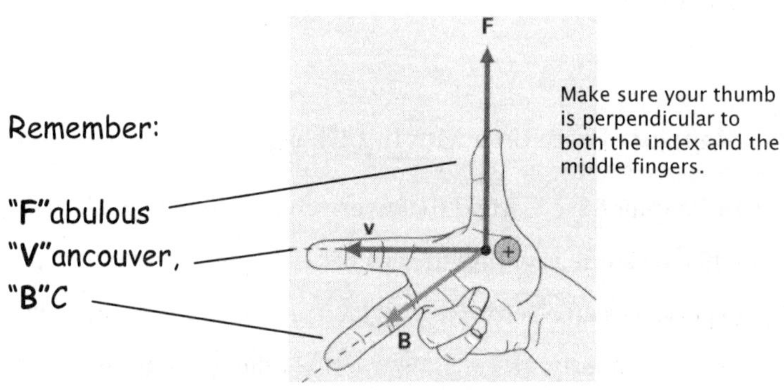

Figure 10-11: Right-hand rule for magnetic force

The magnitude of the magnetic force on the moving charge is given by

$$F_m = |q| v B \sin\theta \quad \dots (10.4)$$

where B is the magnitude of the magnetic field the charge is in (not the one that it generates!) and θ is the angle between that magnetic field and the velocity vector, as indicated in Fig. 10-12. According to Eq. (10.4), the magnetic force becomes a maximum when the charge travels perpendicular to the field lines (that is, when $\theta = 90°$) and becomes zero when it moves parallel to the field lines (that is, when $\theta = 0°$ or $180°$).

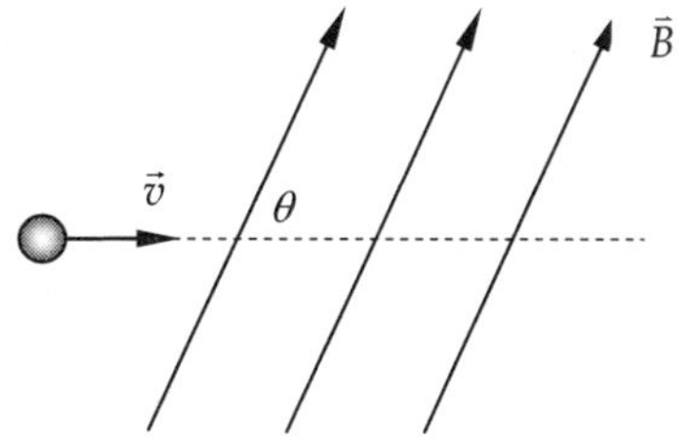

Figure 10-12: Finding angle between velocity and magnetic field

Consider a proton entering a uniform magnetic field at the speed of 2.9×10^5 m/s, as shown in Fig. 10-13. The magnetic field is directed out of the page, and its strength is 0.74 T. According to the figure, the proton's velocity is perpendicular to the magnetic field, so the value of θ in Eq. (10.4) is 90°. Thus, the magnitude of the magnetic force on the proton is

$$F_m = (1.6 \times 10^{-19}\,\text{C})(2.9 \times 10^5\,\text{m/s})(0.74\,\text{T})\sin 90° = 3.4 \times 10^{-14}\,\text{N}$$

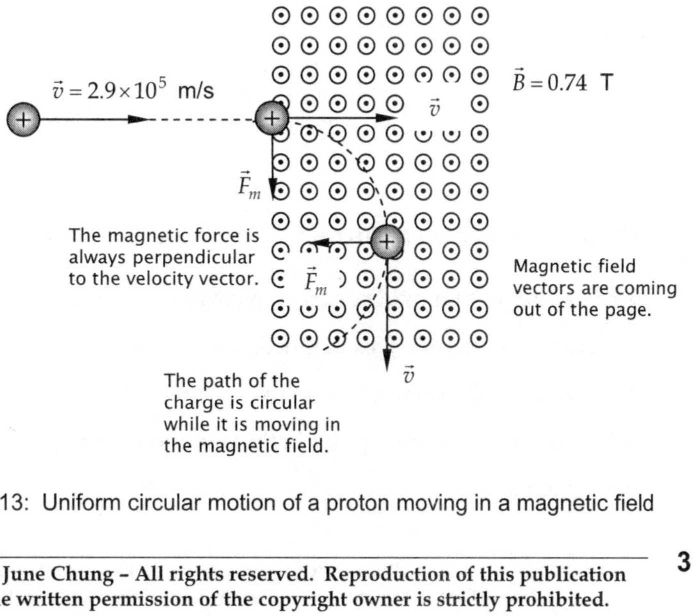

Figure 10-13: Uniform circular motion of a proton moving in a magnetic field

To determine the direction of the magnetic force on the proton just as it enters the magnetic field, use RHR #3, and you'll find that it is downward, as shown in Fig. 10-13. As a result, the proton's path of travel becomes curved, but because the magnetic force remains perpendicular to the proton's velocity, the proton undergoes uniform circular motion while it is moving in the magnetic field.

Recall from Chapter 7 that the equation of motion for an object undergoing uniform circular motion is

$$F_C = m\,a_C = m\frac{v^2}{R} \qquad \ldots (10.5)$$

Since the magnetic force is the centripetal force for the proton in Fig. 10-13, we may write Eq. (10.5) as

sin θ has been assumed to be 1. ⟶

$$|q|\,v\,B = m\frac{v^2}{R} \qquad \ldots (10.6)$$

Then the radius of curvature, R, is expressed as

$$R = \frac{m\,v}{|q|\,B} \qquad \ldots (10.7)$$

⟵ The numerator is the charge's momentum.

Therefore, the proton in Fig. 10-13, upon entering the magnetic field, follows a circular path with the radius equal to

$$R = \frac{(1.67\times10^{-27}\ \text{kg})(2.9\times10^{5}\ \text{m/s})}{(1.6\times10^{-19}\ \text{C})(0.74\ \text{T})} = 4.1\times10^{-3}\ \text{m}$$

In the above example, if it were an electron that entered the magnetic field at 2.9×10^5 m/s, it would experience a magnetic force of the same magnitude as that on the proton, but its path would bend upward. Because of its smaller mass, the radius of its circular path would be much smaller than that for the proton:

$$R = \frac{(9.11\times10^{-31}\ \text{kg})(2.9\times10^{5}\ \text{m/s})}{(1.6\times10^{-19}\ \text{C})(0.74\ \text{T})} = 2.2\times10^{-6}\ \text{m}$$

10-5. Magnetic Force on a Current-Carrying Conductor

Being a stream of moving charges, a current traveling along a conductor placed in a magnetic field experiences magnetic force as well. You can still use RHR #3 to determine the direction of the magnetic force on the conductor, but the index finger in Fig. 10-11 should now point in the direction of the current. The magnitude of the force is equal to

$$F_m = I\,L\,B\sin\theta \qquad \ldots (10.8)$$

where L is the length of the wire segment within the magnetic field and θ is the angle between the direction of the current and the magnetic field. Consider the wire carrying a current of 0.23 A in Fig. 10-14. Placed in a uniform magnetic field directed upward, it experiences a force out of the page, according to RHR #3. The magnitude of the force is

$$F_m = (0.23\,\text{A})(0.36\,\text{m})(1.6\,\text{T})\sin 108° = 0.13\,\text{N}$$

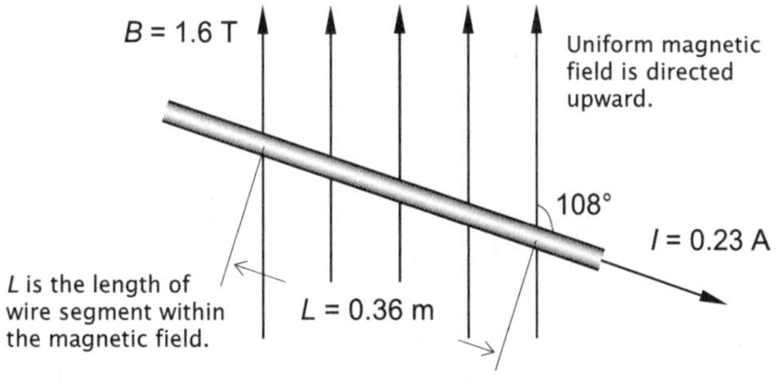

Figure 10-14: A current-carrying wire placed in a magnetic field

Note that B in Eq. (10.8) is the strength of a magnetic field generated by something other than the current in the conductor. The magnetic field created by a moving charge or a current can exert a magnetic force on other moving charges or wires but cannot exert a force on itself. Therefore, charges and currents experiencing any magnetic force must be moving through magnetic fields that have already been created (by another current or magnet).

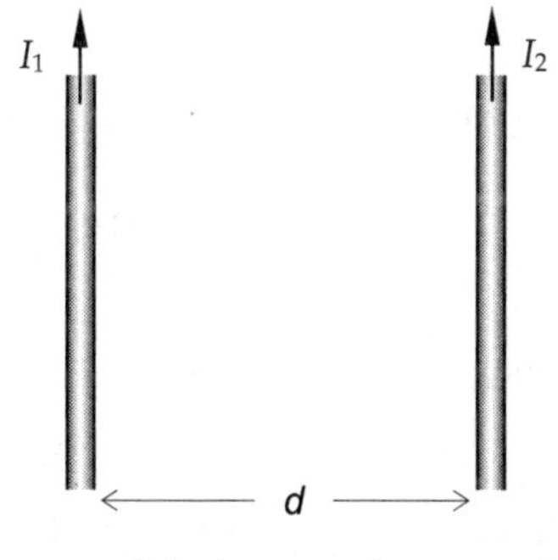

Figure 10-15: Two parallel wires carrying current in the same direction

For example, suppose we have two parallel wires carrying current in the same direction as shown in Fig. 10-15. While each current creates a magnetic field (as described in Section 10-2), the force on each wire is due to the magnetic field generated by the other current. That is, the wire carrying I_1, located to the left of the wire carrying I_2, is in a magnetic field generated by I_2. The direction of that magnetic field is out of the page, and its strength at the left wire is

$$B_2 = \frac{\mu}{2\pi}\frac{I_2}{d} \quad \ldots (10.9)$$

Using RHR #3, we can find that the magnetic force on the left wire is to the right, as shown in Fig. 10-16 (a). The magnitude of this force, per unit length, is

$$\frac{F_1}{L} = I_1 B_2 \sin 90° = \frac{\mu}{2\pi}\frac{I_1 I_2}{d} \quad \ldots (10.10)$$

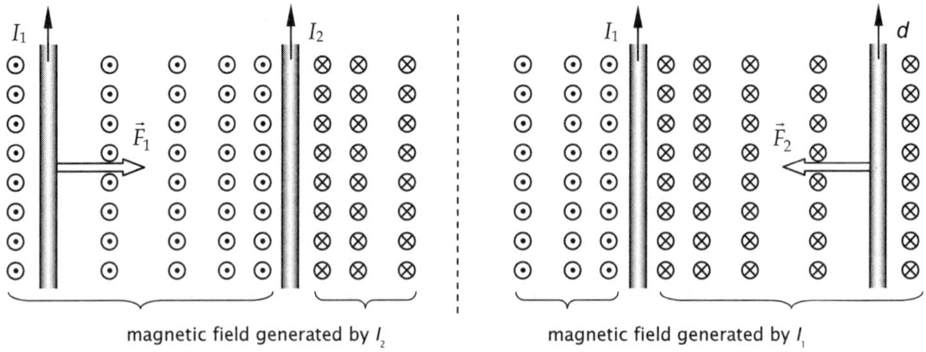

(a) magnetic force on the left wire (b) magnetic force on the right wire

Figure 10-16: Magnetic force on the parallel wires carrying current in the same direction

As for the right wire, it is in a magnetic field going into the page, as shown in Fig. 10-16 (b). The magnetic field strength at the right wire is

$$B_1 = \frac{\mu}{2\pi}\frac{I_1}{d} \quad \ldots (10.11)$$

Then the magnitude of the magnetic force on the right wire, per unit length, is

$$\frac{F_2}{L} = I_2 B_1 \sin 90° = \frac{\mu}{2\pi}\frac{I_1 I_2}{d} \quad \ldots (10.12)$$

The direction of this magnetic force is to the left (from RHR #3), as shown in Fig. 10-16 (b). Note that the two wires experience the same magnitude of force per unit length, and the forces are in opposite directions, toward each other. Indeed, these two forces form an action-reaction pair.

PHYSICS 12 ANALYZED: ELECTROMAGNETISM

10-1. Magnet and Magnetic Field

- A magnet produces a magnetic field around itself.
- To find the direction of the magnetic field at a certain point, place a magnet there. The direction to which the magnet's north pole points is the direction of the magnetic field at that point.
- Properties of magnetic field lines:
  - Magnetic field vectors are tangent to the lines.
  - The closer the lines, the stronger the field.
  - Magnetic field lines do not have starting or ending points.

10-2. The Magnetic Field around Moving Charges or Current-Carrying Wires

- Moving electric charges (and thus electric currents) generate magnetic fields, whose lines form concentric circles in the plane perpendicular to the charge's line of path.
- Finding the direction of the magnetic field: Use the right-hand rule #1 (p. 310)
- Finding the magnetic field strength: For a current-carrying conductor, $B = \frac{\mu_0}{2\pi}\frac{I}{r}$
- The direction of magnetic field lines around a moving <u>negative</u> charge is opposite to that around a moving positive charge.

10-3. Magnetic Field around a Current-Carrying Loop of Wire and a Solenoid

- A current-carrying solenoid, also called an electromagnet, generates a uniform magnetic field inside itself. The magnitude of the field is given by

$$B = \mu_0 \left(\frac{N}{L}\right) I$$

- Finding the direction of the magnetic field: Use the right-hand rule #2 (p. 313)

10-4. Magnetic Force on a Moving Charge

- While an electric charge is moving in an external magnetic field, it experiences a magnetic force. Its magnitude is given by

$$F_m = |q|\, v\, B \sin\theta$$

For negative charges, the direction is the opposite!

- To find the direction, use the right-hand rule #3 (p. 314)

- Since the magnetic force is always perpendicular to the direction of motion, a charge moving in a magnetic field undergoes uniform circular motion. The radius of the circular path is given by

$$R = \frac{m\,v}{|q|\,B}$$

10-5. Magnetic Force on a Current-Carrying Conductor

- A current-carrying wire placed in a magnetic field experiences a magnetic force whose magnitude is given by

$$F_m = I\,L\,B \sin\theta$$

- The direction of this force is determined by RHR #3 (p. 314).

- Two parallel wires carrying current in the same direction attract each other. Two parallel wires carrying current in opposite directions repel each other.

EXERCISES ON CHAPTER 10: ELECTROMAGNETISM

Exercise 10-1

A straight wire is carrying a 1.2-A current into the page, as shown. Find the magnitude and direction of the magnetic field created by the current at point *P*, located 2.5 cm directly above the wire.

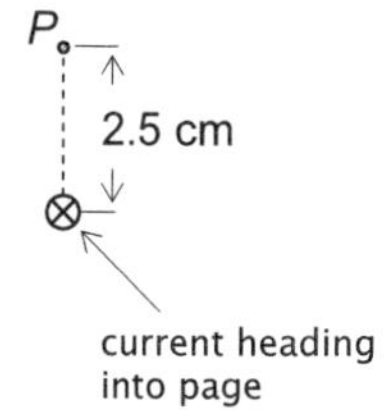

Exercise 10-2

A long straight wire is carrying a 0.34-A current. At what distance from the wire does the magnetic field strength become 2.7×10^{-6} T?

Exercise 10-3

An 8.6-cm long solenoid is connected to a circuit with a 3.5-V battery and a 6.4-Ω resistor as shown. The solenoid has a total of 160 turns of wire. When the switch S is closed, what is the magnitude and direction of the magnetic field inside the solenoid?

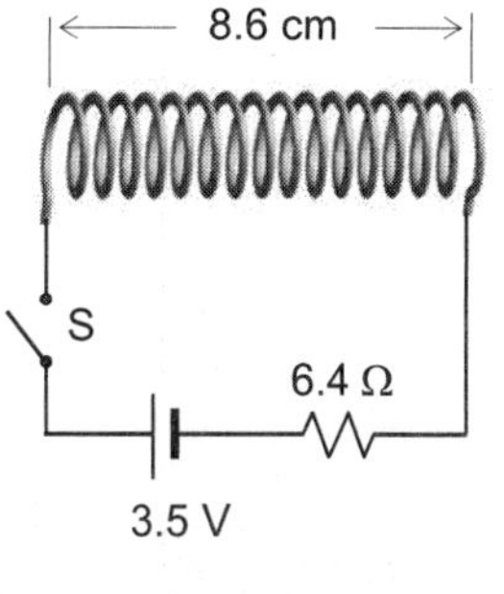

Exercise 10-4

An electron enters a region of uniform magnetic field of 0.74 T. If the electron's speed is 7.3×10^4 m/s, find

(a) the magnitude and direction of the magnetic force on the electron

(b) the radius of curvature of the electron's path

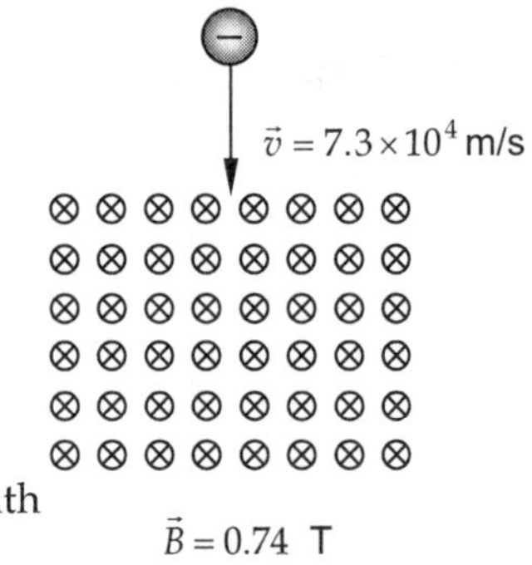

Exercise 10-5

An L-shaped conductor is carrying a 0.82-A current in a magnetic field as shown. The length of each segment, L, is 28 cm. Find the magnitude of the magnetic force on this conductor.

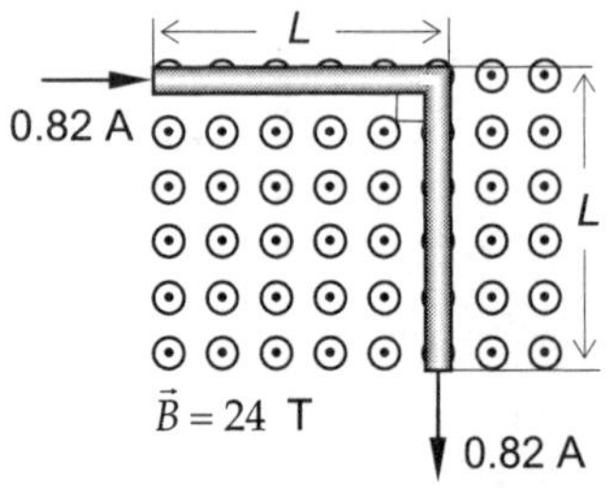

Exercise 10-6

An electron is moving parallel to a conductor carrying a current of 3.6 A as shown. When its speed is 6.3×10^4 m/s, it experiences a magnetic force of 2.2×10^{-20} N. Find

(a) the direction of the magnetic force on the electron

(b) the distance between the electron and the conductor

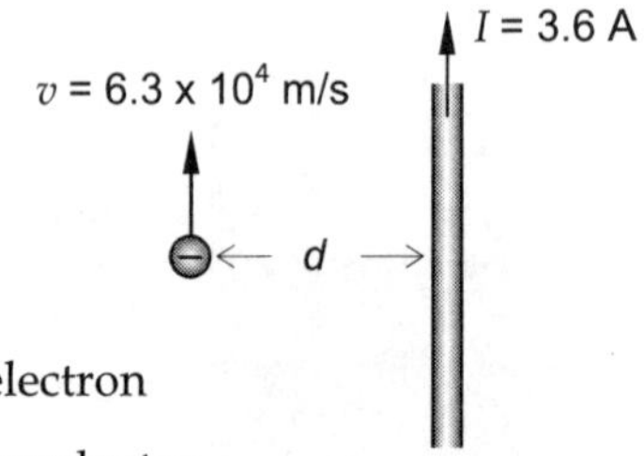

Exercise 10-7

An 3.8-cm long solenoid is connected to a circuit with a 6.0-V battery and a 2.5-Ω resistor as shown. The solenoid has a total of 520 turns of wire.

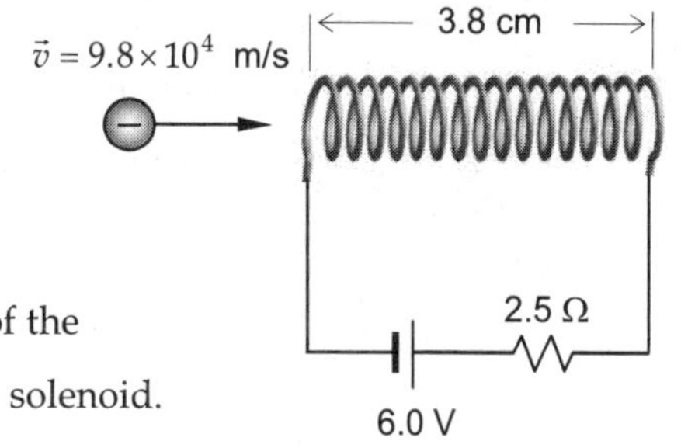

(a) Find the magnitude and direction of the magnetic field generated inside the solenoid.

(b) If an electron enters the solenoid at a speed of 9.8×10^4 m/s, what is the magnitude and direction of the magnetic force on the electron?

Exercise 10-8

An ion with a charge of 3*e* enters a uniform magnetic field of 3.6 T so that its velocity is perpendicular to the magnetic field. If the ion follows a circular path of radius 6.6 cm, what is the magnitude of the ion's momentum?

Exercise 10-9

A proton enters a region of electric and magnetic fields as shown. The potential difference between the plates is 320 V, and the magnetic field strength is 0.67 T. If the proton travels through the region in a straight line, what is its speed?

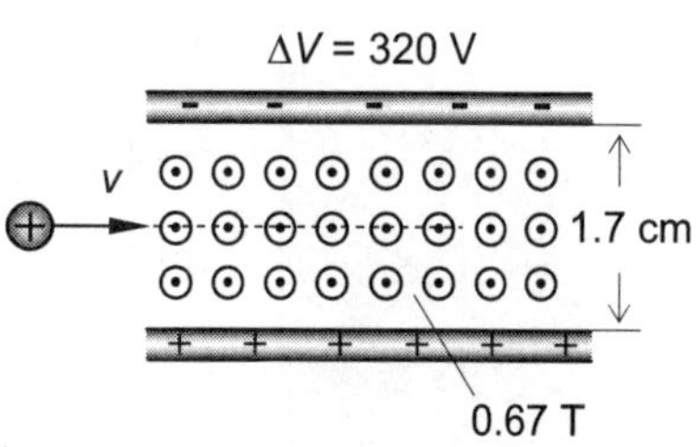

Exercise 10-10

An alpha particle undergoes a uniform circular motion in a magnetic field. If it travels around the circular path of radius 0.85 cm in 5.1×10^{-7} s, what is the strength of the magnetic field?

Exercise 10-11

A loop of wire, carrying a current of 3.7 A, is hanging vertically from a ceiling. The total mass of the loop is 48 g, and its weight is supported by the magnetic force on the loop. Find

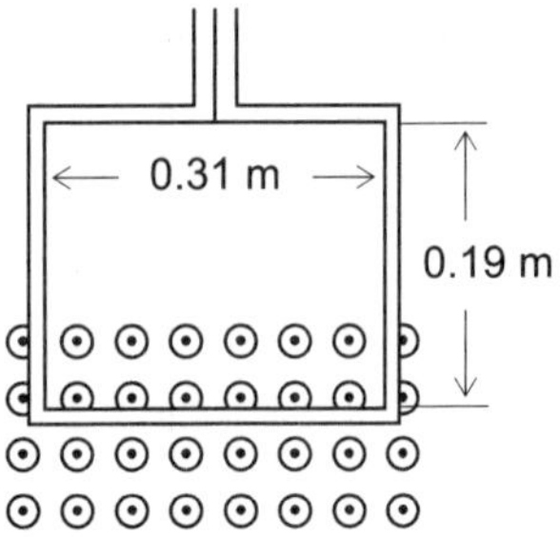

(a) the direction of the current in the loop (i.e., clockwise or counterclockwise)

(b) the magnetic field strength

Exercise 10-12

Two parallel wires are separated by 1.4 m and carry currents in opposite directions as shown. If $I_1 = 2.3$ A and $I_2 = 1.9$ A, find

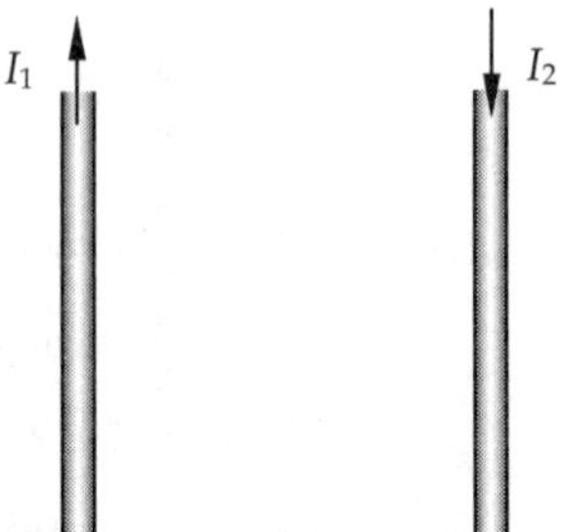

(a) the magnetic field strength felt by each wire

(b) the direction of the magnetic force on each wire

(c) the magnitude of the magnetic force per unit length on each wire

Exercise 10-13

An alpha particle enters a magnetic field at an angle of 73° as shown. If its speed is 3.8×10^3 m/s, find the magnitude and direction of the magnetic force on the particle.

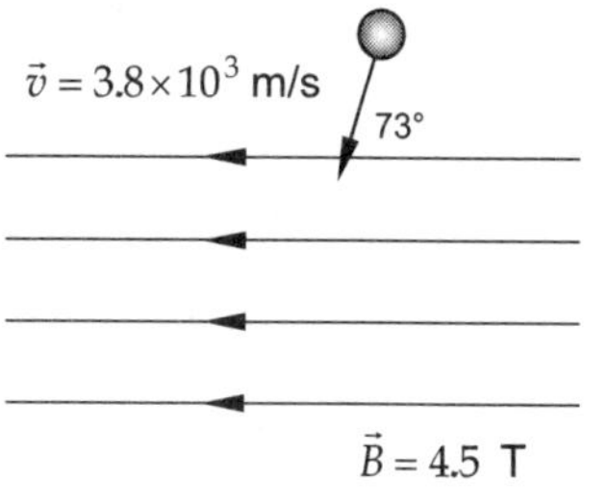

Exercise 10-14

A solenoid with 330 loops wound over 6.3 cm is carrying a current of 2.9 A. If a 3.5-cm segment of wire carrying a 1.1-A current is placed inside the solenoid as shown, what is the magnitude and direction of the magnetic force on that segment?

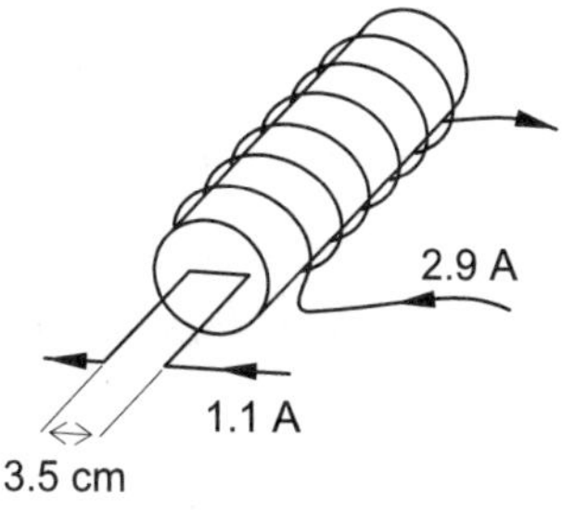

Exercise 10-15

A 45-g conductor is suspended in a magnetic field of 7.2 T by two wires. The conductor, carrying a 0.80-A current, experiences a horizontal magnetic force as shown. If the angle the wires make with the vertical magnetic field lines is 43°, find

(a) the magnitude of the magnetic force

(b) the length of the conductor

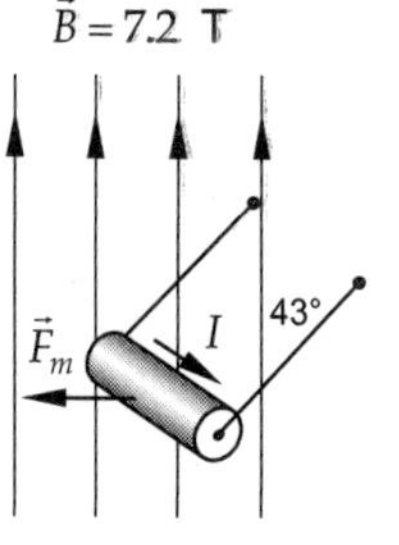

10-6. Magnetic Flux

Electricity and magnetism are closely related. In the first half of this chapter, we have seen an electric charge in motion producing a magnetic field. This allows us to generate a magnetic field in several ways using electricity. We can set a charge in motion, let current flow through a wire, or use a solenoid. It turns out that we can also utilize magnetism to generate electricity, which is the main topic of the remainder of this chapter.

To study how this is done, we first need to learn about what is called **magnetic flux**, which is defined as

The standard unit for magnetic flux is the Weber. ⟶

$$\Phi = BA\cos\theta \quad [\text{T}\cdot\text{m}^2 = \text{Wb}] \qquad \ldots (10.13)$$

where B is the magnetic field strength, A is the area of the loop in the magnetic field, and θ is the angle between the magnetic field vector and the area vector. The area vector is perpendicular to the plane of the loop. Consider the rectangular loops of wire shown in Fig. 10-17. Note that there is no current flowing in either wire, and, therefore, the electric charges in the wires are not generating any magnetic field. The plane of the loop is shaded grey, and the area vector in each figure is perpendicular to the plane. In fact, there are always two possible directions for the area vector. For instance, the area vector may be directed downward for the loop in Fig. 10-17 (a). You may choose any of the two directions, but once you set the direction of the area vector, you have to stick to it throughout the calculations.

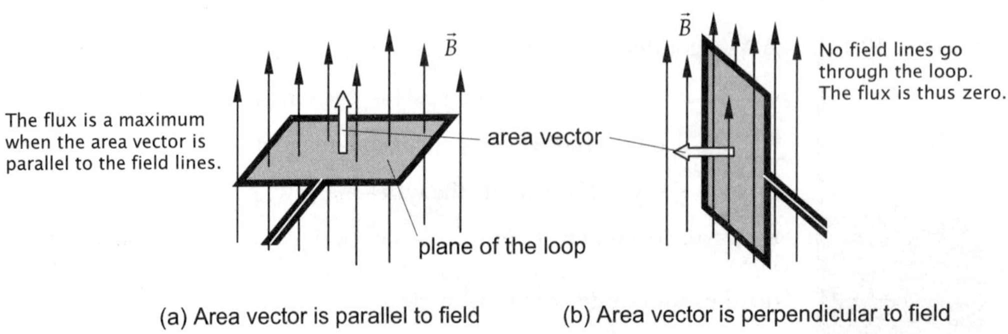

Figure 10-17: Magnetic flux

Suppose in Fig. 10-17 the magnetic field strength is 1.4 T, and the area of the loop is 0.16 m². Then the magnetic flux through the loop in Fig. 10-17 (a) is

$$\Phi = (1.4\,\text{T})(0.16\,\text{m}^2)\cos 0° = 0.22\,\text{Wb}$$

If the area vector is chosen to be downward, the angle between the area vector and the magnetic field would be 180°. Then the magnetic flux becomes

$$\Phi = (1.4\,\mathrm{T})(0.16\,\mathrm{m}^2)\cos 180° = -0.22\ \mathrm{Wb}$$

The magnetic flux through the loop in Fig. 10-17 (b) is zero because the area vector is perpendicular to the magnetic field (cos 90° = 0). It may help you to think of magnetic flux as the number of magnetic field lines going through a loop of wire. Since no magnetic field lines go through the loop in Fig. 10-17 (b), the magnetic flux is zero.

There are three ways to change the magnetic flux:

1. Change the magnetic field strength. (See Fig. 10-18.)
2. Change the area of the loop in the magnetic field. (See Fig. 10-19.)
3. Rotate the plane of the loop. (See Fig. 10-20.)

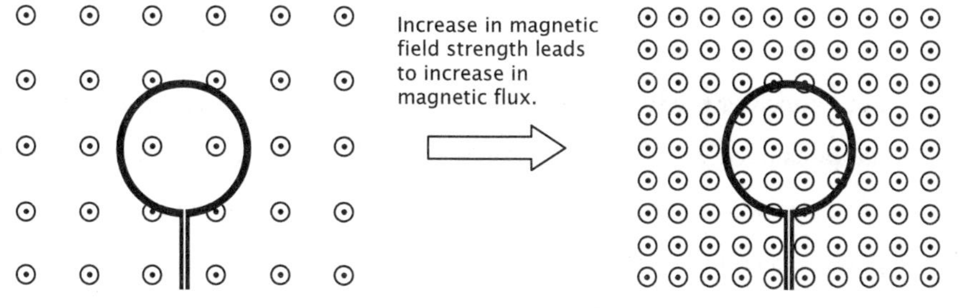

Figure 10-18: Changing magnetic flux by changing the magnetic field strength

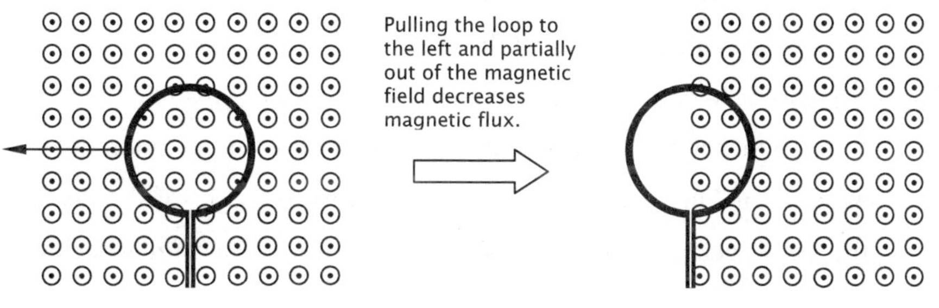

Figure 10-19: Changing magnetic flux by changing area of the loop in the magnetic field

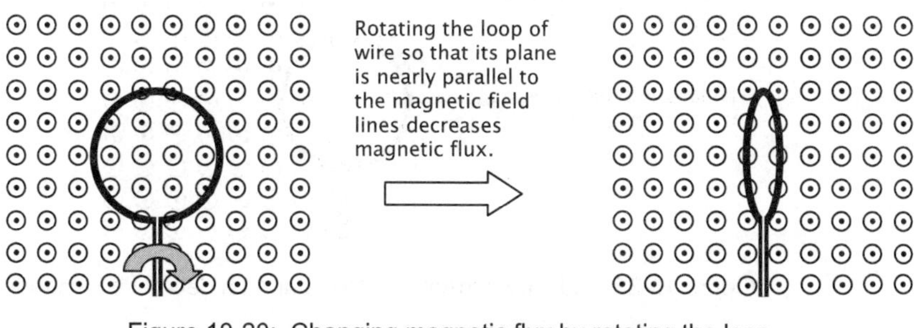

Figure 10-20: Changing magnetic flux by rotating the loop

10-7. Electromagnetic Induction and Lenz's Law

Whenever there is a change in magnetic flux through a loop of wire, an emf is 'induced' and forces electric charges in the wire to move. The resulting current is known as an **induced current**, which flows so that the magnetic field created by it opposes the change in the flux through the loop. The diagram shown in Fig. 10-21 illustrates this process, known as **electromagnetic induction**. In short, a change in magnetic flux triggers all subsequent events, but the induced current attempts to stop the process by canceling that change in magnetic flux.

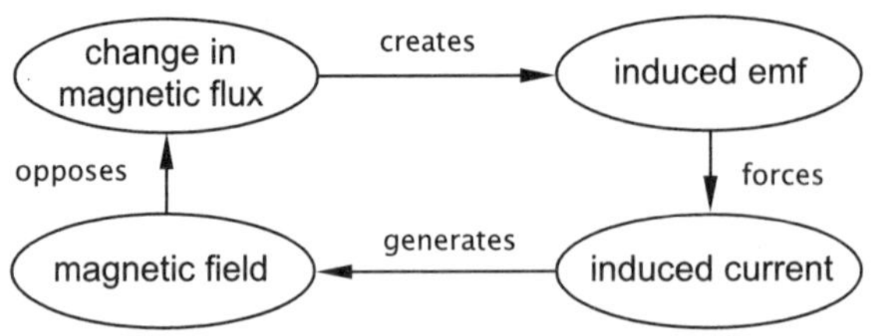

Figure 10-21: Electromagnetic induction

Let's consider the three events shown in Figs. 10-18, 10-19, and 10-20 and see what happens in each case. In Fig. 10-18, the magnetic field lines through the loop increases, and so does the magnetic flux. To cancel these field lines directed out the page, a magnetic field heading into the page is needed, and an induced current flowing in a clockwise direction accomplishes this (see Fig. 10-7 (a), p. 312). On the other hand, the magnetic field lines through the loop decreases in Fig. 10-19 while the loop is being pulled out of the magnetic field. Magnetic field coming out of the page needs to be generated inside the loop in order to maintain the original flux through the loop. Hence, an induced current flows in a counterclockwise direction, as shown in Fig. 10-22. Since the magnetic flux in Fig. 10-20 also decreases while the loop is being rotated, an induced current flows in a counterclockwise direction.

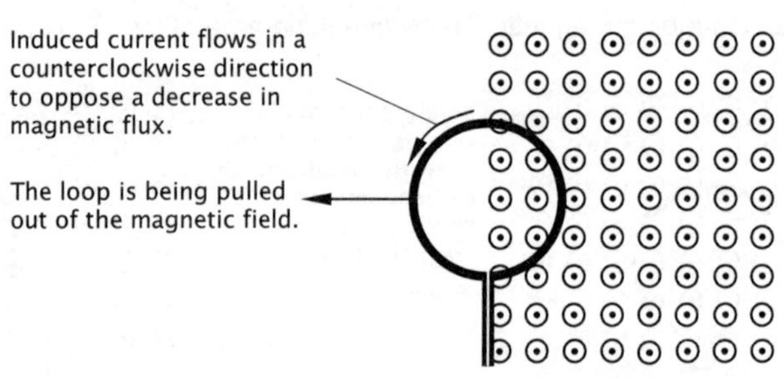

Figure 10-22: Induced current in a loop during a decrease in magnetic flux

Based on such observations, a Russian physicist named Lenz formulated **Lenz's law** in 1834, which states that "the direction of an induced current is such that its magnetic field opposes the change in the magnetic flux through the loop." In other words, an induced current generates a magnetic field so as to maintain the original magnetic flux through the loop. For instance, while the flux through the loop increases, an induced current produces a field directed in the opposite direction to cancel the excess. While the flux decreases, the current produces a field in the same direction to fill the shortage. Thus, when determining the direction of an induced current, you need to pay attention to the original flux through the loop and think about how it has changed.

10-8. Faraday's Law of Induction

Michael Faraday discovered electromagnetic induction in 1831 and formulated a relationship that allows quantitative analysis of electromagnetic induction. The relationship is known as **Faraday's law of induction**, which is expressed as

$$\mathcal{E}_{avg} = -N\frac{\Delta\Phi}{\Delta t} \quad [\text{Wb/s} = \text{V}] \qquad \dots (10.14)$$

induced emf ($\mathcal{E}_{avg}$); number of loops (N); rate of change in magnetic flux ($\frac{\Delta\Phi}{\Delta t}$)

The law states that an **induced emf** is directly proportional to the rate of change in magnetic flux. Thus, the faster the magnetic flux changes, the greater the induced emf will be. The negative sign in the equation is used to determine the direction of the induced current† and has nothing to do with the actual value of the induced emf. This means that, when calculating the induced emf, you should always take the absolute value of the right-hand side in Eq. (10.14).

Once the induced emf is obtained, the amount of induced current can be calculated by applying Ohm's law:

$$I = \frac{\mathcal{E}}{R} \qquad \dots (10.15)$$

induced current (I); induced emf ($\mathcal{E}$); total resistance in the loop(s) (R)

Let's look at a couple of examples and see how Faraday's law is applied in each. In the first example, a rectangular loop of wire is initially in a magnetic field directed

† Using the negative sign in Faraday's law to determine the direction of induced current is not introduced in this course.

into the page. In 0.23 seconds, the direction of the magnetic field is reversed, with the magnetic field strength unchanged, as shown in Fig. 10-23. To find the average induced emf during this time interval, we first need to find the magnetic flux before and after the change. Set the area vector to be into the page. Then

$$\Phi_i = (2.5\,\mathrm{T})[(0.71\,\mathrm{m})(0.93\,\mathrm{m})]\cos 0° = 1.65\ \mathrm{Wb}$$

$$\Phi_f = (2.5\,\mathrm{T})[(0.71\,\mathrm{m})(0.93\,\mathrm{m})]\cos 180° = -1.65\ \mathrm{Wb}$$

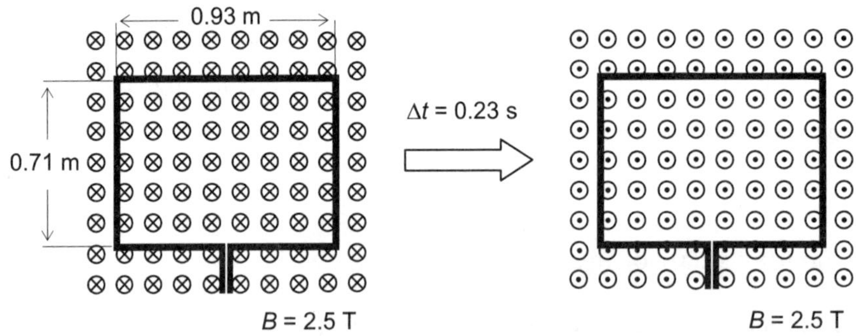

Figure 10-23: Electromagnetic induction example 1

Note that the final magnetic flux is negative because the angle between the area vector (into the page) and the magnetic field lines (out of the page) is 180°. Thus, the induced emf is

$$\mathcal{E}_{avg} = -N\frac{\Phi_f - \Phi_i}{\Delta t} = -(1)\frac{-1.65\,\mathrm{Wb} - 1.65\,\mathrm{Wb}}{0.23\,\mathrm{s}} = +14\ \mathrm{V}$$

Assuming the total resistance in the loop is 8.7 Ω, the amount of induced current is

$$I = \frac{14\,\mathrm{V}}{8.7\,\Omega} = 1.6\ \mathrm{A}$$ ← Using Lenz's law, we can find that this induced current flows clockwise.

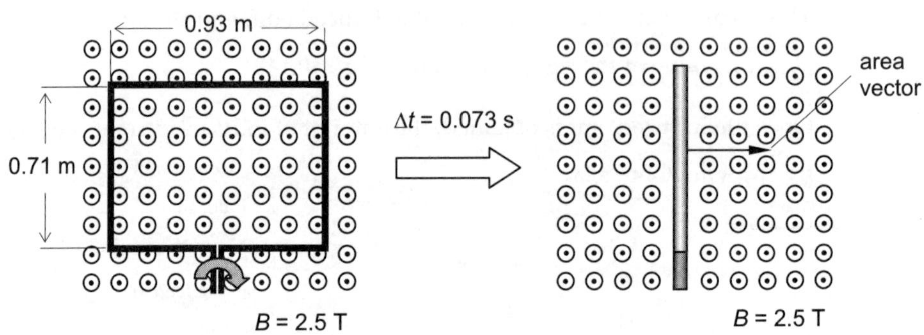

Figure 10-24: Electromagnetic induction example 2

In the next example, 25-loops of wire are rotated 90° in 0.073 seconds in a magnetic field coming out of the page as shown in Fig. 10-24. The dimensions of the loops are the same as those of the loop in the previous example. Again, to find the average induced emf, we first calculate the magnetic flux before and after the rotation. This time, assume the area vector is initially out of the page. The area vector will then rotate with the loops, eventually pointing to the right after turning 90°. Therefore,

$$\Phi_i = (2.5\,\text{T})[(0.71\,\text{m})(0.93\,\text{m})]\cos 0° = 1.65\ \text{Wb}$$

$$\Phi_f = 0\,\text{Wb (because } \cos 90° = 0\text{)}$$

Substitute the values into Eq. (10.14):

$$\mathcal{E}_{\text{avg}} = -N\frac{\Phi_f - \Phi_i}{\Delta t} = -(25)\frac{0\,\text{Wb} - 1.65\,\text{Wb}}{0.073\,\text{s}} = +570\,\text{V}$$

10-9. Electromagnetic Induction for a Conductor Moving in a Magnetic Field

Consider a long conductor initially at rest in a magnetic field. Without any motion, the free electrons on the conductor do not experience any magnetic force and remain stationary. Suppose we now pull the conductor to the right with speed v, as shown in Fig. 10-25. As soon as the conductor starts moving, each electron experiences a downward magnetic force of $F_m = evB$ and begins to accelerate toward the bottom end of the conductor. We have successfully 'induced' the movement of the electrons and created an upward induced current. (Recall that the direction of an induced current is defined as a flow of positive charges, in the opposite direction to electrons' motion.) While this induced current flows toward the top end of the conductor, it experiences a magnetic force of $F_m = ILB$ to the left, which acts to resist the motion of the conductor.

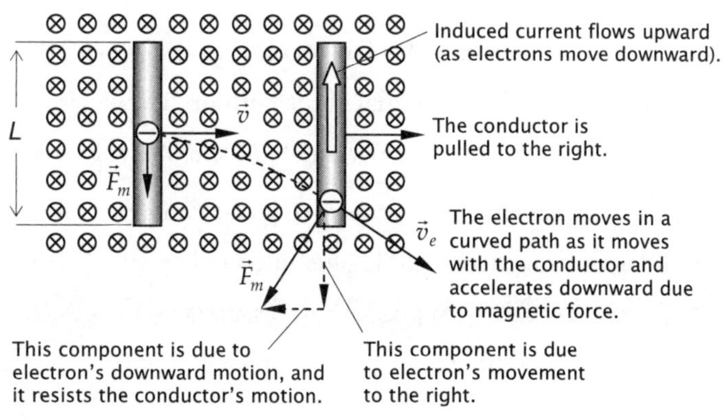

Figure 10-25: Electromagnetic induction for a moving conductor

However, as long as the conductor continues moving to the right, electrons continue moving toward the bottom end of the conductor, and this creates an excess of positive charges at the top and an excess of negative charges at the bottom. The electric field thus produced between the ends exerts an upward electric force of $F_e = qE$ on the electrons, as shown in Fig. 10-26. At first the electric force is not as strong as the magnetic force, but it eventually becomes as strong as the magnetic force as the downward movement of electrons continues. When an equilibrium is reached, the net force acting on an electron becomes zero, which means that

$$qE = qvB \qquad \text{... (10.16)}$$

A potential difference develops between the ends of the conductor.

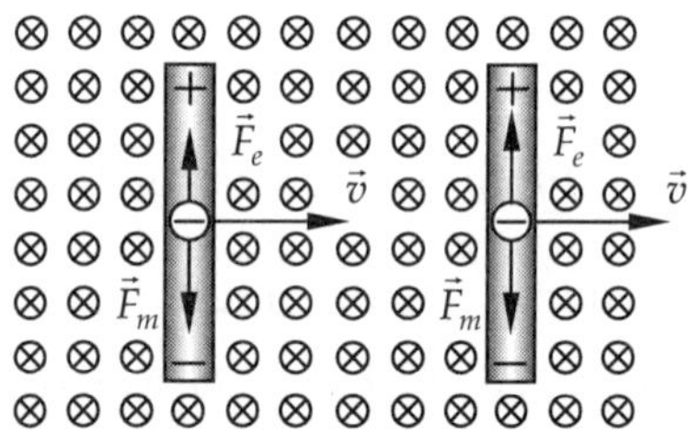

An equilibrium is reached when the electric force on the electron is balanced by the magnetic force.

The upward electric force is small at first but eventually grows stronger as more electrons settle at the bottom end.

Figure 10-26: Electromagnetic induction for a moving conductor

According to Eq. (8.13) (on p. 248), the potential difference between the ends of the conductor, $\mathcal{E}$, is the product of the electric field and the length of the conductor:

$$\mathcal{E} = EL \qquad \text{... (10.17)}$$

Combining Eqs. (10.16) and (10.17), we can express the emf induced between the ends of the conductor as

$$\mathcal{E} = vLB \qquad \text{... (10.18)}$$

At this stage, the conductor has become a battery, with the positive terminal at the top and the negative terminal at the bottom. The electric potential energy that now exists between the ends has originated from the work done on the conductor.

The series of events that occur during electromagnetic induction for a conductor moving in a magnetic field is illustrated in Fig. 10-27. Note that the role of an induced current is to oppose the mechanism that gave rise to it, whether it is created in a loop of wire or in a moving conductor.

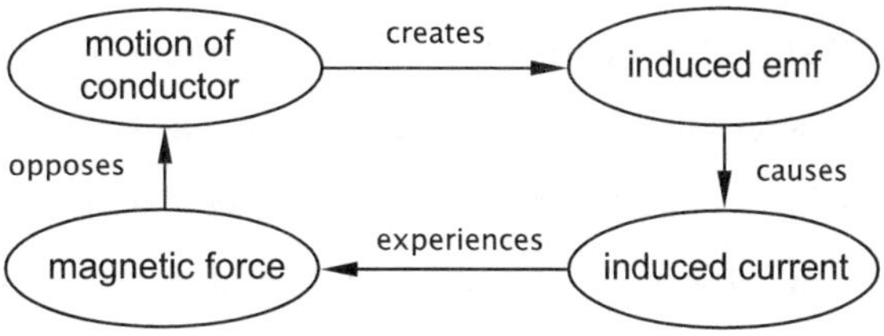

Figure 10-27: Electromagnetic induction for a moving conductor

Let's consider the following example and see how this concept of electromagnetic induction for a moving conductor is applied in an actual problem. In the first example, a rectangular loop of wire is pulled into a magnetic field at 2.8 m/s as shown in Fig. 10-28 (a). While the loop is in motion, an induced emf is developed only across segment *bc*, where the electrons experience an upward magnetic force and move up toward point *b*. The induced emf across *bc* is

$$\mathcal{E} = (2.8\,\text{m/s})(0.40\,\text{m})(3.1\,\text{T}) = 3.5\,\text{V}$$

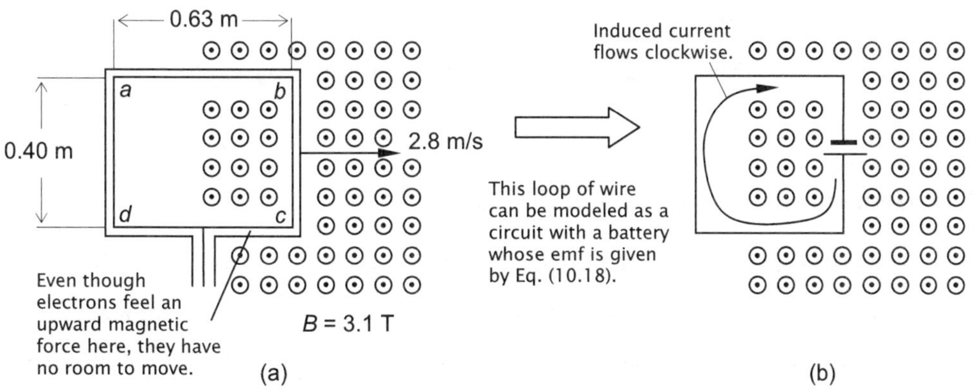

Figure 10-28: A loop of wire being pulled into a magnetic field

In other segments, no emf is induced because either the charges have not yet entered the magnetic field (segment *ad*) or the electrons cannot move upward despite experiencing an upward magnetic force (segments *ab* and *cd*). Therefore, this loop of wire can be thought of as a circuit with a battery facing downward, as shown in Fig. 10-28 (b). The induced current flows clockwise, and if we assume the total resistance of the loop is 8.3 Ω, the amount of induced current becomes

$$I = \frac{3.5\,\text{V}}{8.3\,\Omega} = 0.42\,\text{A}$$

The induced current of 0.42 A flows in the loop until the whole loop is within the magnetic field. Once segment *ad* enters the magnetic field, the induced emf developed across that segment is equal to that across segment *bc*. The loop then can be modeled as a circuit with two batteries facing each other, as shown in Fig. 10-29. Therefore, no current flows in the loop as long as the entire loop is within the magnetic field. We can reach the same conclusion by applying Lenz's law or Faraday's law; as long as the magnetic flux is constant, there is no induced emf or induced current in the loop.

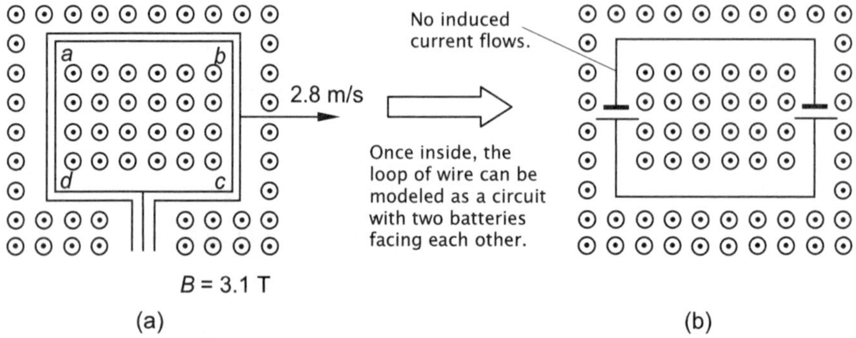

Figure 10-29: A loop of wire moving within a magnetic field

10-10. Generators, Transformers, and Motors

An **electric generator** is a device that transforms mechanical energy into electrical energy by electromagnetic induction. It consists of many loops of wire wound around an armature that is rotated within a magnetic field. Figure 10-30 shows a simple diagram of a generator without the armature. When the axle is turned by some mechanical means, the loops of wire around the armature also turn. Changes in the magnetic flux through the loops generate an induced current in the loop.

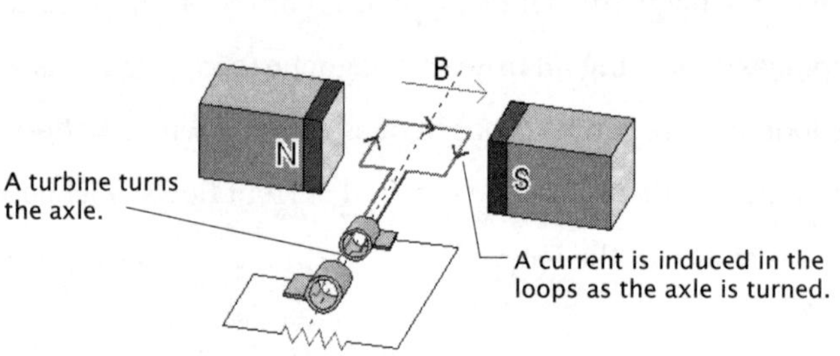

Figure 10-30: An AC generator [Source: http://www.phys.ufl.edu]

The induced current then experiences a magnetic force that resists the motion of the armature, causing it to slow down and eventually turn backwards. When the armature turns the other way, the induced current flows in the opposite direction and exerts a resistive force on the loop again. We can thus imagine the armature of the generator turning back and forth periodically, producing induced current that varies with time as shown in Fig. 10-31. The current shown in the figure is called an **alternating current** because it changes direction periodically.

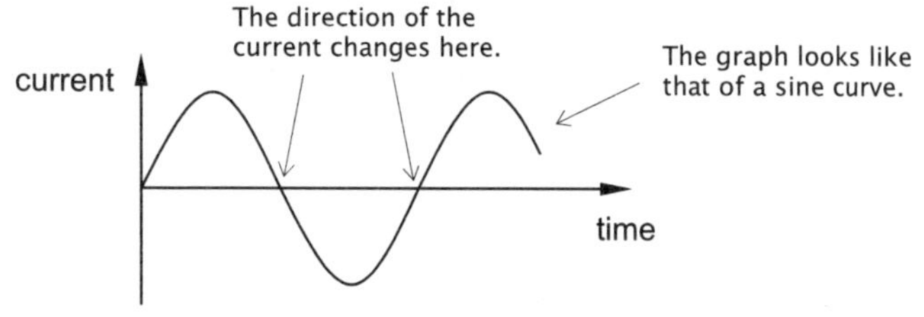

Figure 10-31: Current produced in an AC generator

Electricity thus produced by generators at power plants needs to be sent away at high voltage (usually several hundred kilovolts) in order to minimize energy loss, and this elevated voltage must be lowered before the transmitted electric power can be used safely in homes and factories. The device used for changing the voltage is called the **transformer**, and it does its job while transferring electrical energy from one circuit to another through electromagnetic induction. It consists of two coils of wire known as the **primary** and **secondary coils**, usually linked by a soft iron core as shown in Fig. 10-32. It is designed so that all magnetic flux produced by the current in the primary coil also passes through the secondary coil. When an AC voltage is applied to the primary coil, the changing magnetic field it produces will

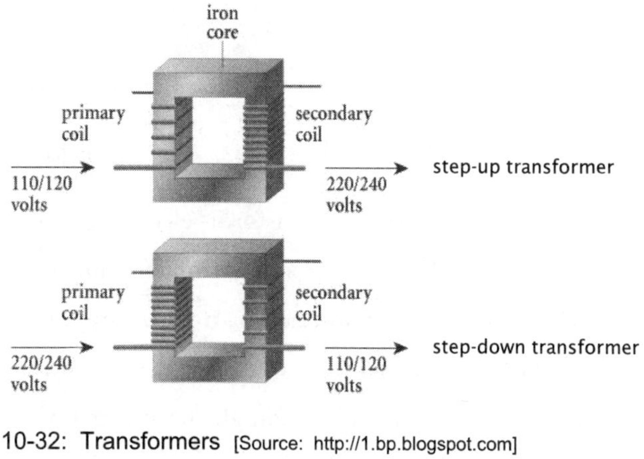

Figure 10-32: Transformers [Source: http://1.bp.blogspot.com]

induce an AC voltage of the same frequency in the secondary coil. However, according to Faraday's law, the voltage is proportional to the number of loops in each coil. Thus,

$$\frac{V_P}{V_S} = \frac{N_P}{N_S} \qquad \ldots (10.19)$$

When a transformer produces a higher voltage in the secondary coil, it is called a **step-up transformer**, and when it lowers the voltage, it is called a **step-down transformer**. Although AC voltage can be manipulated with a transformer, the law of conserva-tion of energy states that the power output must be conserved between the coils. Therefore,

$$V_P I_P = V_S I_S \qquad \ldots (10.20)$$

or

$$\frac{I_P}{I_S} = \frac{N_S}{N_P} \qquad \ldots (10.21)$$

Note that a direct current in the primary coil, unless its amount is made to fluctuate with time, does not produce a changing magnetic flux and therefore induces no emf in the secondary coil. One of the important reasons that AC is preferred to DC is this ease of changing the voltage, which allows efficient distribution of electric power over long distances.

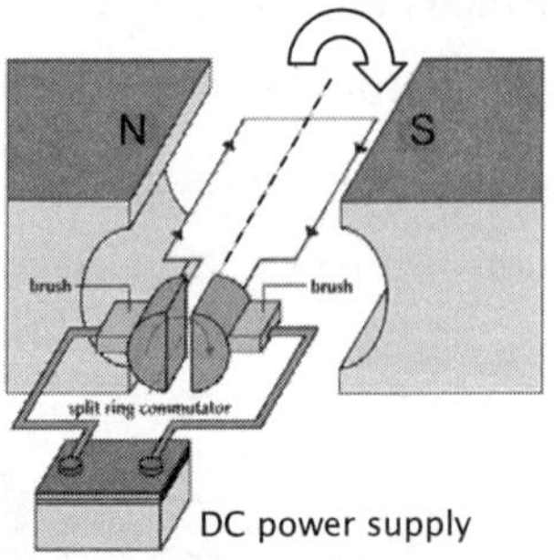

Figure 10-33: A DC motor [http://cache.eb.com]

A **motor** is a device that transforms electrical energy into mechanical energy. As shown in Fig. 10-33, it looks very similar to an electric generator; basically, it is loops of wire wound around an armature and placed within a magnetic field. When a current is supplied to the loops by a power supply, it experiences a magnetic force, which is used to turn the armature. As the armature rotates, however, the magnetic flux through the loops changes, and an induced current is generated (as in a generator) and acts to oppose the motion of the armature. The faster the motor spins, the greater the rate of change in magnetic flux through the loops, and, according to Faraday's law, the greater the induced emf and current.

While the emf in the power supply makes the motor spin and do work, the induced emf developed in the loops resists the motion of the motor. For this reason, this induced emf is called the **counter emf** or **back emf**. Part of the supply voltage, $\mathcal{E}$, is used to overcome this counter emf, and the rest is used to perform work on the load, expressed as the product of the current in the loops and the resistance offered by the armature. That is,

$$\underset{\text{voltage of power supply}}{\mathcal{E}} = \underbrace{V_{\text{back}}}_{\text{back emf}} + \underbrace{I\overset{\text{armature resistance}}{R}}_{\text{voltage used to perform work on the load}} \quad \ldots (10.22)$$

In the above equation, the actual current in the loops, I, is the current supplied by the power supply less the amount of induced current:

$$I = I_{\text{supplied}} - I_{\text{induced}} \quad \ldots (10.23)$$

When a motor is first turned on, the armature starts from rest, and initially there is no induced current in the loops. Hence, the counter emf is zero, and all the supply voltage is used to do work on the load. Then we may re-write Eq. (10.22) as

$$\mathcal{E} = I_{\text{supplied}}\, R \quad \text{when the motor is first started} \quad \ldots (10.24)$$

As the motor gains speed, the counter emf increases, and so does the induced current. This means that less current flows in the loops and less voltage is used to do useful work. (The armature resistance is constant.) Eventually, an equilibrium is reached, and the motor spins at constant speed from then on.

Suppose a 36-V motor is used to lift an object that presents an armature resistance of 12 Ω. When the motor is turned on, the amount of current in the motor is

$$I = \frac{\mathcal{E}}{R} = \frac{36\text{ V}}{12\,\Omega} = 3.0\text{ A} \leftarrow \text{This is the current supplied by the 36-V power supply.}$$

The motor slows down almost immediately and settles to a constant operating speed. Assuming the current in the motor at that speed has been reduced to 1.8 A, we can find the induced current in the loops using Eq. (10.23):

$$I_{\text{induced}} = I_{\text{supplied}} - I = 3.0\text{ A} - 1.8\text{ A} = 1.2\text{ A}$$

The counter emf, according to Eq. (10.22), is

$$V_{\text{back}} = \mathcal{E} - IR = 36\,\text{V} - (1.8\,\text{A})(12\,\Omega) = 14\,\text{V}$$

Knowing the amount of induced current, we could also have obtained the same answer by using Ohm's law:

$$V_{\text{back}} = I_{\text{induced}}\,R = (1.2\,\text{A})(12\,\Omega) = 14\,\text{V}$$

PHYSICS 12 ANALYZED: ELECTROMAGNETIC INDUCTION

10-6. Magnetic Flux

- Magnetic flux is the amount of magnetic field running through a loop of wire. It is defined as $\Phi = B\,A\cos\theta$ [Wb].
- Magnetic flux can be changed in three ways:
  - Change the magnetic field strength.
  - Change the area of the loop in the magnetic field.
  - Rotate the plane of the loop.

10-7. Electromagnetic Induction and Lenz's Law

- Electromagnetic induction: A change in magnetic flux causes electric charges in the loop to move.
- The resulting current is called induced current, and it generates a magnetic field to oppose the change in magnetic flux.
- The direction of an induced current can be determined by Lenz's law, which states that the direction of an induced current is such that its magnetic field opposes the change in the magnetic flux through the loop.

10-8. Faraday's Law of Induction

- Faraday's law of induction states that an induced emf is proportional to the rate of change in magnetic flux:

$$\mathcal{E}_{avg} = -N\frac{\Delta\Phi}{\Delta t} \text{ [Wb/s = V]}$$

- The amount of induced current can be obtained by applying Ohm's law.

10-9. Electromagnetic Induction for a Conductor Moving in a Magnetic Field

- A conductor moving in a magnetic field develops an induced emf given by

$$\mathcal{E} = v L B$$

- The induced current within the moving conductor experiences a magnetic force, which acts to resist the motion of the conductor.

10-10. Generators, Transformers, and Motors

- An electrical generator transforms mechanical energy into electrical energy. The rotating motion of the armature induces an alternating current in the loops of wire wound around the armature.
- A transformer transfers electrical energy from one circuit to another. While doing so, it can increase or decrease an input voltage. Between the primary and secondary coils,

$$\frac{V_P}{V_S} = \frac{N_P}{N_S} = \frac{I_S}{I_P}$$

- A transformer does not operate on direct current unless current changes with time.
- A motor transforms electrical energy into mechanical energy. When it is first turned on, all the supply voltage is used to doing work on the load:

$$\mathcal{E} = I_{\text{supplied}} R$$

- As the motor spins, a counter emf is developed and resists its motion. At full operating speed,

$$\mathcal{E} = V_{\text{back}} + I R$$

EXERCISES ON CHAPTER 10: ELECTROMAGNETIC INDUCTION

Exercise 10-16

A 16-turn circular coil of wire is placed in a uniform magnetic field of 7.4 T. The radius of the loops is initially 8.6 cm but is reduced to 5.6 cm in 0.33 s. Find

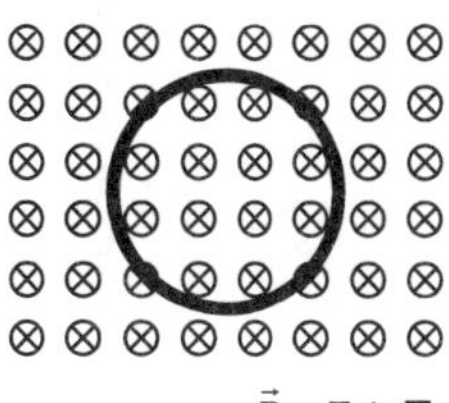

(a) the initial magnetic flux through each loop

(b) the final magnetic flux (after the radius is reduced)

(c) the average emf induced in the coil

(d) the direction of the induced current during the change in the radius

Exercise 10-17

A copper rod is pulled to the right in a uniform magnetic field of 0.38 T as shown. The emf induced across the ends of the rod is 3.4 V. Find

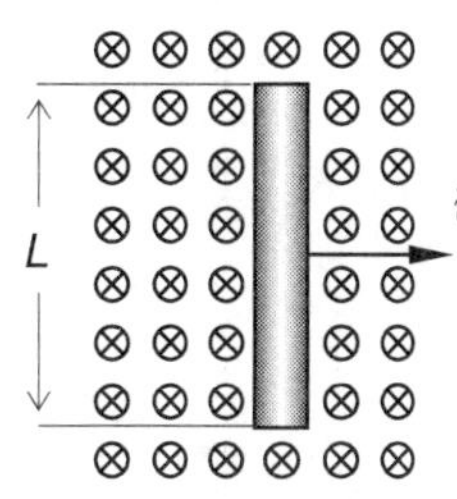

(a) the direction of the induced current

(b) the length of the rod

Exercise 10-18

A DC motor having an armature resistance of 19 Ω draws a 3.2-A current at startup but only a 0.45-A current at normal operating speed. Find

(a) the supply voltage

(b) the back emf at the startup and at normal operating speed

Exercise 10-19

A rectangular coil of wire having 730 loops is placed in a uniform magnetic field of 5.0 T as shown. The magnetic field changes over a time interval of 0.28 s, producing a counterclockwise induced current of 3.9 A. Given that the total resistance of the coil is 86 Ω, find

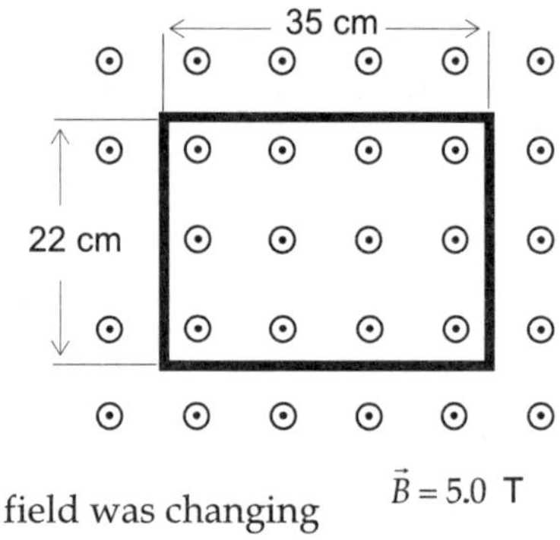

(a) the rate, in T/s, at which the magnetic field was changing

(b) the magnitude and direction of the final magnetic field

Exercise 10-20

A 36-turn square coil of wire is pulled out of a uniform magnetic field of 24 T as shown. Each side of the square is 7.6 cm, and the total resistance of the coil is 0.15 Ω. Find

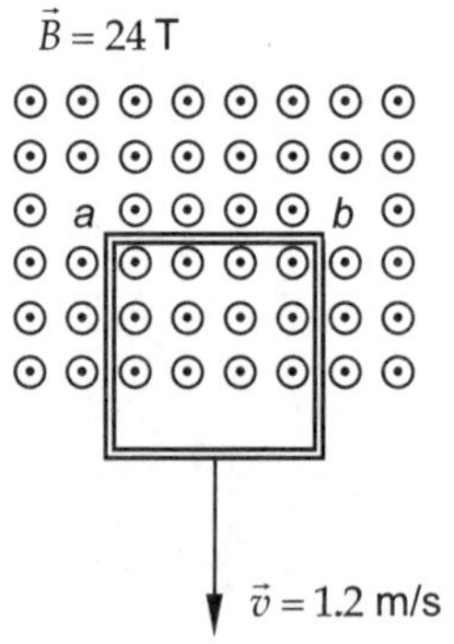

(a) the emf induced between points *a* and *b*

(b) the direction and amount of induced current while the coil is being pulled out of the field

Exercise 10-21

A transformer has 230 turns in the primary coil and 640 turns in the secondary coil. The primary coil draws a 3.1-A current from a 110 V ac. Find

(a) the voltage and the current in the secondary coil

(b) the power being transferred to the secondary coil

Exercise 10-22

A circular coil of wire having 20 loops and the diameter of 33 cm is initially in a uniform magnetic field as shown. The coil is then pulled out of the field within 0.12 s, inducing an average emf of 5.8 V. Find

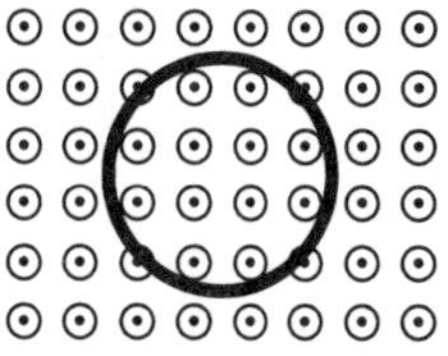

(a) the direction of the induced current while the coil is being pulled out of the field

(b) the strength of the magnetic field

Exercise 10-23

A 3.2-A current flows through a solenoid with 530 turns of wire wound over the length of 21 cm. Inside this solenoid is a 7.8-cm long metal rod, which moves perpendicular to the magnetic field at 1.4 m/s. Find

(a) the magnetic field strength inside the solenoid

(b) the direction of the induced current in the rod

(c) the induced emf between the ends of the rod

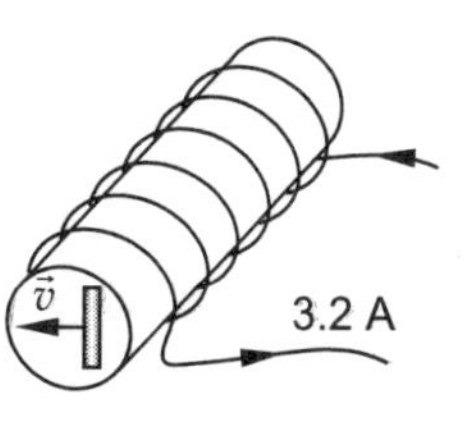

Exercise 10-24

A DC motor, connected to a 35-V power supply, draws a current of 8.5 A when it first starts up. At its normal operating speed, the motor draws a current of 2.2 A. Find

(a) the armature resistance

(b) the back emf at the start-up

(c) the back emf at normal operating speed

Exercise 10-25

A 60-turn rectangular coil of wire is initially placed in a uniform magnetic field of 3.6 T. In 0.29 s, the magnetic field strength increases to 6.3 T, as shown below. During the change in magnetic field strength, find

(a) the direction of the induced current in the coil

(b) the average emf induced in the coil

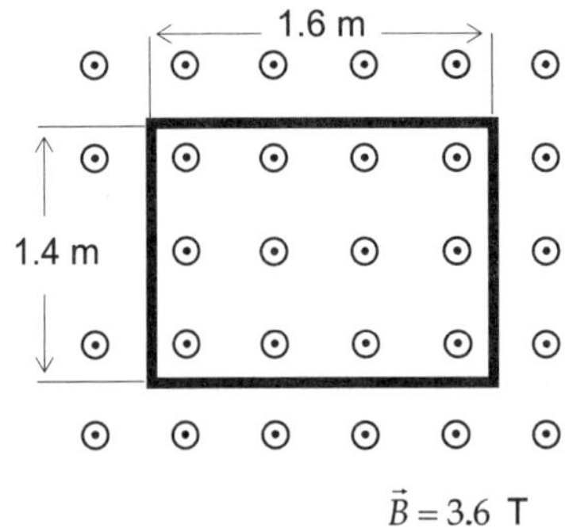

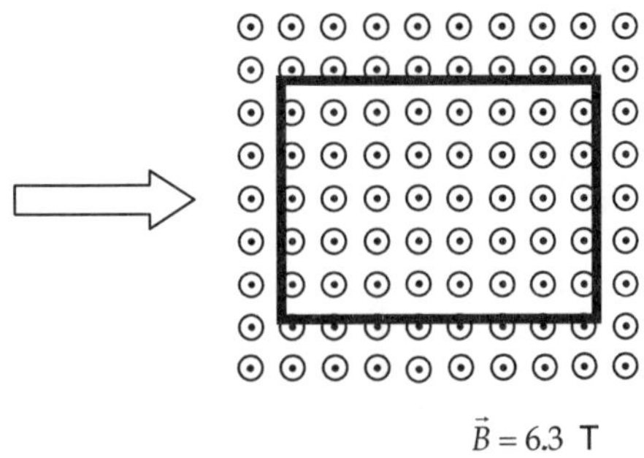

Exercise 10-26

A 30-cm long copper rod slides to the right on metal rails at 0.69 m/s. Find

(a) the emf induced between the ends of the rod

(b) the amount and direction of the induced current in the rod

(c) the magnitude and direction of the magnetic force on the rod

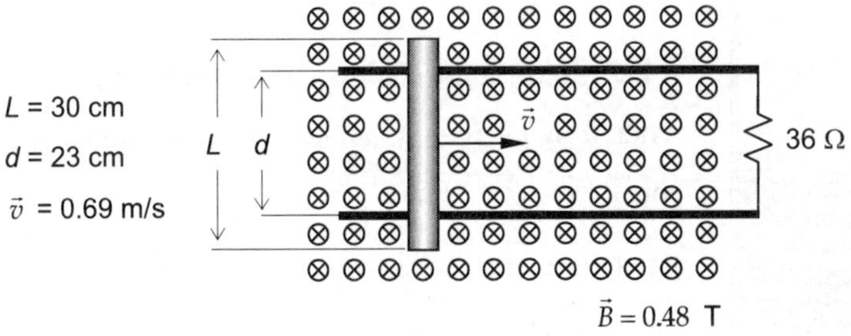

Exercise 10-27

A step-down transformer has 4200 turns in the primary coil, which receives 360 W of power from a 220 V ac. Given that the voltage in the secondary coil is 140 V, find

(a) the number of turns in the secondary coil

(b) the current in the secondary coil

Exercise 10-28

A rectangular loop of wire, initially placed in a uniform magnetic field of 27 T as shown, is rotated about the axis at a rate of 0.20 revolutions per second. Find

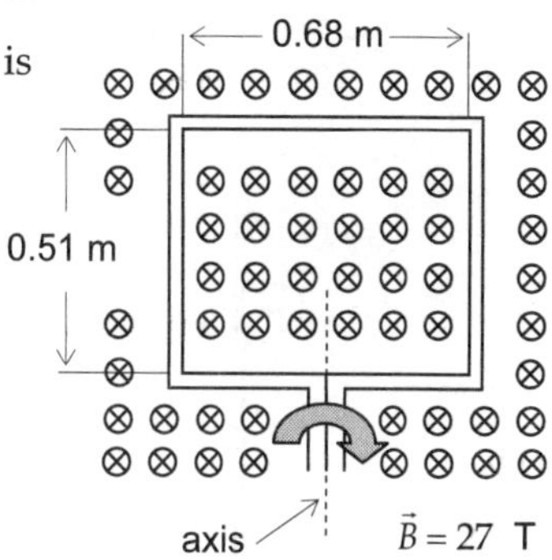

(a) the magnetic flux through the loop at every 0.20 second interval from $t = 0$ s to $t = 1.00$ s

(b) the average emf induced in the loop between $t = 0$ s and $t = 1.00$ s

(c) the average emf induced in the loop between $t = 0.20$ s and $t = 0.40$ s

Exercise 10-29

A square loop of wire with the total resistance of 84 Ω is pulled into a uniform magnetic field of 8.2 T as shown below. If an average current of 0.23 A is produced in the coil, find

(a) the average emf induced in the coil

(b) the direction of the induced current

(c) the speed with which the coil was pulled into the magnetic field

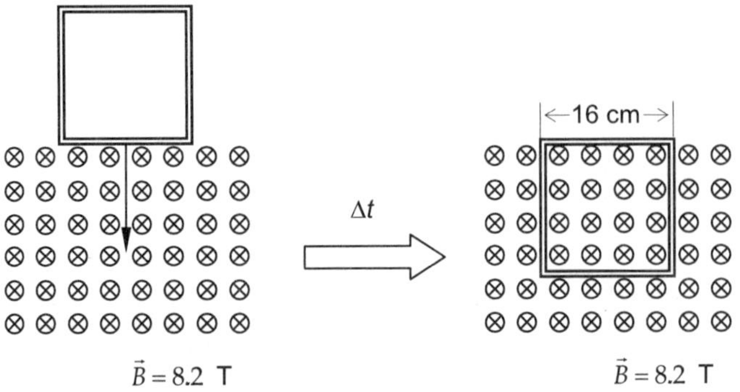

Exercise 10-30

A 12-V power supply is used to operate a DC motor. Running at full operating speed, the motor draws a current of 1.8 A. When it is jammed, the amount of current through the motor increases to 9.0 A. Find

(a) the back emf when the motor is jammed

(b) the armature resistance

(c) the back emf when the motor runs at full operating speed

PROBLEMS ON CHAPTER 10: ELECTROMAGNETISM

1 – 9. Fill in the blanks

1. A moving charge or electric current produces a _________ _________. Its direction can be determined using a right-hand rule.
2. A _________, or electromagnet, is many loops of wire, often wound around a soft iron core to increase the magnetic field strength.
3. _______ _________ is proportional to the number of magnetic field lines through a loop of wire.
4. Whenever _______ _________ changes, an _________ current flows in the loop to cancel that change in _________ _________.
5. _________ law states that an induced _________ flows in a loop of wire so that the magnetic field it generates cancels the change in _________ _________ that gave rise to it.
6. _________ law of induction states that an induced emf is proportional to the rate of change in _________ _________.
7. A _______ is a device that converts mechanical energy to electrical energy.
8. A _______ is a device that converts electrical energy to mechanical energy.
9. A _______ is a device that transfers electrical energy from one circuit to another.

10 – 20. True/False questions

10. _____ Magnetic field lines around a magnet start from its north pole and terminate at its south pole.
11. _____ A proton, when traveling at the same speed as an electron, produces much stronger magnetic field than the electron because of its greater mass.
12. _____ The magnetic field strength around a current-carrying conductor is inversely proportional to the distance from the conductor.
13. _____ When a charge moves parallel to magnetic field lines, it experiences no magnetic force.

14. _____ When two parallel conductors carry currents in the same direction, they repel each other.

15. _____ The magnetic field inside a tightly wound solenoid is uniform.

16. _____ Lenz's law allows us to find the amount of induced current in a loop.

17. _____ Induced emf is directly proportional to the magnetic flux.

18. _____ The magnetic force exerted on a moving conductor due to an induced current always opposes the conductor's motion that caused the induced current to flow.

19. _____ The amount of current within a motor is greatest at start-up.

20. _____ In a step-up transformer, the number of turns in the primary coil is larger than that in the secondary coil.

21 – 46. Multiple choice questions

21. If a bar magnet is placed at point P, in which direction will its north pole point?

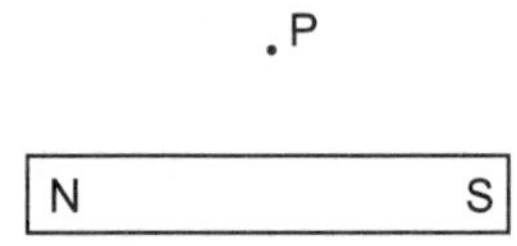

a)  b)  c)  d) 

22. An electron is accelerated by a potential difference of ΔV between two oppositely charged plates. Upon entering a uniform magnetic field, it follows a circular path of radius R. If the potential difference between the plates is quadrupled, the radius of the circular path will

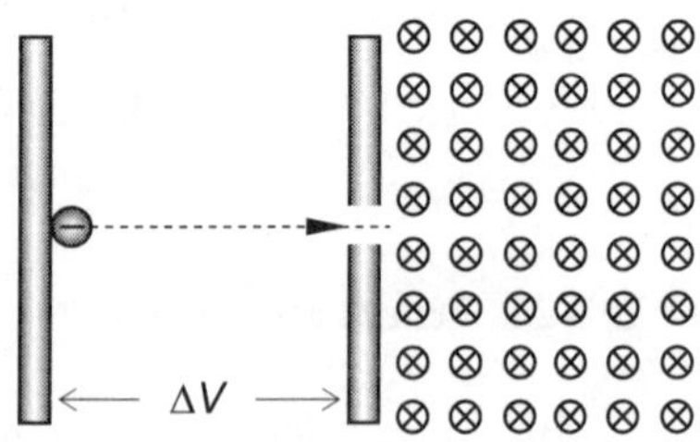

a) increase by a factor of 2.
b) increase by a factor of 4.
c) decrease by a factor of 1/2.
d) decrease by a factor of 1/4.

23. Which of the following is <u>not</u> a unit for magnetic flux?

a) Wb b) $T{\cdot}m^2$ c) J/A d) N·m/V

24. For a moving proton, which of the following fields is present?

1. gravitational field	2. electric field	3. magnetic field

a) 3 only b) 1 and 2 only
c) 2 and 3 only d) 1, 2, and 3

For Problems 25 and 26, refer to the electron placed in a region of both electric and magnetic fields as shown. The electron is initially at rest.

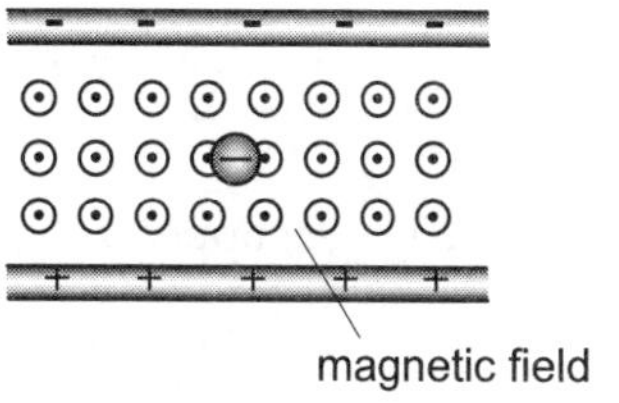

25. The magnitude of the force required to keep the electron at rest is

a) less than the magnitude of the electric force acting on the electron.
b) equal to the magnitude of the electric force acting on the electron.
c) greater than the magnitude of the electric force acting on the electron
d) equal to zero.

26. Which of the following best describes the path the electron takes when it is released from the position shown in the diagram?

a)  b)  c)  d) 

27. A negative charge is traveling horizontally at constant speed, as shown. When the charge is directly below the solenoid, the switch is closed. Which of the following occurs?

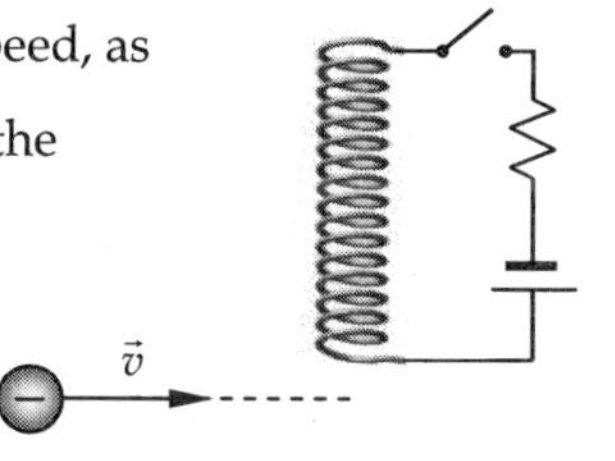

a) The charge comes to a stop immediately.
b) The charge's path is deflected into the page.
c) The charge undergoes a circular motion directly below the solenoid.
d) The charge's motion is not affected.

28. A circular coil of wire is placed in a uniform magnetic field as shown. Which of the following would increase the magnetic flux through the coil?

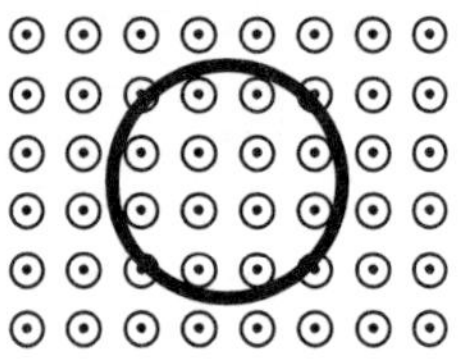

a) Increase the number of loops in the coil.
b) Increase the magnetic field strength.
c) Move the coil to the right.
d) Change the shape of the coil while keeping the area of the loop unchanged.

29. A copper rod, placed in a uniform magnetic field, is to be pulled at constant speed to induced an emf across its ends. In which of the following directions should the rod be moved in order to maximize the induced emf?

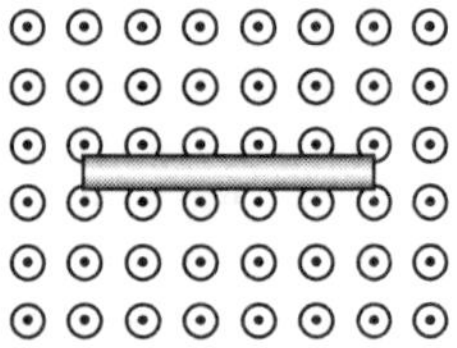

a) upward
b) to the left
c) out of the page
d) to the left and out of the page simultaneously

30. A rectangular coil of wire, placed in a uniform magnetic field and initially at rest, is pulled down as shown. While the coil is in motion,

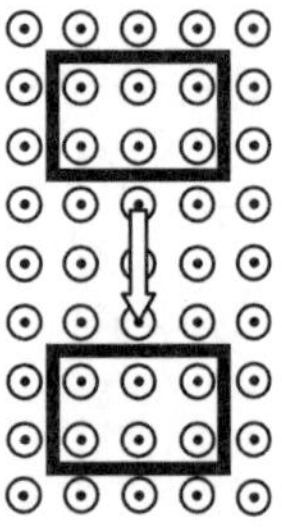

a) the magnetic flux through the coil changes.
b) an induced current flows clockwise.
c) the induced current in the coil increases.
d) the free electrons on the wire experience magnetic force.

31. A bar magnet, with its north pole facing a solenoid as shown, is initially held at rest. Which of the following statements is <u>false</u>?

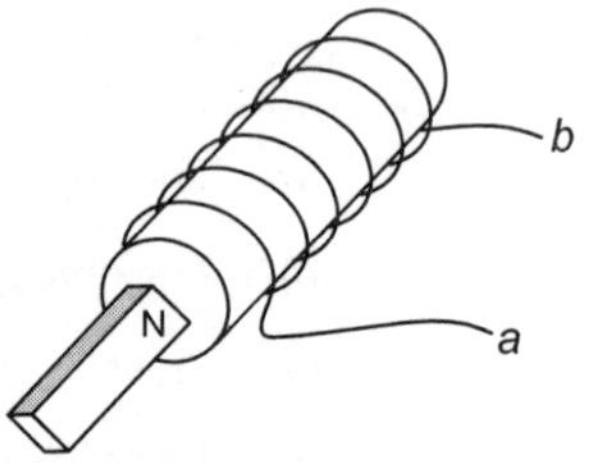

a) As the magnet is pushed into the solenoid, an induced current flows from *a* to *b*.
b) As the magnet is pulled away from the solenoid, the magnetic field is generated inside the solenoid so as to repel the magnet.
c) An induced current flows as long as the magnet is moving.
d) The faster the motion of the magnet into or away from the solenoid, the greater the amount of induced current in the solenoid.

32. An AC generator is used to produce electricity. If the rate at which the armature rotates increases, which of the following does not occur?

a) the rate of change in magnetic flux increases.
b) the amount of induced current increases.
c) the back emf increases.
d) the magnitude of the force opposing the motion of the armature increases.

33. A DC motor produces more heat when jammed than when operating at normal speeds. This increase in heat generated is mainly due to

a) increase in supply voltage.
b) increase in armature resistance.
c) increase in current.
d) all of the above.

34. Two sets of coils are wound around an iron core as shown. The ammeter will detect a current flowing in the secondary coil

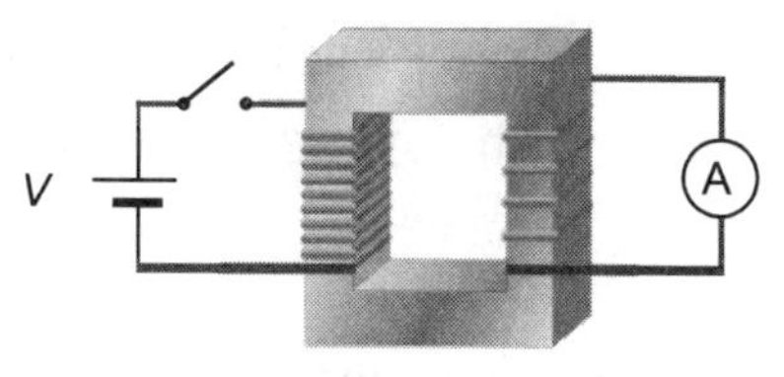

a) as long as the switch remains closed.
b) only while the switch is closed.
c) while the switch is being closed or opened.
d) No current can flow in the secondary coil because the current in the primary coil does not change direction.

35. Two charged objects A and B enter a uniform magnetic field with the same magnitude of momentum and follow the paths as shown. Which of the following statements is true?

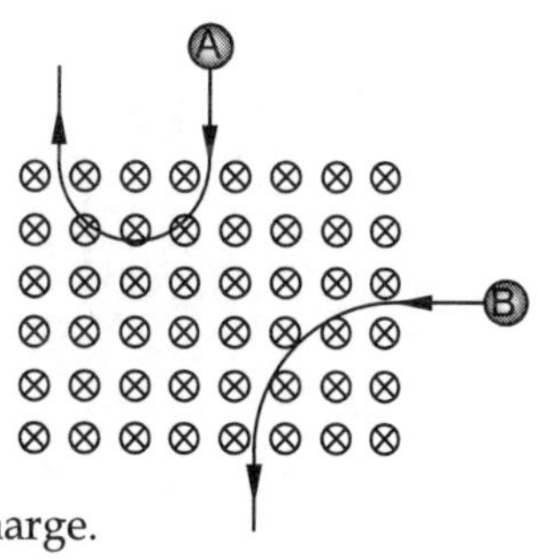

a) Object A is positively charged and has the greater charge.
b) Object A is negatively charged and has the greater charge.
c) Object B is positively charged and has the greater charge.
d) Object B is negatively charged and has the greater charge.

36. Two identical bar magnets are dropped simultaneously from the same height, but the one on the right falls through a copper pipe as shown. Which of the following happens?

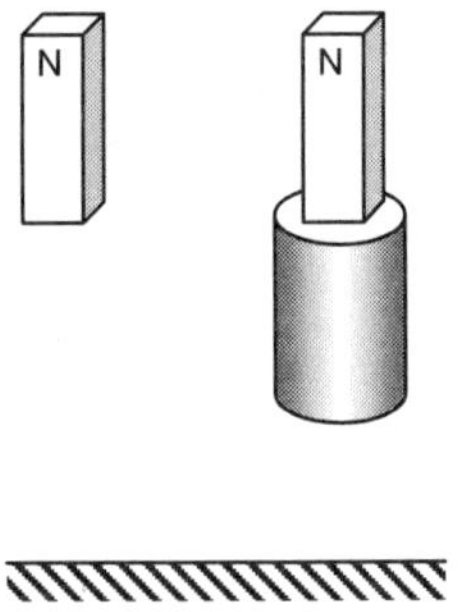

a) The right magnet hits the ground first.
b) Both magnets hit the ground at the same time.
c) The right magnet hits the ground later.
d) The right magnet never reaches the ground.

37. In a loop of wire placed in a uniform magnetic field, an electric current flows counterclockwise as shown. The magnetic force acting on the wire attempts to

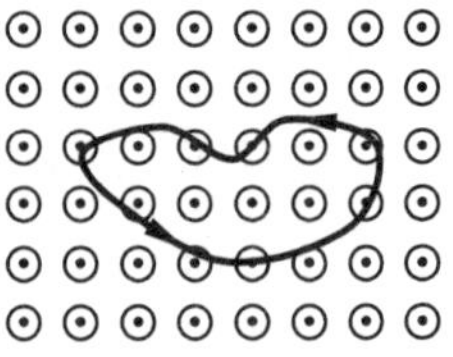

a) decrease the area of the loop.
b) increase the area of the loop.
c) rotate the plane of the loop.
d) slide the loop sideways.

38. A rectangular coil of wire is pulled across a uniform magnetic field as shown. Which of the following graphs best represents the emf induced in the coil versus time?

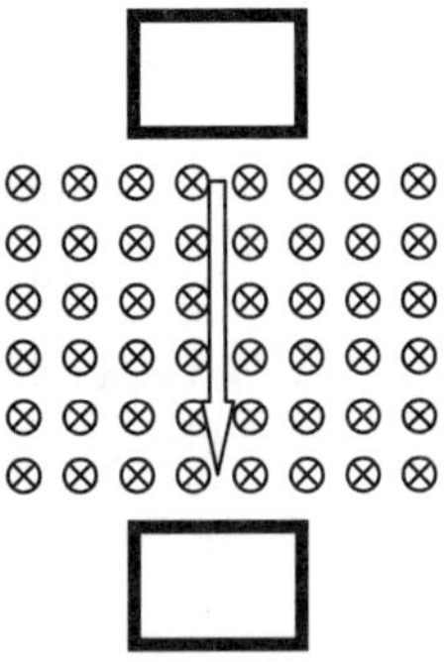

a) 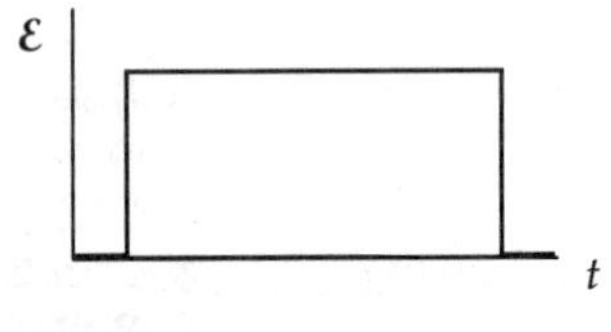

b) 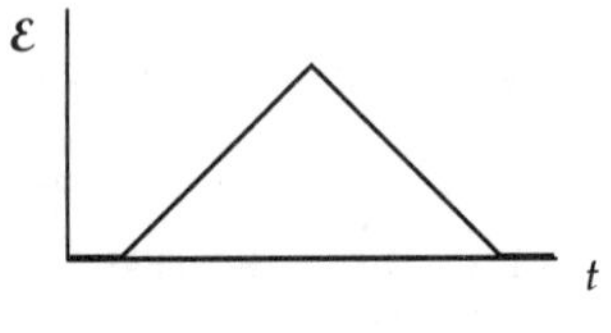

c) 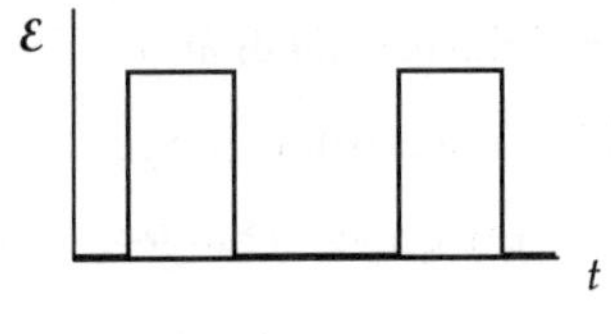

d) 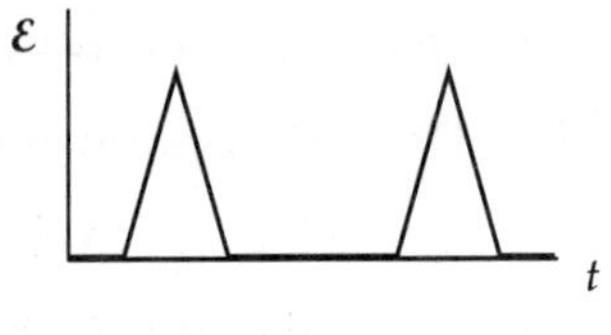

39. Three current-carrying conductors are arranged as shown right. The net force on the middle conductor is

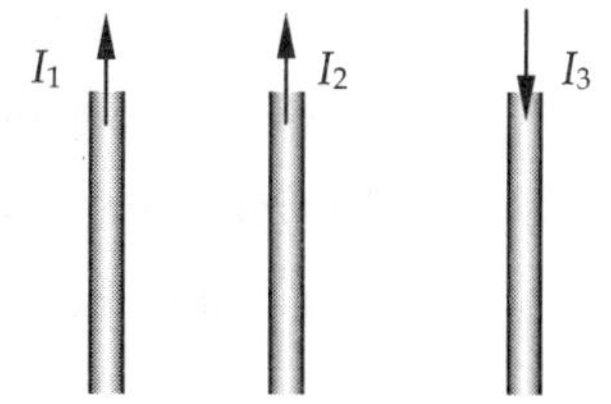

a) to the left. b) to the right.
c) into the page. d) zero.

40. A soft iron core is often used in a solenoid because it

a) absorbs excessive heat generated in the coil.
b) allows the coil to be wound more tightly.
c) increases the magnetic field inside the solenoid.
d) enables the solenoid to generate a magnetic field.

41. A charged object is undergoing uniform circular motion in a magnetic field. Which of the following statements is true?

a) The net force acting on the object is zero.
b) The net work done on the object is zero.
c) The object's momentum remains constant.
d) The radius of the circular path is independent of the magnetic field strength.

42. A charged object, attached to the end of a plastic rod, is rotated in a horizontal circle as shown. If the rotational speed of the rod is constant, where on its path is the magnetic force on the object the greatest and the least?

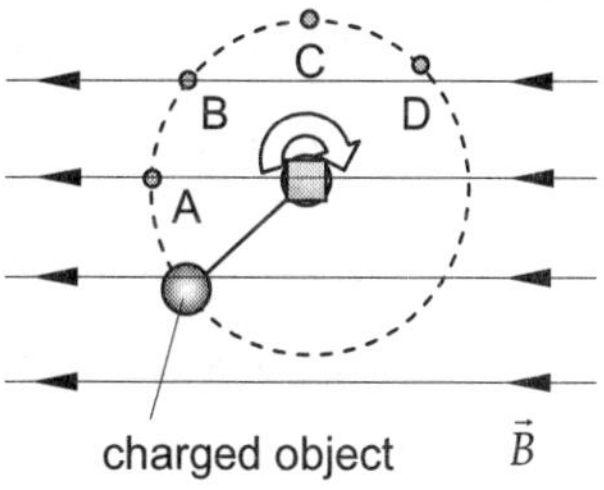

a) greatest at A and least at C
b) greatest at B and least at D
c) greatest at C and least at A
d) greatest at D and least at A

43. The graph shown right is the magnetic flux through a loop of coil plotted versus time. During which time interval is the induced emf within the loop the greatest?

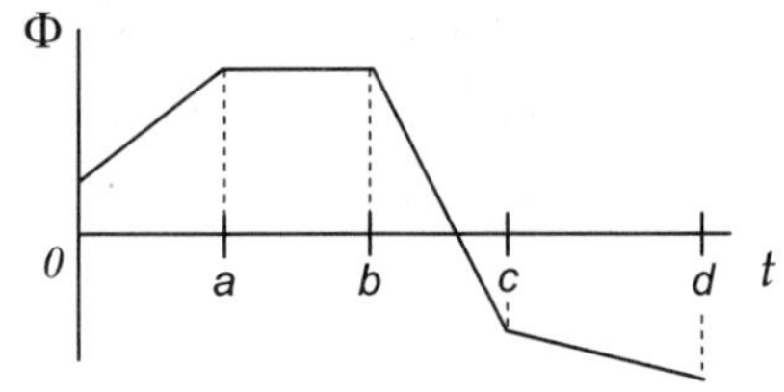

a) between *0* and *a* b) between *a* and *b*

c) between *b* and *c* d) between *c* and *d*

45. Which of the following has least to do with electromagnetic induction?

a) an AC generator b) a DC circuit

c) a DC motor d) a transformer

45. Which of the following statements on a step-up transformer is incorrect?

a) There are more loops in the secondary coil.
b) The voltage is greater in the secondary coil.
c) The secondary coil carries more electric power.
d) Less current flows in the secondary coil.

46. A loop of wire is placed in a magnetic field so that its plane is parallel to the magnetic field lines. If the axle of the loop is turned counter-clockwise as shown, which of the following does not occur?

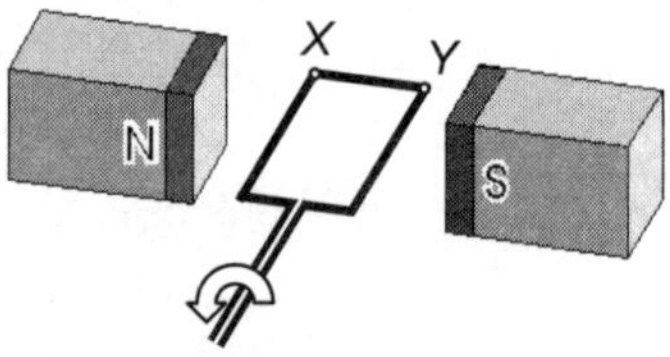

a) Induced current flows clockwise (from *X* to *Y*).
b) The magnetic force acting on the loop opposes the rotation of the loop.
c) An emf is induced in the loop.
d) The faster the loop is turned, the greater the induced current in the loop.

47 – 65. Written-response questions

47. A wire is carrying 5.8-A current out of the page as shown. Draw the magnetic field lines around the wire and find the magnetic field at point P, 4.3 m from the wire.

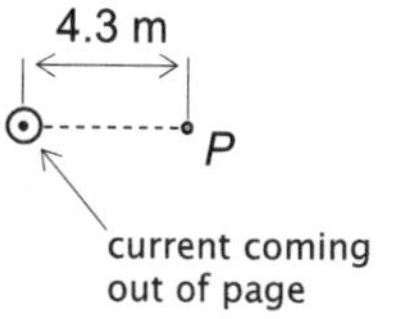

48. Two parallel wires are carrying currents of $I_1 = 2.7$ A and $I_2 = 3.7$ A in the same direction as shown. Find the net magnetic field at point P, 0.47 m from the left wire.

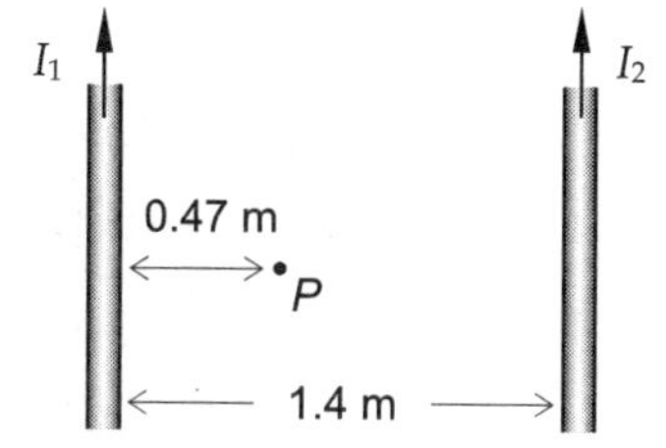

49. A 26-cm long solenoid generates a 3.5×10^{-3} T magnetic field when carrying a 4.1-A current. Find the number of loops in the solenoid.

50. A conducting wire carries a current of 7.7 A in a magnetic field as shown. If the width of the magnetic field is 12 cm, what is the magnitude and direction of the magnetic force on the wire?

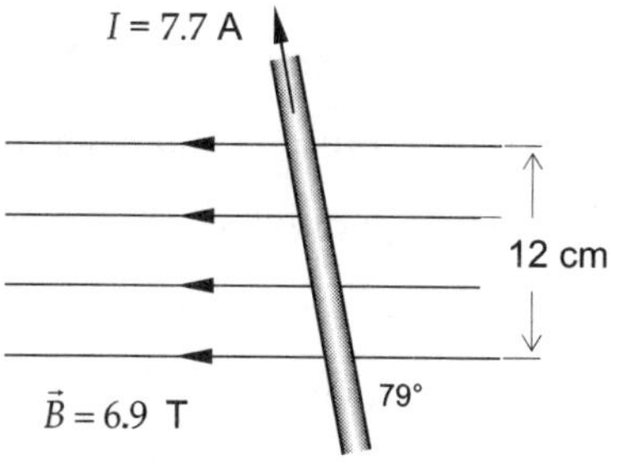

51. A charged object with a charge-to-mass ratio of 0.63 C/kg moves perpendicular to a magnetic field of 0.79 T. Find the period of the object's motion.

52. An electron enters a region of uniform magnetic field, carrying a momentum of 7.6×10^{-25} kg·m/s. If the electron experiences a magnetic force of 2.9×10^{-13} N within the magnetic field, find

a) the strength of the magnetic field

b) the radius of the circular path the electrons follows within the magnetic field

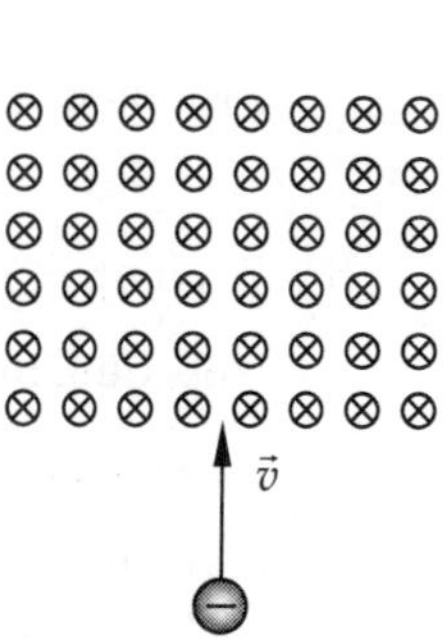

53. A 330-g metal rod carrying a 1.8-A current is suspended in a uniform magnetic field as shown. If the rod is 29 cm long, find the magnetic field strength.

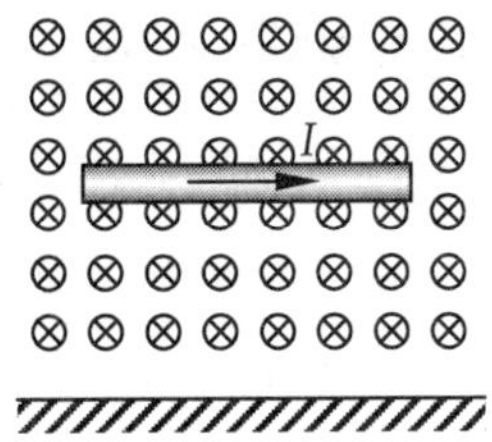

54. An electron enters a region of electric and magnetic fields at $\vec{v} = 6.8 \times 10^5 \text{ m/s}$. The potential difference between the plates is 73 V. If the proton travels through the region in a straight line, what is the magnetic field strength?

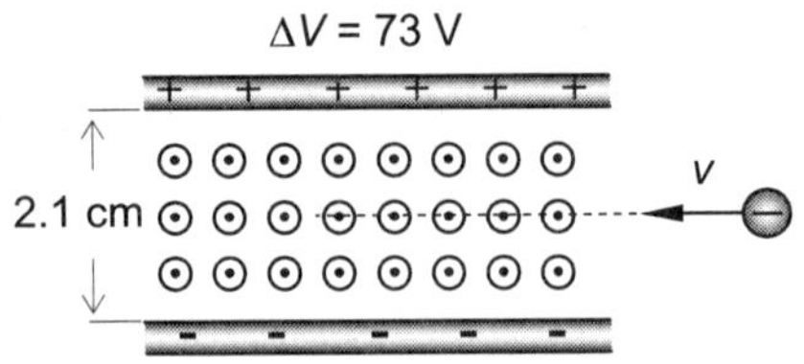

55. Three parallel conductors are carrying currents of $I_1 = 2.6$ A, $I_2 = 3.3$ A, and $I_3 = 6.0$ A, and are separated by $l_1 = 30$ cm and $l_2 = 44$ cm. Find the direction and magnitude of the magnetic force per unit length on each conductor.

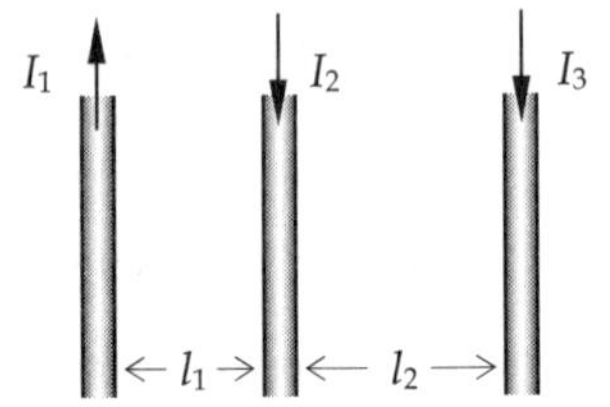

56. A circular coil of wire is placed in a uniform magnetic field so that the plane of the loop is perpendicular to the field lines. The strength of the magnetic field is then doubled with no change in its direction. In order to maintain the initial magnetic flux through the loop, by how many degrees does the loop need to be rotated?

57. A loop of wire is placed within two magnetic fields heading in opposite directions as shown. (The dashed line indicates the boundary between the two magnetic fields.) If a 2.4-A current flows in the loop in a clockwise direction, what is the net magnetic force acting on the loop? (Find both the magnitude and direction.)

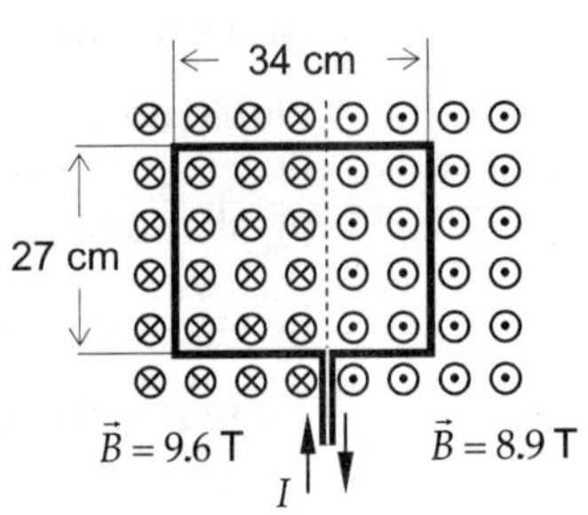

58. Two solenoids, each with 360 loops wound over the length of 17 cm, are arranged as shown, and neither solenoid is producing any magnetic field.

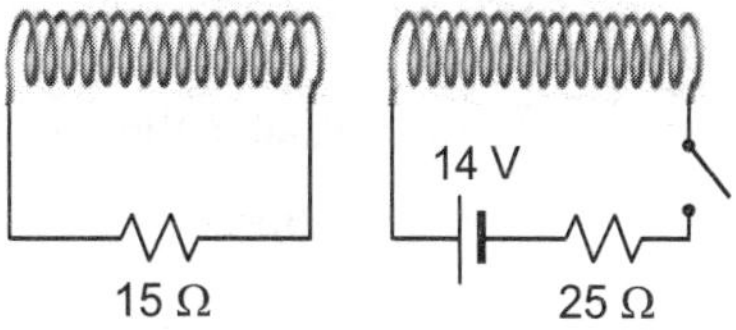

a) Find the magnitude and direction of the magnetic field inside the right solenoid when the switch is closed.

b) Find the direction of the magnetic field inside the left solenoid the moment the switch is closed.

c) How does the answer in (b) change a little after the switch is closed?

59. A 95-turn circular coil of wire is initially placed in a uniform magnetic field directed out of the page. In 0.081 s, the coil changes its shape to a rectangle, and the magnetic field reverses its direction, as shown. Find

a) the magnetic flux before and after the change.

b) the direction of the induced current in the coil.

c) the average emf induced in the coil.

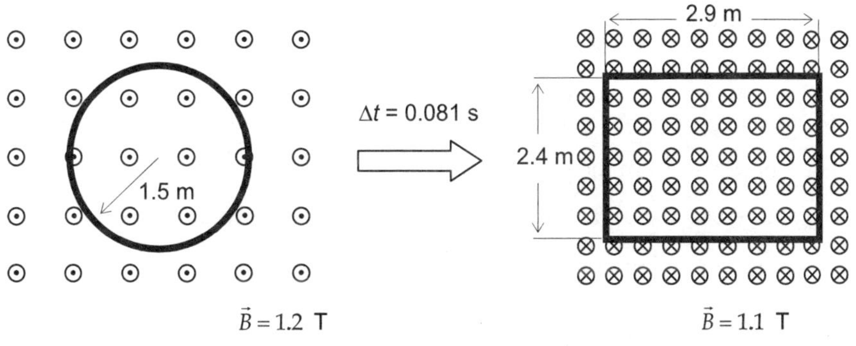

60. A rectangular coil of wire containing 15 loops is pulled out of a uniform magnetic field of 1.6 T at 53 cm/s as shown. Find the direction of the induced current and the emf induced in the loop.

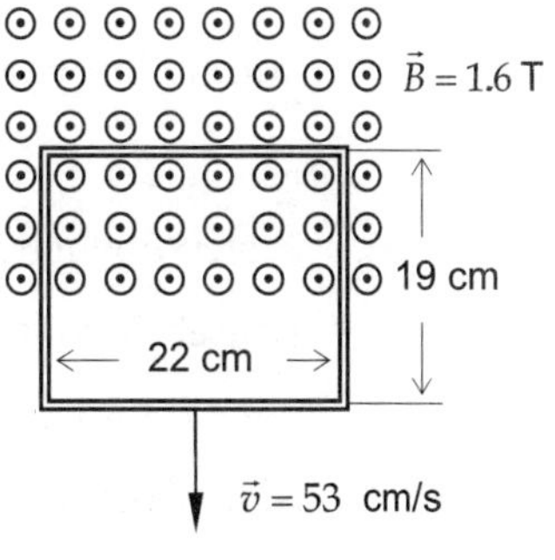

61.* A 490-g aluminum rod is placed on frictionless metal rails and initially at rest.

a) What is the magnitude and direction of the rod's acceleration when the switch is closed?

b) What is the direction of the induced current while the rod is experiencing the magnetic force?

c) What is the maximum speed of the rod? (The rod accelerates and gains speed until the emf induced across the rod becomes 3.0 V.)

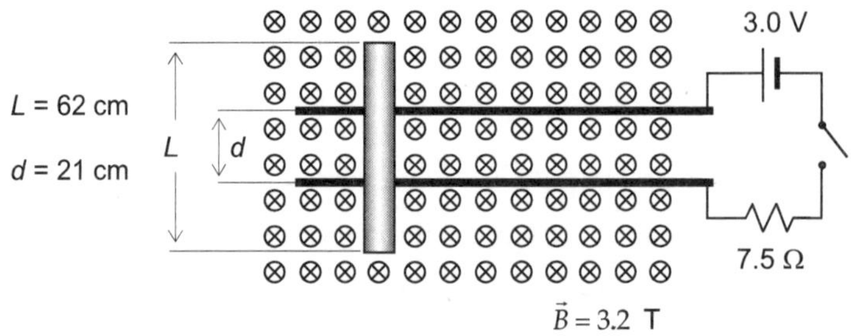

62. A 140-V power supply is used to operate a DC motor. When turned on, the motor draws a 3.1-A current from the power supply. The current quickly reaches a steady-state value of 1.3 A at normal operating speed. Find the armature resistance of the load and the back emf at normal operating speed.

63. A step-up transformer has 240 turns in the secondary coil, which receives 1800 W of power from the primary coil. If the voltages in the coils are V_P = 140 V and V_S = 560 V, find the number of windings in the primary coil and the currents flowing in both coils.

64. A 30-V DC motor, at its full operating speed of 140 rpm, draws a 1.2-A current with the back emf of 22 V. If the motor's speed is slowed to 85 rpm, what is the current through the motor and the back emf?

65. A step-down transformer with an efficiency of 93% is used to lower the household voltage of 120 V to 36 V. If the current in the secondary coil is 0.32 A, what is the current in the primary coil? (The efficiency of a transformer is the ratio of the power in secondary coil to the power in the primary coil.)

APPENDIX A. UNITS

Solving equations is an essential skill in Physics, and a knowledge and understanding of units is extremely useful in this regard. Equations involve quantities with units, and analyzing these units (that is, dimensional analysis) provides a way to verify the equation. If the units in the equation are right, your answer will probably be right as well.

The table below shows all the physical quantities introduced in Physics 12. Review the standard units of those quantities, and become familiar with other equivalent expressions.

Quantity	Symbol	Standard Unit	Equivalent Units
mass	m	kg	
time	t	s	
distance	D	m	
displacement	$\vec{d}$	m	
speed	v	m/s	
velocity	$\vec{v}$	m/s	
acceleration	$\vec{a}$	m/s^2	$1\ m/s^2 = 1\ N/kg$
force	$\vec{F}$	N	$1\ N = 1\ kg{\cdot}m/s^2 = 1\ m{\cdot}A{\cdot}T$
work	W	J	$1\ J = 1\ N{\cdot}m = 1\ kg{\cdot}m^2/s^2$
energy	E	J	$1\ J = 1\ N{\cdot}m = 1\ kg{\cdot}m^2/s^2$ $1\ eV = 1.6\times10^{-19}\ J$
power	P	W	$1\ W = 1\ kg{\cdot}m^2/s^3$ $1\ kw{\cdot}h = 3.6\times10^6\ W$
momentum	$\vec{p}$	$kg{\cdot}m/s$	$1\ kg{\cdot}m/s = 1\ N{\cdot}s$
impulse	$\vec{J}$	$N{\cdot}s$	$1\ N{\cdot}s = kg{\cdot}m/s$
torque	τ	$N{\cdot}m$	$1\ N{\cdot}m = 1\ kg{\cdot}m^2/s^2$
gravitational field	$\vec{g}$	N/kg	$1\ N/kg = 1\ m/s^2$
charge	q	C	$1\ C = 1\ A{\cdot}s = 1\ J/V$
electric field	$\vec{E}$	N/C	$1\ N/C = 1\ N{\cdot}s/A$
electric potential	V	V	$1\ V = 1\ J/C = 1\ A{\cdot}\Omega$
current	I	A	$1\ A = 1\ V/\Omega = 1\ N/(T{\cdot}m)$
resistance	R	Ω	$1\ \Omega = 1\ V/A = 1\ J{\cdot}s/C^2$
electromotive force	ε	V	$1\ V = 1\ J/C = 1\ A{\cdot}\Omega$
magnetic field	$\vec{B}$	T	$1\ T = 1\ N/(A{\cdot}m) = 1\ Wb/m^2$
magnetic flux	Φ	Wb	$1\ Wb = 1\ T{\cdot}m^2 = 1\ J/A$

APPENDIX B. TABLE OF CONSTANTS

Gravitation

Quantity	Value
Gravitational constant (G)	6.67×10^{-11} N·m²/kg²
Gravitational acceleration at the surface of Earth (g)	9.80 m/s²
Mass of Earth	5.98×10^{24} kg
Radius of Earth	6.38×10^{6} m
Orbital period of Earth	3.16×10^{7} s
Earth-Sun distance	1.50×10^{11} m
Earth-Moon distance	3.84×10^{8} m
Mass of the Sun	1.98×10^{30} kg
Radius of the Sun	6.96×10^{8} m
Mass of the Moon	7.35×10^{22} kg
Radius of the Moon	1.74×10^{6} m
Orbital period of the Moon	2.36×10^{6} s

Electromagnetism

Quantity	Value
Constant in Coulomb's law (k)	9.00×10^{9} N·m²/C²
Elementary charge (e)	1.60×10^{-19} C
Mass of electron	9.11×10^{-31} kg
Mass of proton	1.67×10^{-27} kg
Mass of neutron	1.67×10^{-27} kg
Permeability of free space (μ_0)	$4\pi \times 10^{-7}$ T·m/A

APPENDIX C. EQUATIONS

Chapter 1: Vectors

Magnitude of a vector: $R = \sqrt{R_x^2 + R_y^2}$

Direction of a vector: $\theta = \tan^{-1}\left(\frac{R_y}{R_x}\right)$

Chapter 2: Kinematics

Displacement: $\vec{d} = \vec{x}_f - \vec{x}_i = \Delta\vec{x}$

Average velocity: $\vec{v}_{avg} = \frac{\Delta\vec{x}}{\Delta t} = \frac{\vec{d}}{\Delta t}$

Average speed: $v_{avg} = \frac{D}{\Delta t}$

Acceleration: $\vec{a} = \frac{\Delta\vec{v}}{\Delta t} = \frac{\vec{v}_f - \vec{v}_i}{\Delta t}$

Kinematic equations:

$\vec{v}_f = \vec{v}_i + \vec{a}t$	$\vec{d} = \vec{v}_i t + \frac{1}{2}\vec{a}t^2$
$v_f^2 = v_i^2 + 2\vec{a}\cdot\vec{d}$	$\vec{d} = \vec{v}_{avg}\cdot t = \left(\frac{\vec{v}_i + \vec{v}_f}{2}\right)\cdot t$

Relative velocity: ${}_a\vec{v}_b = -{}_b\vec{v}_a$

${}_a\vec{v}_b + {}_b\vec{v}_c = {}_a\vec{v}_c$

Chapter 3: Dynamics

Gravitational force: $F_g = m\,g$

Friction force: $F_f = \mu F_N$

Newton's second law of motion: $\vec{F}_{\text{net}} = m\,\vec{a}$

Newton's third law of motion: $\vec{F}_{\text{on A by B}} = -\vec{F}_{\text{on B by A}}$

Chapter 4: Work and Energy

Work: $W = F\,d\cos\theta$

Kinetic energy: $E_K = \frac{1}{2}mv^2$

Gravitational potential energy: $E_P = mgh$

Mechanical energy: $E_T = E_K + E_P$

Work-energy principle:

$$W_{\text{net}} = \Delta E_K = \frac{1}{2}m\left(v_f^2 - v_i^2\right)$$

$$W_{NC} = \Delta E_K + \Delta E_P = \Delta\left(\frac{1}{2}mv^2 + mgh\right)$$

Power: $P = \frac{\text{Work}}{t} = \frac{F\,d}{t} = F\,v$

Efficiency: $\eta = \frac{W_{\text{out}}}{W_{\text{in}}} = \frac{P_{\text{out}}}{P_{\text{in}}}$

Chapter 5: Momentum

Momentum: $\vec{p} = m\vec{v}$

Impulse: $\vec{J} = \left(\vec{F}_{\text{net}}\right)_{\text{avg}}\Delta t = \Delta\vec{p}$

Conservation of momentum: $\left(m_A\vec{v}_A + m_B\vec{v}_B\right)_{\text{before}} = \left(m_A\vec{v}_A + m_B\vec{v}_B\right)_{\text{after}}$

Conservation of kinetic energy (for perfectly elastic collision):

$$\left(\frac{1}{2}m_Av_A^2 + \frac{1}{2}m_Bv_B^2\right)_{\text{before}} = \left(\frac{1}{2}m_Av_A^2 + \frac{1}{2}m_Bv_B^2\right)_{\text{after}}$$

Chapter 6: Equilibrium

First condition of static equilibrium: $\vec{F}_{\text{net}} = 0$

Second condition of static equilibrium: $\tau_{\text{net}} = 0$ (or $\Sigma\tau_{\text{CW}} = \Sigma\tau_{\text{CCW}}$)

Torque: $\tau = r\,F\sin\theta$

Chapter 7: Circular Motion and Gravitation

Speed: $v = \dfrac{\text{circumference}}{\text{period}} = \dfrac{2\pi R}{T}$

Centripetal acceleration: $a_C = \dfrac{v^2}{R}$

Newton's law of universal gravitation: $F_g = G\dfrac{mM}{r^2}$

Gravitational field: $g = G\dfrac{M}{r^2}$

$\vec{F}_g = m\vec{g}$

Gravitational potential energy: $E_P = -G\dfrac{mM}{r}$

Total mechanical energy of an orbiting body: $E_T = \dfrac{E_P}{2} = -E_K$

Orbital speed: $v = \sqrt{G\dfrac{M}{r}}$

Escape velocity: $v_e = \sqrt{\dfrac{2GM}{R}}$

Chapter 8: Electrostatics

Coulomb's law: $F_e = k\dfrac{|qQ|}{r^2}$

Electric field: $E = k\dfrac{|Q|}{r^2}$

$\vec{F}_e = q\vec{E}$

Electric potential energy: $E_P = k\dfrac{qQ}{r}$

Electric potential: $V = k\dfrac{Q}{r}$

$E_P = qV$

Potential difference between two oppositely charged parallel plates: $|\Delta V| = Ed$

Work-energy principle for a moving charge: $W_{NC} = q\,\Delta V + \dfrac{1}{2}m\left(v_f^2 - v_i^2\right)$

Deflection of cathode ray: $D \propto \dfrac{\Delta V_D}{\Delta V_A}$

Chapter 9: DC Circuits

Current: $I = \dfrac{Q}{t}$

Electric power: $P = I \times \Delta V = I^2 R = \dfrac{(\Delta V)^2}{R}$

Ohm's law: $\left|\Delta V_{\text{across resistor}}\right| = I_{\text{through resistor}} \times R_{\text{of resistor}}$

Kirchhoff's junction rule:	$\sum I_{\text{into junction}} = \sum I_{\text{out of junction}}$
Kirchhoff's loop rule:	$\sum \Delta V_{\text{around a loop}} = 0$

Equivalent resistance: $R_{eq} = R_1 + R_2 + R_3 + \cdots$ (in series)

$$\frac{1}{R_{eq}} = \frac{1}{R_1} + \frac{1}{R_2} + \frac{1}{R_3} + \cdots \quad \text{(in parallel)}$$

Terminal voltage: $V_T = \mathcal{E} - Ir$ (when current flows out of battery)

$V_T = \mathcal{E} + Ir$ (when current is forced into battery)

Chapter 10: Electromagnetism

Magnetic field produced by current: $B = \frac{\mu_0}{2\pi} \frac{I}{r}$

Magnetic field produced by solenoid: $B = \mu_0 \left(\frac{N}{L}\right) I$

Magnetic force on a moving charge: $F_m = |q| v B \sin\theta$

Magnetic force on a current-carrying wire: $F_m = I L B \sin\theta$

Radius of the circular path a moving charge follows in a uniform magnetic field: $R = \frac{m v}{|q| B}$

Magnetic flux: $\Phi = B A \cos\theta$

Faraday's law of induction: $\mathcal{E}_{avg} = -N \frac{\Delta\Phi}{\Delta t}$

Induced emf developed in a moving rod: $\mathcal{E} = v L B$

Transformer equation: $\frac{N_P}{N_S} = \frac{V_P}{V_S} = \frac{I_S}{I_P}$

Back emf: $V_{back} = 0$ at start-up

$V_{back} = \mathcal{E} - I R$ at normal operating speed

APPENDIX D. ANSWER KEY

CHAPTER 1: VECTORS

Test Your Knowledge of Physics 11: Vectors (p. 8)

1. T 2. T 3. T 4. T 5. T 6. d 7. c 8. a 9. d 10. c

11. $\vec{R}, \vec{Q}, \vec{S}, \vec{P}$ 12. 1.6 13. 1.9 km, south 14. 13.0 m 15. 3.9 m

Exercises (p. 16)

1-1. $V_x = 8.33$ m/s, $V_y = 7.63$ m/s
1-2. $V_x = 6.13$ m/s, $V_y = -4.32$ m/s
1-3. 11.6 m/s
1-4. $F_{gx} = -29.4$ N, $F_{gy} = -50.9$ N
1-5. 50.9 N
1-6. 29.4 N
1-7. $F_{1x} = 55.4$ N, $F_{1y} = 46.8$ N
$F_{2x} = 23.0$ N, $F_{2y} = -78.0$ N
$F_{3x} = -59.0$ N, $F_{3y} = -23.0$ N
1-8. 57.6 N, 70.3° below $+x$-axis
1-9. $F_{Tx} = 90.5$ N, $F_{Ty} = 49.1$ N
1-10. $F_N = 52.9$ N, $F_f = -90.5$ N
1-11. $p_{Ax} = 2.10$ kg·m/s, $p_{Ay} = -4.54$ kg·m/s
$p_{Bx} = 6.35$ kg·m/s, $p_{By} = 5.42$ kg·m/s
1-12. 8.50 kg·m/s, 5.95° S of W
1-13. $F_1 = 180$ N, $F_2 = 228$ N
1-14. $F_{1x} = -165$ N, $F_{1y} = 72.4$ N
$F_{2x} = 165$ N, $F_{2y} = 158$ N
$F_{3x} = 0$ N, $F_{3y} = -230$ N
$(F_{net})_x = 0$ N, $(F_{net})_y = 0$ N
1-15. $F_{gx} = -92.4$ N, $F_{gy} = -81.2$ N
$F_{Tx} = -40.7$ N, $F_{Ty} = 81.2$ N
1-16. 133 N, along the rod
1-17. $A_x = 2.4$ m/s, $A_y = -1.2$ m/s
$B_x = -1.3$ m/s, $B_y = 8.0$ m/s
$C_x = 2.9$ m/s, $C_y = 2.1$ m/s
$D_x = -4.1$ m/s, $D_y = -3.7$ m/s
$E_x = 0$ m/s, $E_y = 1.7$ m/s
1-18. 5.9 m/s, 26° above $+x$-axis
1-19. 22 m/s, 78° above $+x$-axis
1-20. 13 m/s, 13° below $+x$-axis
1-21. $a = -0.51$, $b = -0.30$

Problems on Chapter 1: Vectors (p. 21)

1. scalar 2. magnitude, direction 3. resultant 4. components 5. magnitude

6. F 7. T 8. T 9. F 10. F 11. F 12. T 13. F 14. T 15. F

16. c 17. c 18. c 19. c 20. a 21. d 22. c

23. 2.1 m/s
24. -5.9 m, -7.0 m
25. 6.7 N, 58°
26. a) 210 km/h; b) 0.15 h
27. 215 N: $F_x = -110$ N, $F_y = -190$ N
130 N: $F_x = 92$ N, $F_y = -92$ N
28. $F_{gx} = 0$ N, $F_{gy} = -123$ N
$F_{Tx} = -87.8$ N, $F_{Ty} = 23.0$ N
133 N, along the beam (48.7° above horizontal)
29. a) 7.9 m, 18° W of S; b) 12 N, 22° S of E
c) 33.3 m/s, 2.7° above the $+x$-axis
30. 163 paces to the west and 113 paces to the north
31. a) 31.1 m, 54° S of W
b) 98.0 N, 44° S of E
32. A = 192 m/s, B = 49.6 m/s
33. 7.12 km, 47° N of E
34. 178 km, 62° N of E
35. 400 N, 5.8° beneath the rod

CHAPTER 2: KINEMATICS

Test Your Knowledge of Physics 11: Kinematics (p. 32)

1. T 2. F 3. F 4. T 5. T 6. F 7. F 8. T 9. T 10. F

11. d 12. b 13. d 14. b 15. c 16. d 17. c 18. b 19. c 20. c

21. a) 1.5 m, down; b) 0.5 m, up; c) 0 m

23. 0.93 m/s^2

25. a) Object B has greater displacement.
    b) Object B has higher average velocity.
    c) Object A has higher inst. velocity at $t = t_1$.
    d) Object A has greater average acceleration.

27. $V_{\text{after 1.5 s}}$ = 10 m/s, up; $V_{\text{after 3.5 s}}$ = 9.3 m/s, down

29. 25 m/s (or 91 km/h)

31. a) 104 km, east; b) 72 km/h, east

33. 0.42 m/s^2

35. 3.7 m

22. a) displacement = 110 m, up; distance = 170 m
    b) avg. velocity = 55 m/min, up
    avg. speed = 87 m/min

24. -38 m/s, west

26. a) 8.4 m/s^2, right; b) 2.6 s
    c) displacement = 93 m, right; distance = 150 m

28. -200 m/s^2, or -20 g

30. 4.37 s

32. 1.3 km, 21 m/s

34. a) 60.0 m, down; b) 5.9 s; c) 41 m/s, down

Exercises (p. 47)

2-1. a) 2.97 m, up; b) 8.33 m/s

2-3. a) 30 m/s, 33° above horizontal; b) 9.6 s

2-5. 110 m

2-7. a) 6.8 s; b) 60 m/s, 18° below horizontal

2-9. 43 m/s

2-11. A: 4.3 s, B: 2.6 s, C: 2.0 s

2-13. A: 33 m, B: 39 m, C: 27 m
B travels farthest before landing.

2-15. 51 m/s

2-17. 19° S of E

2-19. 168 m/s, 12.5° S of E

2-2. 511 m

2-4. a) 2.1 s; b) 26 m

2-6. a) 0.47 s; b) 17 m

2-8. Its speed is 40 m/s when it's 50 m above ground.

2-10. 0.99 s

2-12. They all have the same speed of 30 m/s.

2-14. a) 0.39 m/s; b) 75 s

2-16. 8.4 m/s, 8.9° N of E

2-18. 87 m/s, 5.6° N of W

Problems on Chapter 2: Kinematics (p. 53)

1. position 2. displacement 3. distance 4. velocity 5. speed

6. acceleration 7. kinematic 8. projectile, trajectory

9. F 10. T 11. T 12. F 13. F 14. F 15. F 16. T 17. F 18. T 19. T

20. d 21. c 22. b 23. c 24. d 25. b

26. at t = 0 s, V_x = 2.7 m/s; at t = 2.5 s, V_x = 2.7 m/s
    slope = 0 m/s^2

28. a) 2.3 m; b) 23 m/s, 51° below horizontal
    c) 1.1 s

30. Baseball will fly over the balloon.

32. 5.0 m/s

34. 12 m/s; Answer does not change.

36. 42° or 63°

38. $_RV_C$ = 18 m/s; $_RV_G$ = 3.4 m/s

40. 11 km/h, 48° N of E

27. slope = -9.8 m/s^2
    Maximum height is reached at t-intercept.

29. 12 m/s

31. 12 m

33. a) 3.8 m/s; b) -6.0 m/s^2; c) 23 m

35. The object will fall off the cliff. a) 51 m; b) 32 m/s

37. a) 44 m/s, 74° above horizontal; b) 9.1 s
    c) 360 m/s; d) 410 m/s, 88° below horizontal

39. 120 m, 68° N of E

CHAPTER 3: DYNAMICS

Test Your Knowledge of Physics 11: Dynamics (p. 65)

1. T 2. T 3. F 4. F 5. F 6. F 7. T 8. F 9. F 10. F

11. d 12. a 13. c 14. c 15. d 16. d 17. a 18. b 19. d 20. b

21. 3.3×10^3 N, up; When he drops the weights, the normal force becomes 910 N, up.

22. 1.5×10^3 N

23. 1.5 m/s^2

24. a_{Liz} = 3.3 m/s^2, away from James
a_{James} = -2.3 m/s^2, away from Liz

25. 7.3 m/s^2, up

26. Tension in top string = 360 N
Tension in lower string = 210 N

27. 59 N

29. 220 N

30. 0.71

31. 31 N

32. 2.5 m/s^2, up

33. b) stopping distance = 32 m.
Yes, the car will stop before entering intersection.

34. 6.1 m/s^2, 43 N

35. 29 N

Exercises (p. 78)

3-1. b) 1.6 N

3-2. 4.9 m/s2, 1.7° N of E

3-3. b) 5.8 m/s^2, down the slope

3-4. b) F_N = 130 N, F_f = 45 N; c) 0.36

3-5. b) 5.0 m/s^2, right

3-6. F_f = 880 N, F_N = 210 N

3-7. b) 910 N; c) 0.058 m/s^2, right

3-8. b) F_f = 210 N, μ = 0.32

3-9. a) 3.5 N, 8.0 N, 14 N; b) 87 N

3-10. b) F_g = 15 N, down, F_T = 9.4 N, up the slope
F_N = 12 N, perpendicular to the slope

3-11. a) 1.1 m/s^2, 210 N; b) 1.1 s; c) 1.2 m/s, up

3-13. 0.29

3-12. 310 N

3-15. 20 N, 0.34 m/s^2, down the slope

3-14. 5.6°

3-17. 29 N, 0.15 m/s^2, to the left

3-16. b) 0.77

3-19. a) 1.1 m/s^2; b) 0.37 m/s^2

3-18. 5.4 m

3-21. 35 N, 1.1 m/s^2, to the left

3-20. a) 0.77 N; b) 5.7 m/s^2, down the slope; c) 13 m

Problems on Chapter 3: Dynamics (p. 85)

1. force 2. gravitational 3. normal 4. friction 5. tension

6. coefficient, friction 7. free-body, diagram 8. inertia 9. Newton's, second 10. Newton's, third

11. F 12. T 13. F 14. T 15. F 16. F 17. F 18. F 19. F

20. b 21. b 22. c 23. c 24. b 25. a 26. d 27. b 28. a 29. c 30. c

31. 15 m

32. 120 N

33. 19 m

34. a) 0.070; b) 6.5 s

35. 0.94 m/s^2, up

36. If M moves up the slope, $F_f = mg - Mg\sin\theta$
If M moves down the slope, $F_f = Mg\sin\theta - mg$

37. 25°

39. a) 20 m; b) 13 m/s

38. 35 N, 0.17 m/s^2, to the left

40. a) 3.0 kg; b) 7.0 m/s^2; c) The accelerations are different because of the change in normal force, and also the friction force.

CHAPTER 4: WORK AND ENERGY

Test Your Knowledge of Physics 11: Work and Energy (p. 96)

1. T 2. F 3. T 4. F 5. T 6. F 7. T 8. F 9. F 10. T

11. b 12. d 13. b 14. c 15. d 16. d 17. d 18. c 19. c 20. c

21. W by F_g = 0 J, W by F_A = 200 J, W by F_f = -24 J
    W_{net} = 170 J

22. -3.6×10^3 J

23. 32 N

24. 61 J

25. a) 23 m/s; b) 9.2 m

26. 6.7 m

27. 1.5 m/s; The speed will be the same.

28. a) 0.87 N; b) 23 J; c) 7.2 W

29. V_A = 11 m/s, V_C = 24 m/s

30. a) 7.3 m/s; b) 52 J

31. 3.3×10^4 W

32. a) 330 W; b) 0.48

33. 0.58

34. P_{out} = 680 W, P_{in} = 1.5×10^3 W

35. $20.30

Exercises (p. 112)

4-1. b) W by F_g = W by F_N = 0 J
     W by F_A = 2.0×10^3 J, W by F_f = -770 J

4-2. a) 7.0 m/s; b) 7.0 m/s

4-3. a) 34 m/s; b) 34 m/s

4-4. a) 1.6×10^6 N; b) 1.6×10^6 N

4-5. a) 0.73; b) 0.73

4-6. a) 1.0 m/s; b) 1.9×10^3 J

4-7. a) 6.8 m; b) 16 J

4-8. a) 320 J; b) 30 m/s

4-9. 1.2 kg

4-10. 32 m/s

4-11. 0.051 m

4-12. 9.3 m/s

4-13. 2.3×10^4 N, against the bullet's motion

4-14. a) 3.5×10^3 J; b) 5.3 m

4-15. a) 140 J; b) 14 N

4-16. 4h

4-17. 0.57 m

4-18. 15 J

4-19. 0.52

4-20. 4.7 m/s

Problems on Chapter 4: Work and Energy (p. 119)

1. work 2. work, energy, principle (or theorem) 3. potential, energy

4. potential, energy, kinetic, energy 5. conservative 6. power 7. efficiency

8. T 9. F 10. F 11. F 12. T 13. T 14. F 15. F 16. T 17. F

18. b 19. d 20. a 21. c 22. c 23. c 24. d 25. c 26. c 27. d 28. a

29. -26 J

30. -10 J

31. 3.2 m/s

32. 37°

33. 2.5 m/s

34. 510 N

35. 420 J must be taken away from the object.

36. a) 2.33 m/s; b) 2.33 m/s

37. a) eq. 1: $24.7 - 0.35\,F_T = 3.6\,v^2$
    eq. 2: $0.35\,F_T = 4.5\,v^2$
    b) 1.7 m/s

38. a) eq. 1: $24.7 - 0.35\,F_T = 3.6\,v^2$
    eq. 2: $0.35\,F_T - 20.7 = 4.5\,v^2$
    b) 0.70 m/s

39. 1.1 m/s

40. 240 W

CHAPTER 5: MOMENTUM

Test Your Knowledge of Physics 11: Momentum (p. 131)

1. T 2. F 3. T 4. T 5. F 6. T 7. T 8. F 9. F 10. T

11. c 12. c 13. a 14. d 15. a 16. c 17. d 18. a 19. b 20. c

21. 31 kg·m/s, up
22. 480 kg·m/s, north
23. 3.5 m/s, south
24. 10 m/s, east
25. 4.5×10^3 N
26. 52 N·s, in the direction of object's motion
27. 56 N·s, up
28. A: 33 kg·m/s, south; B: 33 kg·m/s, north
29. 1.4 m/s
30. 7.3 m/s
31. 1.0 m/s, west
32. 8.7 J
33. 3.4 $(V_x)_{initial}$
34. 5.7 m/s
35. 50 J

Exercises (p. 145)

5-1. a) 6.2 kg·m/s, 75° below horizontal
b) 7.4 N·s, down

5-2. 1.2 N·s, 60° N of W

5-3. a) 240 N·s, north; b) 6.1 m/s, 42° E of N

5-4. 78° E of S for 0.90 s

5-5. 13 m/s

5-6. a) 600 N; b) 13 m/s, 13° above horizontal

5-7. 10 m/s

5-8. a) 3.2 m/s, 23° N of E; b) 36 J

5-9. 2.2 m/s, 77° S of E

5-10. 6.2 kg

5-11. 46 m/s, 85° N of E

5-12. 19 m/s, 81° N of W

5-13. a) 28 m/s, 17° N of E; b) 1.8×10^3 J

5-14. 4.3 m

5-15. a) 12 m/s, 24 m/s; b) 53° N of E

Problems on Chapter 5: Momentum (p. 150)

1. momentum, velocity 2. net, force 3. impulse, momentum 4. impulse

5. perfectly, inelastic 6. perfectly, elastic

7. F 8. F 9. T 10. F 11. T 12. T 13. F 14. T 15. T

16. c 17. d 18. d 19. d 20. b 21. b 22. b 23. a 24. c 25. b

26. a) 1.5 m/s, 39° N of E; b) 7.0 J
27. 120 kg·m/s, right
28. 97 N
29. 8.8 m
30. 3.1 m/s, west and 0.65 m/s, east
31. 240 m/s, 0.27° below horizontal
32. 0.76 kg
33. 25 m
34. 600 kg·m/s, 35° S of W
35. 120°

CHAPTER 6: STATIC EQUILIBRIUM

Exercises (p. 170)

6-1. 42 N, 2.0° S of W

6-3. 8.8 kg

6-5. 1.1 x 10^3 N, 7.5 x 10^2 N

6-7. 42 N

6-9. 1.3 x 10^3 N, 1.3 x 10^3 N

6-11. 1.2 kg

6-13. 52 kg

6-15. F_T = 73 N
F_{hinge} = 61 N, 53° above beam into wall

6-17. 38 kg

6-19. 4.6 N, right

6-21. 2.0 x 10^3 N, up, and 1.1 x 10^3 N, down

6-23. a) 0.80 kg; b) 0.21 m

6-25. 0.15 m

6-27. 0.76 m

6-29. F_g of woman = 480 N, F_g of ladder = 82 N,
F_N by ground = 5.6 x 10^3 N
F_N by wall = F_f = 130 N

6-2. 71°

6-4. 990 N

6-6. 3.8 kg

6-8. 1.9 N, 1.6 N

6-10. 31°

6-12. a) 180 N·m, CW; b) 46 N·m, CW
c) 86 N·m, CCW

6-14. F_T = 71 N, F_{hinge} = 150 N, 62° above ground

6-16. 0.74 m from front wheel

6-18. a) 500 N, up and 280 N, up; b) 1.8 m

6-20. 230 N

6-22. a) 140 N, up; b) 170 N, up

6-24. 36 N

6-26. 5.2 x 10^3 N

6-28. F_T = 700 N, F_{hinge} = 550 N, 36° above horizontal

6-30. 77 N

Problems on Chapter 6: Static Equilibrium (p. 180)

1. equilibrium 2. static, equilibrium 3. rotation 4. lever, arm 5. torque

6. F 7. F 8. F 9. F 10. F 11. T 12. F 13. T 14. T 15. T 16. F

17. a 18. c 19. a 20. b 21. d 22. c 23. a 24. a 25. d 26. c

27. 27°

28. 180 N, 25 kg

29. 210 N, 17°

30. 42 kg

31. 0 N·m

32. 1.2 m

33. All normal forces are 170 N.

34. 72 N, 36 N

35. 40 N, 61 N, 55°

CHAPTER 7: CIRCULAR MOTION AND GRAVITATION

Exercises (p. 196 and p. 215)

7-1. 1.8 N
7-2. 400 N
7-3. a) 5.9 m/s; b) 1600 N
7-4. 0.042 rev/s
7-5. 22 s
7-6. a) 47 N; b) 2.5 m/s
7-7. 7800 N
7-8. 2400 N
7-9. 9.6 m/s
7-10. 380 N
7-11. 2.2 m
7-12. 8000 N
7-13. 250 m/s = 900 km/h
7-14. 28 N
7-15. 180 N
7-16. 0.862 N/kg
7-17. a) 29.8 km/s; b) 0.0027 m/s^2, towards Earth
7-18. 1.8 x 10^{20} kg
7-19. -4.35 x 10^9 J
7-20. 1.3 x 10^{21} kg
7-21. 3.0 x 10^6 m
7-22. 5.7 x 10^{20} kg
7-23. 6.7 km/s
7-24. 1.9 x 10^6 m
7-25. 21 km/s
7-26. 2.6 x 10^4 s
7-27. 3.7 x 10^{14} kg
7-28. 4.5 x 10^{12} J
7-29. 7.7 km/s, -3.4 x 10^{10} J
7-30. 9.3 x 10^7 m
7-31. a) 5.0 x 10^7 m; b) -1.3 x 10^{13} J
7-32. 3.5 km/s
7-33. 3.1 x 10^4 s
7-34. 8.7 x 10^{10} J
7-35. 5.3 x 10^{11} J

Problems on Chapter 7: Circular Motion and Gravitation (p. 220)

1. uniform, circular 2. centripetal 3. universal, gravitation 4. gravitational, field
5. weightlessness

6. T 7. F 8. T 9. T 10. F 11. F 12. T 13. F 14. F 15. F

16. d 17. c 18. d 19. b 20. c 21. b 22. c 23. b 24. a 25. a 26. a 27. d
28. a

29. a/9
30. a) 20 N; b) 2.8 s
31. 11 m/s
32. 3.1 m/s
33. 9.8 m/s
34. 3.2 m/s
35. 8.1 m
36. 27 s
37. 1900 N
38. 3.1 m/s^2
39. a) 7.6 km/s; b) -4.7 x 10^{10} J
40. 170 N
41. 6.8 x 10^3 s
42. 1.94 x 10^{22} kg
43. 9.4 x 10^6 m
44. 4.2 x 10^{25} kg
45. a) -1.3 x 10^{11} J; b) 5.4 km/s
46. 3.9 x 10^7 m
47. 14 km/s
48. 0.72 AU
49. 2.3 x 10^{10} J
50. It can only travel up to 6700 km above the surface.

CHAPTER 8: ELECTROSTATICS

Exercises (p. 238 and p. 253)

8-1. -0.34 μC
8-2. 7.4×10^{-9} N
8-3. 31 N/C, towards the charge
8-4. a) 5.4×10^{-16} N, right; b) 6.0×10^{14} m/s^2, right
8-5. +13 μC
8-6. 9700 N/C, right
8-7. a) 0.15 N, right; b) 0.12 N, left
8-8. 1.0×10^{-7} N/C, up
8-9. -4.0 μC
8-10. 4.4×10^{5} N/C, 53° S of W
8-11. 4.6×10^{10} m/s^2, right
8-12. 60 μC
8-13. 2.2×10^{6} m/s
8-14. a) 3.1×10^{-8} N; b) 2.7×10^{6} m/s; c) 4.9×10^{-16} s
8-15. -150 μC
8-16. -5.2×10^{4} V
8-17. 10 J
8-18. 6.6×10^{-13} C
8-19. -39 μC
8-20. 64 V
8-21. a) -6.9×10^{-16} J; b) 4.6×10^{7} m/s
8-22. 2.5×10^{5} m/s
8-23. -0.89 μC
8-24. a) 7.0×10^{5} m/s
b) Yes, proton reaches right plate. 3.1×10^{5} m/s
8-25. 1.3×10^{-14} J
8-26. 900 V
8-27. 9.8×10^{-16} J
8-28. a) 6.5×10^{6} m/s; b) 5.2×10^{13} m/s^2, down
8-29. 0.11 m
8-30. a) h. def. = 7.6 cm, v. def. = 3.3 cm
b) h. def. = 5.1 cm, v. def. = 3.3 cm

Problems on Chapter 8: Electrostatics (p. 259)

1. elementary, charge 2. Coulomb's, law 3. electric, field 4. electric, field
5. electric, potential 6. equipotential 7. cathode, ray, tube
8. F 9. T 10. T 11. F 12. F 13. T 14. F 15. F 16. T 17. T 18. T 19. T
20. F
21. b 22. d 23. a 24. c 25. d 26. a 27. d 28. c 29. a 30. a 31. b 32. d
33. d 34. c

35. a) 1.0×10^{-67} N; b) 2.3×10^{-28} N; c) 2.3×10^{39}
36. 3.1 kg
37. +33 μC
38. +8.2 μC
39. E = 1.1×10^{6} N/C, 72° S of E, V = 4.8×10^{5} V
40. 7.0×10^{6} m/s
41. +7.0 μC
42. a) 120 m/s; b) 83 m/s
43. 1.7 J
44. a) 7.0×10^{6} m/s; b) 1.8×10^{14} m/s^2, down
45. a) 1.9×10^{-8} s
b) 7.7×10^{6} m/s, 25° below horizontal
46. 1.6×10^{-7} kg
47. 8.5 m/s
48. a) 10 m/s, 2.6 m/s; b) 0 m/s
49. a) falls, a = 7.5 m/s^2; b) 7.1 m/s
c) no collision, They get to 2.2 m apart.
50. 1.9 m

CHAPTER 9: DC CIRCUITS

Exercises (p. 288)

9-1. 2.7 A, right

9-2. 5.7 Ω

9-3. 0.29 Ω

9-4. a) 13 Ω; b) 0.49 A; c) 2.6 V; d) 6.4 V; e) 7.3 C

9-5. a) 5.2 A; b) 6.2 V

9-6. V_b = 7.0 V, V_c = 3.5 V, V_d = -7.0 V

9-7. $I_{2.5\text{-}\Omega}$ = 1.9 A, $I_{5.5\text{-}\Omega}$ = 0.83 A, $I_{4.5\text{-}\Omega}$ = 1.0 A

9-8. a) 3.0 x 10^6 J; b) 2.0 x 10^6 C

9-9. a) 96 W; b) 60 W; c) I_1 = 1.4 A, I_2 = 2.3 A

9-10. a) 12 V; b) 9.4 V; c) 9.4 V; d) 0 V; e) 0.42 V f) 21.6 V

9-11. a) 23 V; b) 10 V

9-13. a) 1.6A, left; b) 3.5 C; c) 1.9 Ω; d) 4.3 W

9-12. a) $I_{33\text{-}\Omega}$ = 0.28 A, $I_{13\text{-}\Omega}$ = 0.71 A, $I_{16\text{-}\Omega}$ = 0.99 A b) $P_{33\text{-}\Omega}$ = 2.6 W, $P_{13\text{-}\Omega}$ = 6.6 W, $P_{16\text{-}\Omega}$ = 16 W

9-15. a) 17 V; b) 3.9 Ω

9-17. a) 11 Ω; b) I_1 = 0.58 A, I_3 = 0.22 A, I_6 = 1.4 A c) P_2 = 1.7 W, P_4 = 1.3 W, P_5 = 0.42 W

9-14. 990 W, 400 W

9-16. a) 1.5 A; b) 16 V; c) 6.3 V

9-19. 3.7 Ω

9-18. a) 2.2 A; b) 18 V; c) 1.7 A, 14 V

9-21. a) 6.5 Ω; b) I_1 = 0.34 A, I_2 = 0.41 A, I_3 = 0.27 A I_4 = 0.34 A, I_5 = 0.13 A, I_6 = 0.75 A c) 4.7 V; d) 0.11 W

9-20. a) 0.54 A, clockwise; b) 6.6 V, 9.1 V

9-22. a) 29 Ω; b) I_1 = 0.66 A, I_2 = 0.44 A, I_3 = 0.43 A I_4 = 0.68 A, I_5 = 1.1 A; c) 30 V; d) 18 V

9-23. a) 5.4 Ω; b) 65 V

9-24. 0.72 Ω, 5.0 A

9-25. 0.34 Ω or 3.6 Ω

Problems on Chapter 9: DC Circuits (p. 297)

1. electric, current 2. resistor 3. junction 4. Kirchhoff's, loop 5. Ohm's

6. electromotive, force 7. internal, resistance 8. terminal, voltage 9. ammeter, voltmeter

10. F 11. T 12. F 13. F 14. T 15. T 16. T 17. F 18. F 19. T 20. T

21. b 22. a 23. c 24. a 25. c 26. c 27. a 28. d 29. a 30. c 31. a 32. a

33. b 34. c 35. c

36. 520 W

37. I_1 = 59 A, right, I_2 = 37 A, right

38. 7.5 V

39. 0.29 Ω

40. a) 5.7 A; b) 82 V

41. a) 14 V; b) 10 Ω; c) 7.3 Ω; d) $P_{R\text{ in parallel}}$ = 1.6 W, $P_{R\text{ in series}}$ = 14 W; e) 3.4 V; f) 11 V; g) 3.24 x 10^{19}

42. a) R_1 = 3.1 Ω, R_2 = 1.6 Ω; b) 4.1 V; c) 7.3 V

44. 230 W

43. a) 0.22 A; b) 6.9 V

46. a) 11 Ω; b) I1 = 0.28 A, I_2 = 1.0 A, I_3 = 0.84 A, I_4 = 0.48 A, I_5 = 1.3 A; c) 14 V; d) 11 V

45. a) $P_{battery} = 16I - 0.18I^2$; b) 86 A or 3.0 A

47. a) 2.9 Ω; b) 5.1 A; c) 0 V

48. a) 2.3 Ω; b) 7.3 V

49. 0.75 Ω, 17 A

50. 76 W

CHAPTER 10: ELECTROMAGNETISM

Exercises (p. 321 and p. 341)

10-1. 9.6×10^{-6} T, right

10-3. 1.3×10^{-3} T, right

10-5. 7.8 N

10-7. a) 0.041 T, left; b) 0 N

10-9. 2.8×10^{4} m/s

10-11. a) clockwise; b) 0.41 T

10-13. 5.2×10^{-15} N into the page

10-15. a) 0.41 N; b) 7.1 cm

10-17. a) upward; b) 1.4 m

10-19. a) -6.0 T/s; b) 3.3 T, out of the page

10-21. a) 310 V, 1.1 A; b) 340 W

10-23. a) 0.010 T; b) down; c) 1.1×10^{-3} V

10-25. a) clockwise; b) 1.3×10^{3} V

10-27. a) 2700 turns; b) 2.6 A

10-29. a) 19 V; b) counterclockwise; c) 15 m/s

10-2. 2.5 cm

10-4. a) 8.6×10^{-15} N, left; b) 5.6×10^{-7} m

10-6. a) left; b) 0.33 m

10-8. 1.1×10^{-19} kg·m/s

10-10. 0.26 T

10-12. a) B felt by $I_1 = 2.7 \times 10^{-7}$ T, B felt by $I_2 = 3.3 \times 10^{-7}$ T
b) repel each other; c) 6.2×10^{-7} N/m

10-14. 7.3×10^{-4} N, up

10-16. a) 0.17 Wb; b) 0.073 Wb; c) 4.8 V; d) clockwise

10-18. a) 61 V; b) 0 V at start, 52 V at normal speed

10-20. a) 79 V; b) 520 A, counterclockwise

10-22. a) counterclockwise; b) 0.41 T

10-24. a) 4.1 Ω; b) 0 V; c) 26 V

10-26. a) 0.099 V; b) 2.8×10^{-3} A, up; c) 3.0×10^{-4} N, left

10-28. a) $\Phi_{t=0.0\,s} = 9.4$ Wb, $\Phi_{t=0.2\,s} = 9.1$ Wb
$\Phi_{t=0.4\,s} = 8.2$ Wb, $\Phi_{t=0.6\,s} = 6.8$ Wb
$\Phi_{t=0.8\,s} = 5.0$ Wb, $\Phi_{t=1.0\,s} = 2.9$ Wb
b) 6.5 V; c) 4.3 V

10-30. a) 0 V; b) 1.3 Ω; c) 9.6 V

Problems on Chapter 10: Electromagnetism (p. 347)

1. magnetic, field 2. solenoid 3. magnetic, flux 4. magnetic, flux, induced, magnetic, flux
5. Lenz's, current, magnetic, flux 6. Faraday's, magnetic, flux
7. generator 8. motor 9. transformer

10. F 11. F 12. T 13. T 14. F 15. T 16. F 17. F 18. T 19. T 20. F
21. b 22. a 23. d 24. d 25. b 26. c 27. b 28. b 29. a 30. d 31. b 32. c
33. c 34. c 35. b 36. c 37. b 38. c 39. a 40. c 41. b 42. a 43. c 44. b
45. c 46. a

47. 2.7×10^{-7} T, up

49. 180 turns

51. 13 s

53. 6.2 T

55. left conductor: 9.9×10^{-6} N/m, left
middle conductor: 1.5×10^{-5} N/m, right
right conductor: 4.8×10^{-6} N/m, left

57. 12 N, left

59. a) $\Phi_{initial} = 8.5$ Wb, $\Phi_{final} = -7.7$ Wb
b) counterclockwise; c) 1.9×10^{4} V

61. a) 0.55 m/s^2, right; b) clockwise; c) 1.5 m/s

63. $N_p = 60$, $I_p = 13$ A, $I_s = 3.2$ A

65. 0.10 A

48. 3.5×10^{-7} T, into the page

50. 6.3 N, out of the page

52. a) 2.2 T; b) 2.2×10^{-6} m

54. 5.1×10^{-3} T

56. 60°

58. a) 1.5×10^{-3} T, right; b) left; c) no magnetic field

60. direction = CCW, induced emf = 2.8 V

62. R = 45 Ω, $V_{back} = 81$ V

64. 13 V, 2.5 A

INDEX

저자 약력

정 준(June Chung)

- 1996년 펜실베니아 주 카네기 멜론 대학에서 기계공학 학사 졸업
- 2002년 일리노이 대학 항공우주 공학과 박사 졸업
- 2008년 현재 캐나다 밴쿠버 소재 88 아카데미 원장

Patricia Grindlay

- 1961년 영국 맨체스터 대학에서 Bachelor of Science
- 1970년 영국 런던 대학에서 Diploma of Education 취득
- 1985년~2006년까지 밴쿠버 명문 사립고 Little Flower Academy에서 물리 및 화학 과장 역임

ACING PHYSICS 12

2009년 7월 20일 초판 인쇄
2009년 7월 25일 초판 발행

저 자 : 정 준, Patricia Grindlay
펴낸이 : 천승배
펴낸곳 : 도서출판 유한문화사

주소 : (157-801) 서울시 강서구 가양동 146-63
전화 : 2668-2055~6
팩스 : 2668-2565
http://www.yuhansa.com
E-mail : yuhansa@paran.com
등록 : 제 5-31호. 1979. 3. 6.

값 18,000 원

ISBN : 978-89-7722-118-5 93420